Metal
Building
Systems

Metal Building Systems

Design and Specifications

Alexander Newman

McGraw-Hill

New York San Francisco Washington, D.C. Auckland Bogotá
Caracas Lisbon London Madrid Mexico City Milan
Montreal New Delhi San Juan Singapore
Sydney Tokyo Toronto

Library of Congress Cataloging-in-Publication Data

Newman, Alexander, date.
   Metal building systems / Alexander Newman.
      p.   cm.
   Includes index.
   ISBN 0-07-046379-4 (hc)
   1. Building, iron and steel.   I. Title.
TA684.N435   1997
693'.7—dc20   96-49570
                                     CIP

McGraw-Hill

A Division of The McGraw-Hill Companies

   ` 3 4 5 6 7 8 9 0 DOC/DOC 9 0 1 0 9 8 7

ISBN 0-07-046379-4

The sponsoring editor for this book was Larry Hager, the editing supervisor was David E. Fogarty, and the production supervisor was Suzanne W. B. Rapcavage. It was set in Century Schoolbook by Renee Lipton of McGraw-Hill's Professional Book Group composition unit.

Printed and bound by R. R. Donnelley & Sons Company.

McGraw-Hill books are available at special quantity discounts to use as premiums and sales promotions, or for use in corporate training programs. For more information, please write to the Director of Special Sales, McGraw-Hill, 11 West 19th Street, New York, NY 10011. Or contact your local bookstore.

Contents

Preface

There was a time when architectural decision making concerning pre-engineered metal buildings was confined to a choice among shades of beige. Those days are gone. Now metal building systems can accommodate a variety of floor plan requirements and are offered in a rainbow of exterior finishes. More and more, architects and engineers retained directly by prospective owners conceive and conceptually design these buildings before manufacturers become involved. According to a recent survey, architects are engaged in over one-half of all metal building contracts.

These professionals need reference material. Indeed, according to another survey, many architects, contractors, and building owners inexperienced with metal building systems seek information about the systems' performance specifications, energy efficiency, and compliance with building codes. Unfortunately, design schools and textbooks tend to ignore this type of construction, apparently assuming that it belongs in the manufacturer's domain. True, the manufacturers and their trade associations do produce facility planning guides, publish codes of standard practice for the industry, and disseminate a wealth of other literature. These publications are useful for people in the field and those outside it who have an adequate grasp of the critical issues, plus the time and patience to struggle through dozens of three-ring binders filled with glossy brochures, proprietary product names, and specialized jargon. For the rest of us, it may seem that all metal building suppliers are essentially the same and that they would just as soon avoid our meddling or any owner-imposed special criteria.

The stubborn fact remains, however, that as with any other type of construction, metal buildings can be poorly conceived, specified, designed, and erected, which may result in problems with excessive deflections of structural members, cracking of finishes, building vibrations, and damage from excessive snow accumulation or violent hurricanes. Furthermore, the proprietary nature of the field makes misunderstanding of the design intent by a supplier difficult to detect, unless the specifiers know which questions to ask. Quite often, the manufacturer who offers a seemingly irresistible quote on a project may not end up being selected under more rigorous criteria that consider factors other than price alone and compare "apples to apples."

This book is written by a practicing structural engineer not affiliated with any suppliers, dealers, or their trade groups. He can thus provide an impartial and objective overview of the industry, its strengths, weaknesses, and tremendous potential.

The book's principal objective is to present architects, engineers, construction specifiers, facility managers, and building officials with enough information to enable them to make intelligent decisions in their design and evaluation activities. It contains plenty of data to help design professionals accomplish tasks ranging from feasibility studies to preparation of construction documents. It should also prove helpful to contractors seeking to undertake construction utilizing metal building systems.

The book is arranged by chapters dealing with various design aspects of metal building systems. Although some chapters are devoted to rather technical matters intended mainly for structural engineers, every attempt has been made to keep the presentation readable and enjoyable.

In addition to explanatory and reference data, the book contains a wealth of tips on how to avoid potential pitfalls when specifying these deceptively simple structures. Most of the tips are distilled from "learning experiences" that caused some embarrassment for the designers, claims for "extras," and even litigation. The readers are invited to learn from these mistakes to be better prepared to participate in the success of this extraordinary industry.

The proprietary nature of metal building systems has necessitated substantial reliance on information and illustrations provided by the manufacturers. The author fervently hopes that proper credit has been given to all sources of the material presented here and wishes to express his sincere thanks and appreciation to everyone who helped in the production of this book. He is especially grateful to his colleagues and friends at Maguire Group Inc. who encouraged and assisted him in this endeavor.

Alexander Newman

Metal
Building
Systems

1

Metal Building Systems: Yesterday and Today

1.1 The Origins

1.1.1 What's in a name?

There is confusion—even among some design professionals—about what a metal building system is. "Just what kind of a building is it? Is it a modular building? Or prefabricated? Or maybe panelized? Is it the same as a pre-engineered building?"—you might hear a lot. Though all of these terms involve some sort of structure designed and partially assembled in the shop by its manufacturer, they refer to quite different concepts. Before proceeding further, the distinctions need to be sorted out.

Modular buildings consist of three-dimensional plant-produced segments that are shipped to a site for erection and final assembly by a field contractor. One of the most popular materials for modular buildings is wood, and such factory-produced units are common in housing construction. Another common application involves precast concrete formed into modular stackable prison cells that are completely prewired and prefinished. These modules are composed of four walls and a ceiling that also serves as a floor for the unit above. Modular steel systems, consisting of three-dimensional column and joist modules bolted together in the field, were marketed in the 1960s and 1970s, with limited success. Modern metal building systems, however, cannot be called modular.

Panelized systems include two-dimensional building components such as wall, floor, and roof sections, produced at the factory and field-assembled. In addition to the "traditional" precast concrete, modern exterior wall panels can be made of such materials as metals, brick, stone, and composite assemblies known as EIFS (exterior insulation and finish system). While the exterior "skins" of metal buildings generally employ panels, the term "panelized" does not capture the essence of metal building systems and should not be used to describe them.

Prefabricated buildings are made *and* substantially assembled at the factory. While the metal building industry has its roots in prefabricated buildings, this type today includes mostly small structures transported to the site in one piece, such as toll booths, kiosks, and household sheds. Modern metal buildings are not prefabricated in that sense.

As we shall see, the changes in terminology parallel the evolution of the industry itself.

1.1.2   The first metal buildings

The first building with an iron frame was the Ditherington Flax Mill constructed in Shrewsbury, England, in 1796.[1] Cast-iron columns were substituted for the usual timber in a calico mill constructed in nearby Derby 3 years earlier. These experiments with iron were prompted by frequent devastating fires in British cotton mills of the time. Once the fire-resistive properties of metal in buildings had been demonstrated, wrought-iron and cast-iron structural components gradually became commonplace.

In the middle of the nineteenth century, experimentation with rolling of iron beams finally culminated in construction of the Cooper Union Building in New York City, the first building to utilize hot-rolled steel beams. In 1889, Rand McNally Building in Chicago became the first skyscraper with all-steel framing.[2]

Prefabricated metal buildings first appeared at about the same time. As early as the mid-nineteenth century, "portable iron houses" were marketed by Peter Naylor, a New York metal-roofing contractor, to satisfy housing needs of the California Gold Rush fortune seekers; at least several hundred of those structures have been sold.[1] Eventually, of course, California's timber industry got established and Naylor's invention lost its market.

In the first two decades of the twentieth century, prefabricated metal components were mostly used for garages. Founded in 1901, Butler Manufacturing Company developed its first prefabricated building in 1909 to provide garage space for the ubiquitous Model T. That curved-top building used wood framing covered with corrugated metal sheets. To improve fire resistance of its buildings, the company eventually switched to all-metal structures framed with corrugated curved steel sheets. The archlike design, inspired by cylindrical grain bins, influenced many other prefabricated metal buildings.[3]

In 1917, the Austin Company of Cleveland, Ohio, began marketing 10 standard designs of a factory building that could be chosen from a catalog. The framing for these early metal buildings consisted of steel columns and roof trusses which had been designed and detailed beforehand. The Austin buildings were true forebears of what later became known as pre-engineered construction, a new concept that allowed for material shipment several weeks earlier, because no design time needed to be spent after the sale. Austin sold its buildings through a newly established network of district sales offices.[4]

In the early 1920s, Liberty Steel Products Company of Chicago offered a prefabricated factory building that could be quickly erected. The LIBCO ad pictured the building and boasted: "10 men put up that building in 20 hours. Just ordinary help, and the only tools needed were monkey wrenches...."[1]

By that time, steel was an established competitor of other building materials. The first edition of Standard Specification for the Design, Fabrication and Erection of Structural Steel for Buildings was published by the newly formed American Institute of Steel Construction in 1923.

Several metal-building companies were formed in the 1920s and 1930s to satisfy the needs of the oil industry by making buildings for equipment storage; some of these companies also produced farm buildings. For example, Star Building Systems was formed in 1927 to meet the needs of oil drillers in the Oklahoma oil boom. Those early metal buildings were rather small—8 by 10 or 12 by 14 ft in plan—and were framed with trusses spanning between trussed columns. The wall panels, typically 8 by 12 ft in size and spanning vertically, were made of corrugated galvanized sheet sections bounded by riveted steel angles.

1.1.3   The war years and after

During World War II, larger versions of those metal buildings were used as aircraft hangars. Their columns were made of laced angles, perhaps of 6 by 4 by $\frac{3}{8}$ in in section, and roof structure consisted of bowstring trusses. Military manuals such as NAVFAC DM-2 were used for design criteria. These buildings, unlike their predecessors, relied on intermediate girts for siding support.

The best-known prefabricated building during World War II was the Quonset hut, which became a household word. Quonset huts were mass-produced by the hundreds of thousands to meet a need for inexpensive and standardized shelter (Fig. 1.1). Requiring no special skills, these structures were assembled with only hand tools, and—with no greater effort—could be readily dismantled, moved, and re-erected elsewhere.

Quonset huts followed GIs wherever they went and attested to the fabled benefits American mass production could bestow. Still, these utterly utilitarian, simple, and uninspiring structures were widely perceived as being cheap and ugly. This impression still lingers in the minds of many, even though quite a few Quonset huts have survived for over half a century.

The negative connotation of the term "prefabricated building" was reinforced after the war ended and the next generation of metal buildings came into being. Like the Quonset hut, this new generation filled a specific need: the postwar economic boom required more factory space to satisfy the pent-up demand for consumer products. The vast sheet-metal industry, well-organized and efficient, had just lost its biggest customer—the military. Could the earlier sheet-metal prefabricated buildings and the Quonset hut, as well as the legendary Liberty Ship quickly mass-produced at Kaiser's California shipyard, provide a lesson for a speedy making of factory buildings? The answer was clearly: "Yes!"

In the new breed of sheet-metal-clad buildings, the emphasis was, once again, on rapid construction and low cost, rather than aesthetics. It was, after

Figure 1.1 Quonset hut, Quonset Point, R.I. (*Photo: David Nacci.*)

all, the contents of these early metal structures that was important, not the building design. Using standardized sheet-metal siding and roofing, supported by gabled steel trusses and columns—a 4:12 roof pitch was common—the required building volumes could be created relatively quickly. In this corrugated, galvanized environment, windows, insulation, and extensive mechanical systems were perceived as unnecessary frills. The sheer number of these prefabricated buildings, cloned in the least imaginative mass-production spirit, was overwhelming.

Eventually, the economic boom subsided, but the buildings remained. Their plain appearance was never an asset. As time passed and these buildings frayed, they conveyed an image of being worn out and out of place. Eventually, prefabricated buildings were frowned upon by almost everyone. The impression of cheapness and poor quality that characterized the Quonset hut was powerfully reinforced by the "boom factories." This one-two punch knocked respectability out of "prefabricated buildings" and may have forever saddled the term with negative connotations.

The metal building industry understood the problem. It was looking for another name.

1.1.4   Pre-engineered buildings

The scientific-sounding term *pre-engineered buildings* came into being in the 1960s. The buildings were "pre-engineered" because, like their ancestors, they relied upon standard engineering designs for a limited number of off-the-shelf configurations.

Several factors made this period significant for the history of metal buildings. First, the improving technology was constantly expanding the maximum clear-span capabilities of metal buildings. The first rigid-frame buildings introduced in the late 1940s could span only 40 ft. In a few years, 50-, 60-, and 70-ft buildings became possible. By the late 1950s, rigid frames with 100-ft spans were made.[5] Second, in the late 1950s, ribbed metal panels became available, allowing the buildings to look different from the old tired corrugated appearance. Third, colored panels were introduced by Stran-Steel Corp. in the early 1960s, permitting some design individuality. At about the same time, continuous-span cold-formed Z purlins were invented by Stran-Steel, the first factory-insulated panels were developed by Butler, and the first UL-approved metal roof appeared on the market.[1]

And last, but not least, the first computer-designed metal buildings also made their debut in the early 1960s. With the advent of computerization, the design possibilities became almost limitless. All these factors combined to produce a new metal-building boom in the late 1950s and early 1960s.

As long as the purchaser could be restricted to standard designs, the buildings could be properly called "pre-engineered." Once the industry started to offer custom-designed metal buildings to fill the particular needs of each client, the name "pre-engineered building" became somewhat of a misnomer. In addition, this term was uncomfortably close to, and easily confused with, the unsophisticated "prefabricated buildings," with which the new industry did not want to be associated.

Despite the fact that the term *pre-engineered buildings* is still widely used, and will be often found even in this book, the industry now prefers to call its product *metal building systems*.

1.2   Metal Building Systems

Why "systems"? Is this just one more application of the cyber-speak indiscriminately applied to describe everything made of more than one component? Nowadays, even the words "paint system" or "floor cleaning system" do not provoke a smile.

In all fairness, metal building system satisfies the classical definition of a system as an interdependent group of items forming a unified whole. In a modern metal building, the components such as walls, roof, main and secondary framing, and bracing are designed to work together. A typical assembly of a metal building is shown in Fig. 1.2. In addition to a brief discussion here, the roles played by various metal building components are examined in Chap. 3.

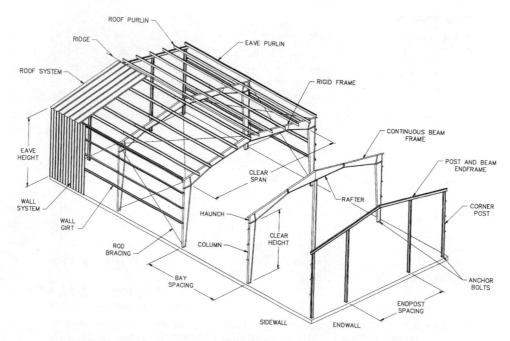

Figure 1.2 Typical components of a metal building system. (*Varco-Pruden Buildings.*)

A building's first line of defense against the elements consists of the wall and roof materials. These elements also resist structural loads, such as wind and snow, and transfer the loads to the supporting secondary framing. The secondary framing—wall girts and roof purlins—collects the loads from the wall and roof covering and distributes them to the main building frames, providing them with valuable lateral restraint along the way. The main structural frames, which consist of columns and rafters, carry the snow, wind, and other loads to the building foundations. The wall and roof bracing provides stability for the whole building. Even the fasteners are chosen to be compatible with the materials being secured and are engineered by the manufacturers.

The systems approach, therefore, is clearly evident. The term *metal building system* is proper and well-deserved. Over time, it will undoubtedly displace the still-common name, *pre-engineered buildings*.

1.3 Some Statistics

Today, metal building systems dominate the low-rise nonresidential market. According to the Metal Building Manufacturers Association (MBMA), pre-engineered structures comprised 65 percent of all new one- and two-story buildings with the areas of up to 150,000 ft^2 in 1995. The 1995 metal building sales of MBMA members totaled $2.21 billion; 355 million ft^2 of space was put in place. Large industrial buildings with areas of over 150,000 ft^2 added another 34.3 million ft^2 of new space.[6]

Figure 1.3 Metal building system in a commercial application. (*Photo: Bob Cary Construction.*)

Metal building systems are found in many applications. "Commercial" uses accounted for 36 percent of total 1995 sales. This category includes warehouses, small office buildings, garages, supermarkets, and retail stores (Figs. 1.3, 1.4, and 1.5). Another 40 percent of the sales were dedicated to "manufacturing"—factories, material recycling facilities, automotive and chemical plants (Figs. 1.6, 1.7, and 1.8). Some 12 percent of the metal buildings were used for "community" purposes: schools, town halls, and even churches (Fig. 1.9). The catch-all "miscellaneous" use includes everything else and, notably, agricultural buildings such as grain storage facilities, farm machinery, sheds, storage buildings, and livestock shelters.

1.4 The Advantages of Metal Building Systems

Most metal buildings are purchased by the private sector, which seems to appreciate the advantages of proprietary pre-engineered buildings more readily than the public entities. What are these advantages?

■ *Faster occupancy.* Anyone who has ever tried to assemble a piece of furniture can remember the frustration and the amount of time it took to comprehend

Figure 1.4 Office building of pre-engineered construction. (*Photo: Varco-Pruden Buildings.*)

the various components and the methodology of assembly. The second time around, the process goes much faster. A similar situation occurs at a construction site when a stick-built structure is being erected. The first time it takes a little longer..., but there is no second time to take advantage of the learning curve. With standard pre-engineered components, however, an experienced erector is always on familiar ground and is very efficient.

By some estimates, the use of metal building systems can save up to one-third of construction time. This time is definitely money, especially for private clients who can reap considerable savings just by reducing the duration of the inordinately expensive construction financing. It is not uncommon for small (around 10,000 ft^2) metal building projects to be completed in 3 months. By this time, many stick-built structures are just coming out of the ground.

- *Cost efficiency.* In a true systems approach, well-fitting pre-engineered components are assembled by one or only a few construction trades; faster erection means less expensive field labor. In addition, each structural member is designed for a near-total efficiency, minimizing waste of material. Less labor and less material translate into lower cost. The estimates of this cost efficiency vary, but it is commonly assumed that pre-engineered buildings are 10 to 20 percent less expensive than conventional ones. However, as is

Figure 1.5 Auto dealership housed in a metal building. (*Photo: Metallic Building Systems.*)

demonstrated in Chap. 3, some carefully designed stick-built structures can successfully compete with metal building systems.

■ *Flexibility of expansion.* Metal buildings are relatively easy to expand by lengthening, which involves disassembling bolted connections in the end-wall, removing the wall, and installing an additional clear-spanning frame in its place. The removed endwall framing can often be reused in the new location. Matching roof and wall panels are then added to complete the expanded building envelope.

■ *Low maintenance.* A typical metal building system, with prefinished metal panels and standing-seam roof, is easy to maintain: metal surfaces are easy to clean, and the modern metal finishes offer a superb resistance against corrosion, fading, and discoloration. Some of the durable finishes available on the market today are discussed in Chap. 6.

■ *Single-source responsibility.* The fact that a single party is responsible for the entire building envelope is among the main benefits of metal building systems. At least in theory, everything is compatible and thought through. The building owner or the construction manager does not have to keep track of many different suppliers or worry about one of them failing in the middle

Figure 1.6 A modern manufacturing facility made possible with metal building systems. (*Photo: Bob Cary Construction.*)

of construction. Busy small building owners especially appreciate the convenience of dealing with one entity if anything goes wrong during the occupancy. This convenience is a major selling point of the systems.

1.5   Some Disadvantages of Metal Buildings

An objective look at the industry cannot be complete without mentioning some of its disadvantages. As with any type of construction, metal building systems have a negative side that should be clearly understood and anticipated to avoid unwanted surprises.

- *Variable construction quality.* Most people familiar with pre-engineered buildings have undoubtedly noticed that all manufacturers and their builders are not alike. Major manufacturers tend to belong to a trade asso-

Figure 1.7   Material recycling facilities often utilize metal building systems. (*Photo: Maguire Group Inc.*)

ciation that promotes certain quality standards of design and manufacture. Some other suppliers might not accept the same constraints, and occasionally they provide buildings that are barely adequate, or worse. In fact, a structure can be put together with separately purchased metal-building components, but without any engineering—or much thought—involved. Such pseudo-pre-engineered buildings are prone to failures and give the industry a bad name. It is important, therefore, to know how to specify a certain level of performance, rather than to assume that every manufacturer will provide the quality desired for the project.

- *Lack of reserve strength.* The flip side of the fabled efficiency of the metal building industry is the difficulty of adapting existing pre-engineered buildings to new loading requirements. With every ounce of "excess" metal trimmed off to make the structure as economical as possible, any future loading modifications must be approached with extreme caution. Even the relatively small additional weight imposed by a modest rooftop HVAC unit or by a light monorail can theoretically overstress the structure designed "to the limit," unless structural modifications are considered.

- *Possible manufacturer's unfamiliarity with local codes.* When a metal building is shipped from a distant part of the country, its manufacturer might not

Figure 1.8 A large manufacturing facility housed in pre-engineered building. (*Photo: Varco-Pruden Buildings.*)

be as familiar with the nuances of the applicable building codes as a local contractor. While most major manufacturers keep a library of national and local building codes and train their dealers to communicate the nuances of the local codes to them, a few smaller operators might not. Owners should make certain that the building they purchase complies in all respects with the governing building codes, a task that requires some knowledge of both code provisions and manufacturing practices. (To be sure, some local codes might be based on obsolete editions of the model codes; for such cases the *MBMA Manual* could be a better design guide.)

The many advantages of metal building systems clearly outweigh a few shortcomings, a fact that helps explain the systems' popularity. Still, specifying pre-engineered buildings is not a simple process; it contains plenty of potential pitfalls for the unwary. Some of these are described in this book.

Figure 1.9 This community building utilizes a metal building system. (*Photo: Metallic Building Systems.*)

References

1. "MBMA: 35 Years of Leading the Industry," a collection of articles, *Metal Construction News,* July 1991.
2. Leslie H. Gillette, *The First 60 Years: The American Institute of Steel Construction, Inc. 1921–1980,* AISC, Chicago, Ill., 1980.
3. *Design/Specifiers Manual,* Butler Manufacturing Company, Roof Division, Kansas City, Mo., 1995.
4. *Metal Building Systems,* 2d ed., Building Systems Institute, Inc., Cleveland, Ohio, 1990.
5. "Ceco Celebrates 50th Year...," *Metal Construction News,* February 1996.
6. "1995 Metal Building Systems Sales Top $2 Billion in Low-Rise Construction," *Metal Construction News,* July 1996.

2

Industry Groups and Publications

2.1 Introduction

Metal building systems dominate the low-rise nonresidential building market, as demonstrated in Chap. 1. For dozens, if not hundreds, of manufacturers offering proprietary framing systems, profiles, and materials, a level-field competition is impossible without a common thread of standardization and uniformity. This unifying role has been traditionally handled by the industry organizations and trade associations.

A designer who is seriously involved in specifying metal building systems and who wants to become familiar with common industry practices will at some point need to review the design manuals and specifications promulgated by these groups. The specifier might even wish to follow the latest industry developments by becoming a member of the trade association or by subscribing to some of its publications. Sooner or later the designer will be faced with a question about the availability of a certain metal-building component, or about the feasibility of some nontraditional design approach—the questions that can be answered only by an industry expert. Also, were any disagreements to arise during construction, manufacturers and contractors would probably reference the trade literature to the specifiers to support their position.

For all these reasons, it is important to become familiar with the industry groups and their publications. While our list is not intended to be all-inclusive—in this dynamic industry, new organizations are being formed every year—it should prove helpful to anyone seeking further information about metal building systems.

2.2 Metal Building Manufacturers Association (MBMA)

2.2.1 The Organization

In the 1950s, the metal building industry was still disorganized and faced a host of problems ranging from building-code and insurance restrictions to union conflicts. The idea that the fledgling industry needed a trade organization was conceived by Wilbur Larkin of Butler Manufacturing Co., who invited his competitors to a meeting. MBMA was founded in Oct. 1, 1956, with 13 charter members. The charter members were Armco Drainage, Behlen Manufacturing, Butler Manufacturing, Carew Steel, Cowin & Company, Inland Steel, Martin Steel, Metallic Building, Pascoe Steel, Soule Steel, Steelcraft Manufacturing, Stran-Steel Corp., and Wonder Building. Wilbur Larkin of Butler Manufacturing was elected MBMA's first chairman.[1] The new group set out to forge the industry consensus on dealing with such common issues as code acceptance, design practices, safety, and insurance.

Today, MBMA consists of about 30 members, representing the best-known metal building manufacturers and over 9000 builders. Together, the group members account for about 9 out of 10 metal building systems built in this country. Recently, MBMA has opened its membership to industry suppliers, who may join as associate members to have greater access to MBMA programs and information.[2]

One of the most important roles played by the Association is providing engineering leadership to the manufacturers. Before MBMA was formed, each building supplier was using its own engineering assumptions and methods of analysis, a situation that resulted in a variable dependability of metal buildings. Development of the engineering standards was among the first steps taken by the new organization. Indeed, the MBMA Technical Committee was formed at the Association's first annual meeting on Dec. 4, 1956. Throughout the years, the MBMA's director of research and engineering served as the main technical representative of the industry. The Association's technical efforts, which have become especially intensive since the 1970s, were directly responsible for the increased sophistication of the manufacturers' engineering departments.

During its adolescence, the industry had to contend with a lack of building-code information about the behavior of one- and two-story buildings under wind loading. Since the low-rise buildings were the main staple of metal building manufacturers, new research was desperately needed. MBMA has risen to the occasion and, by teaming up with the American Iron and Steel Institute (AISI) and the Canadian Steel Industries Construction Council, sponsored in 1976 a wind-tunnel research program at the University of Western Ontario (UWO), Canada.[3] Wind-tunnel testing had been in existence for decades, but this program under the leadership of Dr. Alan G. Davenport was its first extensive application to low-rise buildings. The results of this testing have been incorporated into the 1986 edition of the *MBMA Manual*[5] and have contributed to the development of the wind-load provisions in ANSI 58 (now ASCE 7), Standard Building Code, and other codes around the world.

While most of the research work at the UWO was conducted between 1976 and 1985, wind studies there continue to this day. A case in point is a program called Random City, which is being conducted at the UWO Boundary Layer Wind Tunnel Laboratory. The Random City is a miniature model of a typical industrial town being subjected to a hurricane; the objective is to measure the wind forces acting on a typical low-rise building.[4] Similar studies are conducted at Clemson University, where standing-seam roof panels are being tested for dynamic wind forces, and at Mississippi State University, the site of experimental load simulation by electromagnets.

Research on snow loading also gets a share of MBMA's attention. For example, the effects of unbalanced snow loading on gable buildings are being studied at Rensselaer Polytechnic Institute with MBMA's sponsorship.

Another area of the MBMA-sponsored research concerns the thermal effects that solar radiation produces in metal roof and wall assemblies. This research has been conducted mainly at the University of Idaho between 1979 and 1983. Its results are summarized in two reports, "Part I: Extreme Surface Skin Temperatures Induced by Solar Radiation and Re-Radiative Cooling," and "Part II: Structural Response of Roof Systems." Both of these reports help in better understanding of the temperature-induced loads, and in a more realistic approach to building in the sunny locales.[3]

Since metal building systems are not inherently fireproof, the establishment of UL-listed assemblies involving the system components is critical to acceptance of the industry by building officials. MBMA has facilitated the progress on this front by sponsoring the fire-rating tests of the tapered steel columns and of metal-roof assemblies.

In addition to its role in the development of engineering standards for low-rise buildings, MBMA serves as a promotional arm of the metal building industry. The group publishes the *MBMA Fact Book* and the *Annual Market Review* and offers videos, slide presentation shows, and other promotional materials explaining the benefits of metal building systems.

The Association has been instrumental in expanding the scope of the Quality Certification Program, administered by the American Institute of Steel Construction (AISC), to include metal building manufacturers. Originally, the program was intended to certify structural steel fabricators by assuring consistently high quality throughout the entire production process. The new certification category MB (Metal Building Systems) is applicable to manufacturers of pre-engineered buildings "that incorporate engineering services as an integral part of the fabricated end product." The program objectives include evaluation of the manufacturer's design and quality assurance procedures and practices, certification of those manufacturers who qualify, periodic audits of the certified companies, and encouragement of others to adopt it. A certification by the well-known agency obviously enhances the manufacturer's image and facilitates acceptance of its system by local building officials.

The Metal Building Manufacturers Association is located at 1300 Sumner Avenue, Cleveland, OH 44115-2851; the telephone number is (216) 241-7333.

2.2.2   MBMA's Low Rise Building Systems Manual

Since its first edition in 1959, the *Manual*[5] has been a desktop reference source for metal building manufacturers and their engineers and builders. This 300-page book is of value to anybody involved with metal building systems. As an introduction to the first section, "Design Practices," states, the *Manual* reflects three decades of research applicable to all low-rise buildings.

The "Design Practices" along with its "Commentary" provides information on design criteria and structural loads, explains the methods of load application, and lists a number of standard load combinations. Section 11, "Design Examples," illustrates how to apply snow, wind, and seismic loads acting on low-rise buildings.

Another major part of the MBMA *Low Rise Building Systems Manual,* "Common Industry Practices," includes a diverse range of topics dealing with sale, design, fabrication, delivery, and erection of metal building systems, and with some insurance and legal matters. The specifiers of metal buildings should pay particularly close attention to Section 2, "Sale of a Metal Building System," that spells out in detail which parts and accessories are included in a standard metal building system package, and which are normally excluded.

The next section of the *Manual,* "Guide Specifications," is intended to be used as a guide in preparing contract specifications. The *Manual* also includes bibliography, nomenclature, and appendixes, which are quite useful for those who seek a deeper understanding of the *Manual*'s philosophy.

It is important to keep in mind that, while the *Manual* is widely used and respected, the information in it is presented from the standpoint of the manufacturers and is primarily intended to guide *them*. The outside specifiers have every right to modify the *Manual*'s criteria whenever those are in conflict with the project's requirements, and whenever there is a strong technical justification for doing so.

2.3   American Iron and Steel Institute (AISI)

The American Iron and Steel Institute has evolved from the American Iron Association, which was founded in 1855. Throughout the years, the Institute was instrumental in development of design codes and standards for a variety of steel structural members, occasionally crossing its ways with American Institute of Steel Construction (AISC). To avoid duplication, the two institutes have agreed to divide the applicability of their standards. Presently, the *AISC Manual* covers the design of hot-rolled structural steel members, which include the familiar wide-flange beams, angles, and channels. These members are cast and roll-formed to their final cross-sectional dimensions at steel mills, while still at an elevated temperature. The *AISC Manual* also covers plate girders fabricated from plates with thicknesses generally greater than $\frac{3}{16}$ in.

In contrast, steel members produced without heat application, or cold-formed, are in the AISI domain. Today, AISI is a recognized authority in the field of cold-formed construction. Cold-formed framing is made from steel sheet, plates, or flat bars by bending, roll forming, or pressing, and is usually confined to thinner materials. Some examples of cold-formed shapes used in construction include metal deck, siding, steel studs, joists, and purlins—the "meat and bones" of pre-engineered buildings. These structural members are usually less than $\frac{3}{16}$ in in thickness and are known as "light-gage" framing.

Most components of a typical metal building system, such as secondary members and wall and roof covering, are likely to be governed by the AISI provisions; the main steel frames, by AISC specifications.

While the *AISC Manual*[6] can be found on the bookshelf of most structural engineers, the *AISI Manual* is less known, perhaps because cold-formed structures have been traditionally designed outside of consulting engineering offices. Indeed, most consulting engineers deal predominantly with stick-built structures that utilize familiar off-the-shelf hot-rolled members.

The heart of the *AISI Manual* is Specification for the Design of Cold-Formed Steel Structural Members.[7] The Specification was first published in 1946 and has been frequently revised since, often drastically, reflecting the rapidly developing state of knowledge in cold-formed design. The original Specification was developed largely from AISI-funded research at Cornell University under Dr. George Winter and at other institutions in the late 1930s and early 1940s. The current code development is in the hands of the Committee on Construction Codes and Standards.

The Specification includes design procedures for various stiffened and unstiffened light-gage structural members, provides detailed design criteria for connections and bracing, and describes the required tests for special cases. The Specification's equations are used by manufacturers and fabricators of pre-engineered buildings, steel deck, siding, and steel studs, and are utilized in numerous nonbuilding applications such as steel vessels and car bodies. Some Specification provisions are discussed in Chap. 5.

The *AISI Manual* also contains a commentary and design examples explaining and illustrating the Specification.

In addition to publishing the *Manual*, AISI is involved in technical education efforts and promotional activities. The Institute's network of regional engineers is ready to answer technical questions from the specifiers and code officials. The AISI Construction Marketing Committee is actively promoting targeted areas of steel construction. A major marketing program undertaken by the committee that included direct mail, presentations at construction conventions, and one-on-one marketing was largely responsible for the huge success of metal roofing systems.

The Institute is also engaged in many other activities such as representing all of the steel industry before the lawmakers and the executive branch.

American Iron and Steel Institute is located at 1101 17th Street, NW, Suite 1300, Washington, DC 20036-4700; the telephone number is (202) 452-7100.

2.4 System Builders Association (SBA)

In contrast with MBMA and AISI, System Builders Association represents contractors and erectors of metal buildings. Formed in 1968 as Metal Building Dealers Association (MBDA), SBA now includes representatives from nearly every corner of the industry. The Association offers several membership categories for builders, independent erectors, metal roofing contractors, light-gage metal framers (engaged mostly in residential work), suppliers, and design professionals. Most of its activities take place at local chapters, where competitors by day join each other in the evening to discuss common problems and exchange information. According to SBA's brochures, its chapters have been instrumental in solving a variety of local zoning issues.

At the national level, SBA offers its members several standard legal forms, including "Standard Form of Agreement" between contractor and client, "Subcontract Agreement," and "Proposal-Contract." In addition, members in the builder and erector categories may receive free legal help on the matters dealing with contracts, liens, collection problems, and the like. System Builders Association works with Associated Builders and Contractors on a joint certification program for builders and erectors. SBA is in the award business, too, recognizing its annual Building of the Year, maintaining a Hall of Fame, and bestowing other honors. SBA-sponsored annual trade shows include exhibits, seminars, and social events. The Association publishes a magazine, *Systems Building Review,* which is mailed to prospective clients.

Systems Builders Association's address is 28 Lowry Drive, P.O. Box 117, West Milton, OH 45383-0117; the telephone number is 1-800-866-NSBA.

2.5 North American Insulation Manufacturers Association (NAIMA)

North American Insulation Manufacturers Association represents major manufacturers of fiberglass, rock wool, and slag wool insulation. NAIMA, which traces its roots to one of its predecessor organizations established in 1933, seeks to disseminate information on proper application, performance, and safety of insulation products. Like other similar trade groups, NAIMA conducts both technical-education and promotional affairs.

Since the group's interests go well beyond metal building systems, it is NAIMA's Metal Building Committee that sets performance standards and establishes testing programs for insulation products used in pre-engineered buildings.

Among the most valuable NAIMA's publications applicable to metal building systems are:

- Understanding Insulation for Metal Buildings
- ASHRAE 90.1 Compliance for Metal Buildings
- NAIMA 202 Standard

North American Insulation Manufacturers Association is located at 44 Canal Center Plaza, Suite 310, Alexandria, VA 22314; the telephone number is (703) 684-0084.

2.6 Building Systems Institute (BSI)

The Building Systems Institute, formed in 1984, is an umbrella organization comprising three major associations discussed above—Metal Building Manufacturers Association, American Iron and Steel Institute, and System Builders Association. In addition to regulatory issues, BSI produces educational and marketing materials for the metal building industry. Perhaps the most important of those is *Metal Building Systems,* a 232-page book that covers, among other topics, the origin and growth of the industry, building systems nomenclature, general design principles, energy considerations, and life-cycle costing. Since some of those subjects are beyond the scope of our book, the readers are well advised to read the BSI's publication. Since 1995, BSI coordinates an annual forum of all the industry organizations to improve and foster communication between the groups.

At present, Building Systems Institute shares its address and telephone number with MBMA.

2.7 Metal Construction Association (MCA)

Established in 1983, MCA was formed mainly for promoting the wider use of metal in construction.[3] MCA's best-known contribution to this goal is its annual *Metalcon International,* a major trade show that represents the entire metal building industry from around the world. MCA has its own Merit Award Program, bestowing honors on the projects it judges noteworthy, publishes a newsletter, and conducts market research.

MCA's market research activities include gathering and disseminating information on emerging and growing market segments and on promising new uses of metal components. The group's annual Metal Roof and Wall Panel Survey tracks use of metal panels by installed weight and square footage. To discuss a few specific areas of interest to only some of its members, MCA sponsors its Industry Councils—Light Frame, Construction Finishes, and Architectural Products/Metal Roofing and Siding. The membership is open to any person or company involved in the manufacture, engineering, sale, or installation of metal construction components.

Metal Construction Association is located at 11 S. LaSalle Street, Suite 1400, Chicago, IL 60603; the telephone number is (312) 201-0101.

2.8 National Roofing Contractors Association (NRCA)

Membership of this 100-year-old organization consists mostly of roofing contractors but also includes manufacturers, suppliers, consultants, and specifiers of roofing. NRCA offers a variety of educational programs, tests, and evaluations of new and existing roofing materials and disseminates technical information to its members. Rather than develop its own design standards or performance requirements, NRCA prefers to support other standard-writing bodies. Of particular interest to the specifiers of metal building systems is its *Commercial Low-Slope Roofing Materials Guide* that contains, among other useful data, a wealth of information about metal roofing manufacturers and their products. NRCA compiles this information by incorporating data submitted by the manufacturers and does not independently verify its validity. The guide's format greatly facilitates side-by-side product comparison. Another NRCA publication, *Residential Steep-Slope Roofing Materials Guide,* deals with the likes of asphalt shingles and clay tile. NRCA also publishes a monthly magazine, *Professional Roofing.*

NRCA is located at 10255 W. Higgins Road, Suite 600, Rosemont, IL 60018-5607; the telephone number is (708) 299-9070.

2.9 Light Gage Structural Institute (LGSI)

Most metal building systems manufacturers produce their own cold-formed metal building components, but there are also independent producers of roof purlins, eave struts, and wall girts used in pre-engineered buildings. For years, these producers felt underrepresented by the existing trade organizations. As was already mentioned, light-gage cold-formed construction is governed by rather complex and often-changing AISI Specification. After a Specification revision in 1986, several producers of light-gage framing felt the need to work together to address the major changes in Specification provisions. In 1989, they formed the Light Gage Structural Institute.

The main engineering result of the Institute's activities was a publication of its *Light Gage Structural Steel Framing System Design Handbook,*[8] which contains tables of design properties and allowable load-bearing capacities for typical C and Z steel sections produced by LGSI members. This information is quite valuable, as we shall see in Chap. 5.

Apart from producing technical information, LGSI is active in promoting quality of light-gage-framing manufacturing. Manufacturing plants of the member companies receive up to four unannounced annual inspections by LGSI's representatives. The inspectors verify thickness and material properties of the steel used by the manufacturer and perform product measurements for compliance with LGSI guidelines; a special sticker is affixed to each inspected steel bundle.

Light Gage Structural Institute can be contacted by writing to P.O. Box 866301, Plano, TX 75086-6301; the telephone number is (214) 618-3977.

2.10   Modern Trade Communications Inc.

Modern Trade Communications is best known for publishing three magazines that serve different segments of the metal building industry:

- *Metal Architecture,* of interest to architects and other specifiers of metal building systems

- *Metal Construction News* (formerly *Metal Building News*), the first tabloid-size industry magazine intended mostly for builders, manufacturers, and suppliers

- *Metal Home Digest,* dealing with residential applications of metal building systems

These three publications, especially *Metal Architecture,* should be of value to anyone interested in staying abreast of the latest industry developments.

Modern Trade Communications Inc. is located at 109 Portage Street, Woodville, OH 43469; the telephone number is (419) 849-3109.

References

1. "MBMA: 35 Years of Leading the Industry," a collection of articles, *Metal Construction News,* July 1991.
2. "MBMA Opens Membership to Industry Suppliers," *Metal Construction News,* February 1996.
3. "Industry's Associations Playing an Important Role...," *Metal Architecture,* 1994.
4. "MBMA Research Impacts Building Codes and Standards," *Metal Architecture,* May 1993.
5. *Low Rise Building Systems Manual,* 1986 ed. with 1990 Supplement, Metal Building Manufacturers Association, Inc., Cleveland, Ohio.
6. *Manual of Steel Construction, Allowable Stress Design,* American Institute of Steel Construction, Inc., Chicago, Ill., 1989.
7. Specification for the Design of Cold-Formed Steel Structural Members, American Iron & Steel Institute, Washington, D.C., 1986, with 1989 Addendum.
8. *Light Gage Structural Steel Framing System Design Handbook,* LGSI, Plano, Tex., 1995.

3

Selection of
Structural System

3.1 The Decision Time

How and when do we, architects and engineers, decide whether or not to specify metal building systems for the project? How and when do we compare the systems with the other available types of framing? These are not idle questions. Specified too early by enthusiastic designers, before the project requirements are fully established, metal building systems may end up being stretched beyond their optimum range of applications. Specified too late in the design process, the systems might prove incompatible with the project items that have already been designed.

Let us look briefly at the milestones of a typical building project, which starts when we are invited to help. Whether called by the client directly or responding to a formal Request for Proposals, we learn that the client has a problem to solve. The problem could be anything ranging from a lack of operating space to a need for new equipment.

During the first phase, *programming,* the problem is studied and analyzed. The program report summarizes the designers' recommendations on the amount of new space actually needed and establishes basic requirements for the proposed building. At this stage, it is too early to discuss structural systems, unless the only solution is already obvious.

During *conceptual design* and *preliminary design,* the program requirements are translated into a proposed layout, size, and mass of the building; various building code aspects are studied; and a preliminary cost estimate is prepared. This is the best time to get the structural engineers involved. Unfortunately, all too often the engineers are not brought on board until the preliminary design is completed, and an opportunity to influence the design decisions dealing with shape and clear span of the building is missed. Moreover, some large clients prefer to perform schematic design in-house.

Eli Cohen, one of the most respected engineers of our time, when asked about lessons he had learned, replied, "You have to spend more time in the conceptual design, because with the first 10 percent of your time you can save 25 percent of the cost of the building."[1]

At this point, the project can go in one of three directions:

1. *Conventional delivery.* The building is designed by an outside architect-engineer and later constructed by a general contractor selected via public bidding or negotiated process.

2. *Design-build.* The building is designed and constructed by a single entity that includes both designers and constructors.

3. *Pre-engineered construction.* A local builder, acting as a dealer for the metal building system manufacturer, contracts directly with the owner, who may or may not be assisted by an architect.

Obviously, selection of the third method indicates that a metal building system has been already chosen for the job. If, however, one of the first two delivery methods is pursued, a decision whether to use metal building systems, and of what type, will be made during the next design phase, *design development.* At that time, armed with the information about the building from the preliminary drawings, and after the building code research, structural engineers will determine the design loads on the structure and evaluate various framing alternatives.

To understand that process better, we have to refresh our structural engineering knowledge a bit; we refer to the structural basics explained in the next section throughout the rest of the book.

3.2 Structural Loads

The types of structural loads (or loads, for simplicity) listed below are commonly encountered in the design of metal building systems.

3.2.1 Dead load

Dead load is the weight of all permanent construction materials, such as roofing, framing, and other structural elements. Being well defined and known in advance, dead load is assigned a relatively low factor of safety in the ultimate (load factor) design. *Collateral* or *superimposed dead load* is a specific type of dead load that includes the weight of any materials other than the permanent construction. It may account for the weight of mechanical ducts, pipes, sprinklers, electrical work, future ceilings, and reroofing.

Collateral load is usually specified as an allowance of 5 to 10 lb/ft² (psf). It represents a maximum probable additional weight distributed over a relatively large area, and is therefore most useful in the design of roof girders and main frames. Roof purlins, however, should be carefully evaluated for the actual effect of any suspended loads, rather than for an averaged allowance. For example, if a sprinkler pipe is supported every 10 ft on a roof with a 5-ft purlin spacing, every other purlin will be subjected to a concentrated load, while its neighbors will not be. What is "average" about that?

Some argue exactly the opposite point, that collateral load is most suitable for purlins and may be excessive for the design of main framing and foundations. The building codes, however, do not provide any criteria for a reduction of collateral load for large tributary areas (i.e., the total areas supported by the member). Furthermore, it would seem that cold-formed purlins should not be designed for collateral load in excess of 5 lb/ft^2, if that, because it is very difficult to economically support heavy loads from the light-gage steel members and still satisfy deflection criteria.

Another variant of dead load is *equipment load,* which accounts for the weight of each specific piece of equipment supported by the roof or floor. The weight of any HVAC rooftop unit heavier than 50 lb, for example, is best represented by a concentrated downward force in the design of the supporting purlins. However, that same load could be "averaged out"—converted to a uniform collateral load—for the main framing design.

3.2.2 Live load

Live load is produced by building occupants, furniture, storage items, portable equipment, and partitions (some codes consider partitions as dead load). Owing to the fact that live load is relatively short-term, not easily predictable or quantifiable, it carries large factors of safety (uncertainty, really) in the ultimate design methods. Other sources of live load that arise during construction, repair, or maintenance of the building are even more difficult to predict and quantify.

To deal with this uncertainty, building codes have enacted conservative values for live loads—the framing must be designed to resist the loads which might occur only once or twice in the lifetime of the structure, if at all. For example, office buildings are normally designed for the live load of 50 lb/ft^2, while the actual weight of all the people and furniture in a typical office probably does not exceed 15 lb/ft^2.

It is quite probable that the design live load will occur in a relatively small area of the building at some time or another; it is much less probable that the whole floor will ever see that load. To reflect this reality, building codes set forth the rules governing the *live load reduction* for members supporting relatively large floor or roof areas. For metal building systems, *roof live load,* essentially an allowance for the roof loading during its construction and maintenance, is the load being reduced. With live load reduction, larger uniform loads are assigned to secondary members supporting limited roof areas than to primary structural framing.

The magnitude of roof live load is often compared to snow load and the larger value used in the design.

3.2.3 Snow load

Snow load represents the maximum weight of snow likely to collect on the roof. Unlike live load, snow load is independent of the building occupancy but is highly dependent on location. Local building codes specify the magnitude of "basic" snow load to be considered, but this is only the beginning.

This "basic" or "ground" snow load needs to be modified for roof configuration and exposure. Many codes specify the design roof snow load to be 70 percent of the ground snow load because some roof snow is removed by melting and blown off by wind. There are circumstances when the opposite is true: *more* snow may collect on a superinsulated and sheltered roof than on the ground. A judgment of experienced structural engineers should be sought for such cases.

When applicable, two other snow-related factors often prove critical: snow sliding and snow drift. Most people living in northern climates have watched snow sliding down a smooth pitched roof; this snow can slide on an adjacent roof below and add to the snow load on it.

Roof snow drifts against walls and parapets are another familiar sight for northerners. The amount of this *additional* snow load depends on the roof size, wall or parapet height, and other factors (Fig. 3.1). The extra weight from sliding and drifting snow is highly concentrated and cannot be averaged out over the whole roof. It follows that some elements of the roof structure must resist higher snow loads than others. Indeed, the roof areas adjacent to walls and high parapets are often designed for up to three times the snow loading elsewhere.

Another design condition that should be considered is unbalanced snow on gable roofs. The design provisions of various codes vary in this regard; the *MBMA Manual*[3] requires, for example, that all roofs with slopes greater than 3:12 but not more than 12:12 be designed for snow loading applied only to one slope.

Some codes specify yet another loading to be added to all other roof snow load: *rain-on-snow* load, which is intended to approximate the effects of a heavy rain following a major snowstorm, or of partial melting and accumulation of snow.

Incidentally, heavy rains can result in roof *ponding* conditions, which may become critical for flexible near-flat roofs with parapets. This is one of the rea-

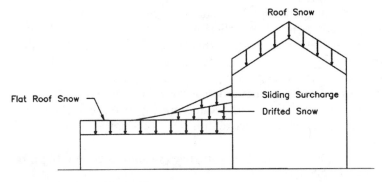

Figure 3.1 Snow load on buildings.

sons why metal building roofs are traditionally free-draining and without high parapets.

3.2.4   Wind load

Ever since being told about the sad experience of the three little pigs, most of us have an appreciation of the wind's destructive power. Several recent hurricanes, such as Hugo, Andrew (1989), and Iniki (1992), have highlighted our vulnerability to this common natural disaster. The property losses attributed to wind are enormous.

To design wind-resisting structures, the engineers need to know how to quantify the wind loading and distribute it among various building elements. Unfortunately, the wind effects on buildings are still not perfectly understood; the continuing research results in frequent building code revisions.

Most modern building codes contain maps specifying *design wind speed* in miles per hour for various locales. Design wind speed used to be defined as the fastest-mile wind speed measured at 33 ft above the ground and having an annual return probability of 0.02. The 1995 edition of ASCE 7,[2] however, defines it as the maximum three-second gust, reflecting a new method of collecting data by the National Weather Service. By using the code-provided formulas, it is possible to translate wind speed into a corresponding *velocity pressure* in pounds per square foot. From the velocity pressure, the design wind pressure on the building as a whole can be determined as a function of height and *exposure* category that accounts for local ground surface conditions.

Hurricane damage investigations reveal that local failures of walls and roofs occur most often near the building corners and roof eaves. The secondary members and covering in those areas should be designed for much higher wind loads—both inward and outward—than those in the rest of the building. The actual formulas for such an increase vary among the building codes and are not reproduced here, but the basic definition of the "salient corner" areas subjected to the higher wind loads is similar. Figure 3.2 illustrates the approach of ASCE 7[2] and *MBMA Manual.*[3]

Winds can damage buildings in four basic ways:

1. *Component damage,* when a part of the building fails. Some examples include a roof being blown off, wall siding torn out, or windows shattered.

2. *Total collapse,* when lack of rigidity or proper attachments causes the building to fall apart like a house of sticks.

3. *Overturning,* when the building stays in one piece and topples over, owing to insufficient weight and foundation anchorage.

4. *Sliding,* when the building stays in one piece but loses its anchorage and slides horizontally.

For a long time, engineers considered wind to be a strictly horizontal force and computed it by multiplying the velocity pressure by the projected area of the

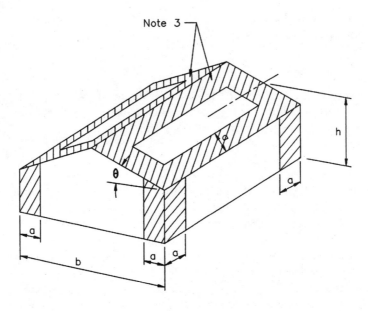

Notes:
1. The dimension "a" ("The Salient Corner" distance) is defined as the smaller of 0.1b or 0.4h (but not less than 0.04b nor 3 feet)

2. The dimension "h" is taken as mean roof height (when $\theta < 10°$, eave height may be used)

3. Areas adjacent to the ridge are included only when $10° < \theta \le 45°$

Figure 3.2 Areas of high localized wind loading for low-rise buildings.

building (Fig. 3.3a). As wind research progressed, often pioneered by the metal building industry, a more complex picture of the wind force distribution on gable buildings gradually became acknowledged (Fig. 3.3b). In the current thinking, the wind is applied perpendicular to all surfaces; both pressure and suction on the roof and walls are considered, as are internal and external wind pressures. Sorting out the various permutations of all these wind load components takes some practice and should be delegated to experienced professionals.

3.2.5 Earthquake load

Earthquake damage makes front-page news; even if not witnessed firsthand, devastating effects of the earth shaking appear uninvited on our living room TV screens, accompanied by familiar commentaries about the limitations of scientific knowledge in this area. Can we design seismically resistant buildings? The answer is yes, and we are learning how. As the forces of nature become better understood, building codes prescribe increasingly sophisticated methods of earthquake analysis. Still, the most basic notions of seismic design do not change, and it is worthwhile to review some of them.

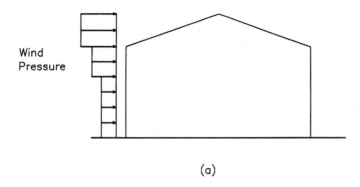

(a)

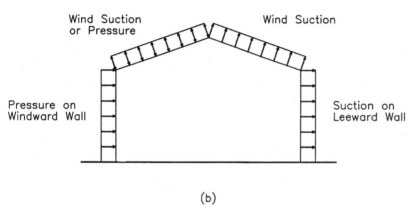

(b)

Figure 3.3 Wind load on gable buildings. (*a*) Old method of wind load application on projected area (obsolete); (*b*) current wind load application (normal to surface).

The first classic theory holds that the majority of earthquakes originate when two segments of the earth crust collide or move relative to each other. The movement generates seismic waves in the surrounding soil that are perceived by humans as ground-shaking; the waves diminish with the distance from the earthquake epicenter. The wave analogy explains why earthquakes are cyclical and repetitive in nature.

The second seismic axiom states that, unlike wind, earthquake forces are not externally applied. Instead, these forces are caused by inertia of the structure that tries to resist ground motions. As the earth starts to literally shift away from the building, it carries the building base with it, but inertia keeps the rest of the building in place for a short while. From Newton's first law, the movement between two parts of the building creates a force equal to the ground acceleration times the mass of the structure. The heavier the building, the larger the seismic force that acts on it.

Factors affecting the magnitude of earthquake forces on the building include the type of soil, since certain soils tend to amplify seismic waves or even turn to a liquidlike consistency (the liquefaction phenomenon). The degree of the building's rigidity is also important. In general terms, the design seismic force is inversely related to the fundamental period of vibration; the force is also affected by the type of the building's lateral load-resisting system.

The notion of ductility, or ability to deform without breaking, is central to modern seismic design philosophy. Far from being just desirable, ductility is fundamental to the process of determining the level of seismic forces. The building codes may not explicitly state this, but a certain level of ductility is required in order for the code provisions to be valid. Without ductility, the design forces could easily have been four or five times larger than those presently specified. The systems possessing ductile properties, such as properly detailed moment-resisting frames, may be designed for smaller seismic forces than those with less ductility, such as shear walls and braced frames. Why?

To answer this question, one needs to examine the goals of seismic design in general. Most building codes agree that the structures designed in accordance with their seismic code provisions should resist minor earthquakes without damage, moderate earthquakes without structural damage, but with some nonstructural damage, and major ones without collapse. Since the magnitude of the actual earthquake loads is highly unpredictable, the goal of collapse avoidance requires the structure to deform but not to break under repeated major overloads. The structure should be able to stretch well past its elastic region in order to dissipate the earthquake-generated energy.

To achieve this goal, the codes are filled with many prescriptive requirements and design limitations; a particular attention is given to the design details, since any disruption of the load path destroys the system.

It is important to keep in mind that real-life seismic forces are *dynamic* rather than static, even though their effects are commonly approximated in practice by a so-called equivalent static force method. This method is used partly for practicality, as dynamic analysis methods are quite cumbersome, and partly for comparison of the results to those of wind-load analysis and using the controlling loading to design against overturning, sliding, and other modes of failure discussed in Sec. 3.2.4.

The actual formulas for determination of seismic forces differ widely among the building codes and even among the various code editions. In general, these formulas start with the weight of the structure and multiply it by several coefficients accounting for all the factors discussed above.

3.2.6 Crane load

This type of loading is produced by cranes, monorails, and similar equipment. We consider crane loads in detail in Chap. 15.

3.2.7 Temperature load

This often-ignored and misunderstood load occurs whenever a steel member with fixed ends undergoes a change in temperature. A 100-ft-long piece of structural steel, free to move, expands by 0.78 in for each 100°F rise in temperature and similarly contracts when the temperature drops. If something prevents this movement, the expansion or contraction will not occur, but the internal stresses within the "fixed" member will rise dramatically. A basic formula for thermal stress increase in a steel element with fixed ends is

$$\text{Change in unit stress} = E \cdot \epsilon \cdot t$$

where E = modulus of elasticity of steel (29,000,000 lb/in^2)
ϵ (epsilon) = coefficient of linear expansion (0.0000065 in/°F)
t = temperature change (°F)

For example, if the temperature rises 50°F, a steel beam 50 ft long that is restrained from expansion will be under an additional stress of 29,000,000 × 0.0000065 × 50 = 9425 lb/in^2, a significant increase.

Temperature loads seldom present a problem either in conventional bolted-steel construction or in metal building systems. Indeed, the author is not aware of any building failures caused by temperature changes, although he has investigated a metal building system damaged by heat buildup due to fire—a separate issue. Thermal loading is insignificant and may often be ignored in the design of primary framing for small pre-engineered buildings, or of buildings located in the areas with relatively constant ambient temperatures, or of climate-controlled buildings. It may be more important for unheated fixed-base large-span structures with small eave heights located in climates with large temperature swings, or for cold-storage facilities. Thermal contraction of steel frames seems to be more dangerous than thermal expansion, because during contraction the steel is under tensile stress which, if large enough, can damage the connection bolts or welds. In any case, effects of temperature acting on exposed metal roofing seem to be more important than those of temperature acting on primary framing.

Many building codes, including the *MBMA Manual,* are silent on temperature loads. How much of a temperature change to assume in the design? The answer depends on the climate, building use, and insulation levels. If this loading is included at all, thermal stresses due to at least a 50°F rise or fall (100°F total variation) from the probable temperature at the time of the erection should be considered.

3.3   Methods of Design and Load Combinations

The loads discussed above need not be lumped together indiscriminately. It is highly improbable, for instance, that a once-in-a-lifetime hurricane will occur at the same time as a record snowfall. The odds of the roof live load, an allowance for infrequent roof maintenance and repair effects, being present during a major earthquake are similarly slim. To produce a realistic picture of combined loading on the structure, two approaches have been traditionally taken, reflected in the ultimate and the allowable stress designs methods.

3.3.1   Ultimate design method

In this method, also known as the strength design method, the loads are added together in various combinations, using *load factor* multipliers for each load and modifying the total by a "probability factor." The resulting combined load is then compared to the "ultimate" capacity of the structure. The load factors reflect a degree of uncertainty and variability of the loads, as was already mentioned. For steel design, this method is followed in the Load and Resistance Factor Design (LRFD) Specification for Structural Steel Buildings published by the American Institute of Steel Construction,[4] that contains a list of load combinations similar to those of ASCE 7.

The LRFD method of structural analysis provides a more uniform reliability than the allowable stress design discussed below and may take over the pre-engineered industry in the future. As of this writing, however, it does not yet have a widespread acceptance in the design community, and therefore it is not covered here in greater detail.

3.3.2   Allowable stress design method

In this method, some fractions of loads that represent perceived probabilities of the simultaneous load occurrence are added together in various combinations. The total stress level from the loads in each combination is then computed and compared with the allowable stress value (expressed as a function of yield stress for steel members). The allowable stress can usually be increased by one-third for wind or earthquake loading.

There is no universal agreement or a single best way to combine the loads acting on the building. Specifiers should follow the provisions of the governing building code, or, if not available, of a nationally recognized standard such as ASCE 7[2] or the *MBMA Manual*[3] modified for project conditions.

For single-story metal building systems, the following "basic" load combinations might be appropriate:

Dead + snow (or roof live load)

Dead + wind (or earthquake)

Dead + snow + earthquake

Dead + ½ wind + snow

Dead + wind + ½ snow

The dead load in these combinations should include collateral load if that increases the total effect. For example, the collateral load should be ignored for the uplift determination in the "dead + wind" combination but included when the wind acts downward. Thermal loading, not included in the above "basic" combinations, should be considered when appropriate, as discussed above. Both balanced and unbalanced snow loading should be considered in all the loading combinations involving snow.

Buildings with crane equipment require some additional load combinations:

Dead + crane

Dead + snow + crane

Dead + ½ wind (or full seismic) + crane

The MBMA recommends that the roof snow load in the combination of "dead + snow + crane" be taken as

- Zero for design snow loads less than 13 lb/ft^2
- One-half of design snow loads when their value is between 14 and 31 lb/ft^2
- Three-quarters of design snow loads greater than 31 lb/ft^2

Occasionally, projects may require that some nonstandard load combinations be considered, whether based on the local code provisions or on engineering judgment. In this case, the specifiers should bring the manufacturers' attention to this requirement early—at the bidding or negotiating stage—and be prepared to persevere in the face of some resistance to alter routine practice.

3.4   How Metal Buildings Work Structurally

3.4.1   Some building anatomy

Having refreshed our knowledge of structural loads, let us take the next step and briefly examine how building structures actually function. Far from attempting to conduct a comprehensive review of available structural systems, we focus on our main topic—metal building systems—and then look at some competing framing types. The goal of this chapter, after all, is to illustrate how and when to make an informed judgment about the suitability of pre-engineered buildings for a particular project.

A typical single-story metal building system is supported by *main frames* forming a number of bays (Fig. 1.2). *Bay size* is the space between frame centerlines measured along the sidewall. In the perpendicular direction, *frame clear span* is the clear distance between frame columns. At the roof level, *metal roof panels* form a weathertight enclosure and carry structural loads to *purlins,* the secondary structural members spanning between the main frames. Metal building systems can have a variety of wall materials, the original and still the most popular being metal siding, supported by sidewall or endwall *girts.*

Endwalls are commonly framed with *endwall columns,* which provide support for the girts and therefore are spaced at the intervals dictated by the girt's structural capacity. The endwall columns carry roof beams spanning from column to column, as in post-and-beam framing. If a future building expansion is planned, a regular main frame can be used instead of the endwall framing; the only function of the endwall columns then is girt support. During the future expansion, the columns are removed and one or more bays added.

3.4.2   Lateral stability of buildings in the frame direction

A building lacking lateral stability against wind and earthquake loads will not be standing for long. The most popular pre-engineered structure, *rigid frame,* relies on its own moment-resisting ability to laterally support the building (Fig. 3.4). Other frame systems, such as the familiar post-and-beam construction, do not possess such rigidity and may collapse like a house of cards if pushed laterally (Fig. 3.5a). Thus the second way to achieve lateral stability of the building is to provide *braced frames,* as shown in Fig. 3.5b. Proper bracing not only resists lateral loads but also stiffens the building in general, especially against the crane-induced loads, minimizes vibrations, and helps during building erection.

A typical design solution for metal building systems is to provide moment-resisting frames spanning the short direction of the building and braced frames in the exterior walls. The bracing located in the endwalls acts primarily in resisting lateral loads acting in the direction parallel to the frames, while the sidewall bracing resists the loads in the perpendicular direction. The *roof diaphragm,* usually a system of horizontal braces, distributes the loads between the lateral load-resisting elements (Fig. 3.6). A typical metal building roof diaphragm is made of diagonal steel rods and also includes roof purlins and beams designed for axial loading (Fig. 3.7). For large buildings and severe loads, steel pipes or tubes might be needed instead of the rods.

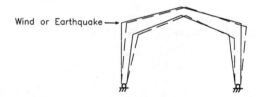

Figure 3.4   Rigid frame's moment-resisting ability.

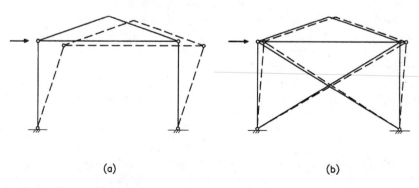

(a)   (b)

Figure 3.5   Post-and-beam frames. (*a*) Unbraced (unstable); (*b*) braced (stable).

When rigid frames are used in combination with endwall bracing, the roof diaphragm plays a relatively minor role in resisting wind or earthquake loads acting parallel to the frames, since its span is only one bay—between the frames. However, the roof diaphragm plays a critical role in buildings with non-rigid frame types such as a simple-span beam and bar joist system, where the diaphragm spans the distance between the endwalls.

While roof bracing represents the usual type of roof diaphragm found in metal building systems, the same result can be achieved by the rigidity of roof decking made of steel, wood, or concrete. Corrugated metal roof deck is probably the most common diaphragm used in conventional construction; it has its place in metal building systems as well. Through-fastened metal roofing operates on the same principle as metal deck, although it possesses a lesser degree of rigidity owing to the thinner metal gages.

Another method of resisting lateral loads—dispensing with bracing, rigid frames, and diaphragms altogether—takes advantage of the flagpole principle: Each column is designed as a cantilever fixed in the ground. The foundations and columns designed that way tend to become rather expensive, as will be discussed in the next chapter. This method is only infrequently used to replace the rigid frames but is often encountered in its *wind post* incarnation. Wind posts, exterior wall columns fixed at the base, can be used in lieu of wall bracing, where the latter is not desirable because of openings or appearance. Wind posts generate bending moments in the foundations and should be approached with caution.

3.4.3 Building stability in the direction perpendicular to frames

Rigid frames offer little or no lateral resistance normal to their plane, unless fixed at the base—an infrequent and often undesirable solution. Instead, stability in that direction is typically provided by sidewall bracing, spaced as

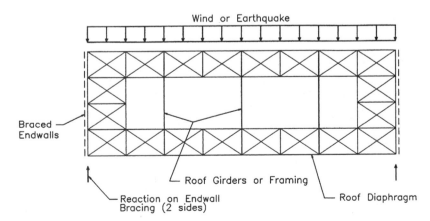

Figure 3.6 Roof diaphragm distributes lateral loads to braced endwalls.

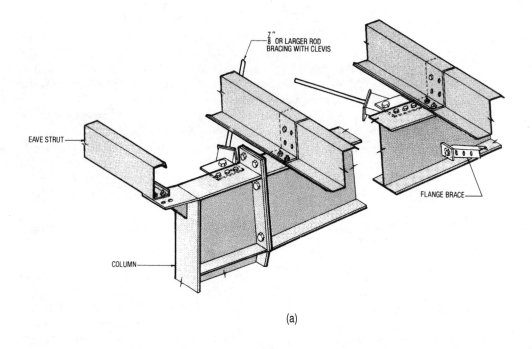

(a)

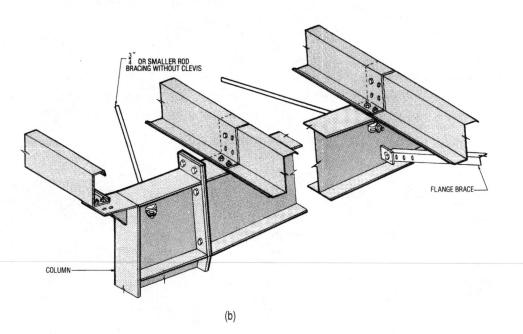

(b)

Figure 3.7 Typical roof diaphragm details. (*a*) With clevis, used with $^{7}/_{8}$ in or larger rods; (*b*) without clevis, with $^{3}/_{4}$ in or smaller rods. (*Star Building Systems.*)

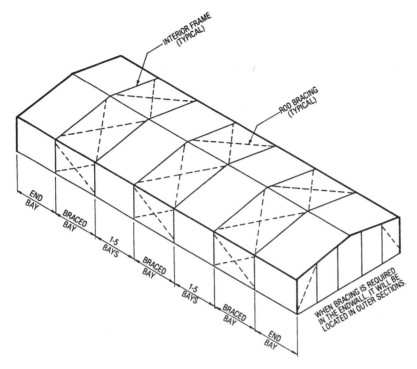

Figure 3.8 Typical bracing locations. (*Star Building Systems.*)

shown in Fig 3.8. A typical sidewall bracing bay consists of steel rod or cable diagonals, eave strut and columns on each side. The details of diagonal rod and cable bracing used by the industry are shown in Fig. 3.9. Essentially, concentrated loads from the bracing are transferred via hillside washers directly into the column webs.

Despite their widespread use, these details are not always appropriate, since the thin unreinforced webs are rarely checked for local bending from the point loads and may not survive the real load application. In fact, during a Feb. 17, 1994 Northridge earthquake in California, five out of six such connections in one building failed. The failure mechanism involved fracture of the hillside washers and, in some cases, subsequent pull-through of the rods.[5] (MBMA points out that, despite these anchorage failures, none of the buildings suffered more than trivial damage, while some buildings of other construction were severely damaged.) Until a better off-the-shelf product is developed, we recommend that for critical applications a custom-designed detail be used. One solution could consist of a steel reinforcing plate under a tapered washer. The plate would be field-cut and fitted between the column flanges, and welded to the column on three sides. Another solution is to utilize a separate foundation clip to which the bracing is attached (Fig. 12.25).

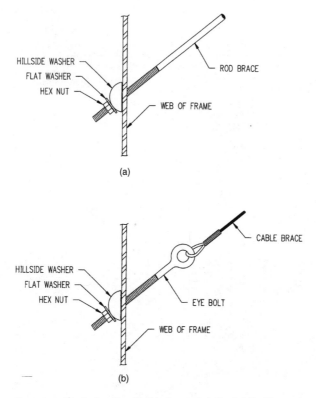

HILLSIDE WASHER
FLAT WASHER
HEX NUT
ROD BRACE
WEB OF FRAME

(a)

HILLSIDE WASHER
FLAT WASHER
HEX NUT
CABLE BRACE
EYE BOLT
WEB OF FRAME

(b)

Figure 3.9 Typical rod and cable brace details. (*a*) Rod brace to
frame detail; (*b*) cable brace to frame detail. (*Metallic Building
Systems.*)

If the bracing is architecturally undesirable in some bays, this fact should be
communicated to the manufacturer, who will propose other bracing options
such as *portal frames*. A portal frame is a small rectangular rigid frame that
fits between, and is attached to, the building columns (Fig. 3.10). A fixed-base
wind post approach, mentioned above, can also be taken. Various types of wall
bracing should not be combined in the same wall, unless a detailed relative-
rigidity analysis is first made.

Alternatively, concrete or masonry *shear walls* that actually possess higher
rigidity than the bracing may be used to provide lateral stability (Fig. 3.11).
While expensive to construct specifically for the bracing purpose, shear walls
cost very little in buildings with masonry or precast exteriors.

3.4.4 Load path

In a properly functioning building, structural loading is transferred between
various building elements, like a ball in a football game, until it is absorbed by
the soil or otherwise extinguished. This system of load transfer is known as the

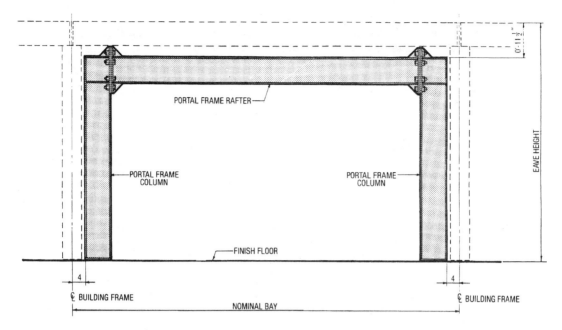

Figure 3.10 Portal frame in a sidewall. (*Star Building Systems.*)

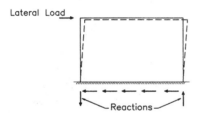

Figure 3.11 Shear wall.

load path. To illustrate its function, let's trace the path of a wind loading acting on a pre-engineered building's roof.

As Fig. 3.3b indicates, wind acts normal to the roof, either toward the surface (pressure) or away from it (uplift or suction). When wind pressure occurs, the roofing panels, the building's first line of defense, are pushed against the purlins and transfer the load by bearing. During wind uplift, the panels are pulled away from the roof; the fasteners holding them in place, if improperly designed, may fail and let the roofing fly. If the fasteners hold, the purlins get into action, transferring the load by flexure into the primary frames. Again, the connections must be adequate, or the whole assembly of the roofing and purlins will be in the air.

The primary frames, in turn, resist the load by bending and might also fail if either their strength or connections are deficient. If the frames hold, and the uplift load is not overcome by the weight of the structure, the load travels to the anchor bolts attaching the frames to the foundations. And finally, if the anchor bolts hold, the wind load is transferred to the foundation, which, hopefully, has sufficient weight to counteract the wind uplift. Otherwise the whole building might be lifted up like a giant tree with shallow roots.

The final load transfer occurs between the anchor bolts and the foundation and is typically not the responsibility of the metal building manufacturer. This leaves the outside architect or engineer to complete the final link of the load path and to design the foundations for the most critical loading effect, a task discussed in Chap. 12.

3.4.5   Bracing for stability of compression flange

The bracing discussed so far was for lateral resistance of *buildings*. Each flexural *member*—purlin, frame, truss, or joist—needs to be stable under load as well. It is a well-known phenomenon that the compression flange of members in bending tends to buckle laterally and must be restrained from doing so by proper bracing . Compression-flange bracing for primary framing members is usually provided by roof purlins, while the purlins, in turn, rely on the purlin bracing or through-fastened roofing.

To be effective, this type of bracing should be attached to the compression flange or near it. While the top-bearing purlins are certainly in the right place for this task, the common purlin bracing consisting of sag rods or sag angles is often attached to the purlin's web, some distance away from the compression flange. Whether this presents a problem or not depends on the size of the member, level of stress, and web thickness. A 5-ft-deep moment frame section may easily tolerate a 4-in eccentricity, while an 8-in-deep purlin may not.

The bracing should be designed for a compression load required to restrain the compression flange from buckling. The force to be resisted by the bracing is usually taken as 2 percent of the compressive flange force in simple-span members and is sometimes increased to 4 percent for continuous members. Brace stiffness should be carefully evaluated, since a deflected brace is essentially useless. Bracing of primary framing and purlins will be revisited in the following two chapters.

3.5   The Competition of Metal Building Systems

Our study of metal building systems will not be complete without a cursory review of the competition. Even the die-hard enthusiasts of pre-engineered buildings can benefit from an objective comparison with other framing systems, as there is no single most economical framing solution for all circumstances.

3.5.1 Open-web steel joists

One of the most economical contemporary framing systems consists of open-web steel joists carrying galvanized metal deck and supported on joist girders or wide-flange steel beams (Fig. 3.12). Open-web steel joists, popularly known as bar joists, are commonly made of double-angle chords (top and bottom horizontal members) and round bar diagonals. The joists are designed and built by their manufacturers in accordance with Steel Joist Institute Specifications,[6] often using proprietary steel design software. Utilizing high-strength steel, the open-web joists offer an exceptional strength-to-weight ratio. The joist system is ideal for roof framing supporting uniformly distributed loads, suspended ceilings, and mechanical ducts. The open-web design saves space by allowing passage of piping, conduits, and even small HVAC ducts. Concentrated loads present a problem, however, and should be applied only at the panel points, because the chord sections are usually rather weak in local bending and the joist shear capacity may be insufficient. Heavy point loads may require special joist design.

The joists are unstable during erection, and SJI specifications require several rows of bridging for lateral bracing. Once the bracing is properly secured and roof deck attached, the system is stable. The steel deck, together with perimeter steel beams, forms a horizontal roof diaphragm serving the same function as horizontal roof bracing in metal building systems. Joist spacing is governed by the roof deck's capacity.

Historically, bar joists have been used with relatively flat roofs, since large slopes present some difficulties in making sloped bearing seats, and require a careful structural analysis. Many engineers avoid this system when roof slope exceeds 45°.

The open-web joist system is very economical for spans ranging from 25 to 50 ft. Long-span steel joists are cost-effective for even longer spans, from 60 to 144 ft. For a rectangular column layout (a 30- by 40-ft grid is considered by many to be the most economical bay size), the joists normally run in the long direction (40 ft in a 30- by 40-ft bay).

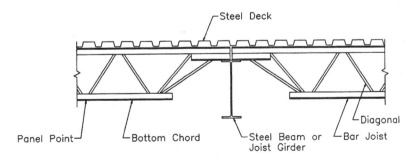

Figure 3.12 Open-web joist and steel deck system.

Open-web joists can be supported by either hot-rolled wide-flange beams or joist girders. The joist girders function on the same principle as bar joists—as a minitruss—but are commonly made of heavier all-angle sections. The panel points of joist girders coincide with bar joist locations. Joist girders are often preferred to wide-flange sections, especially for longer spans and for larger projects where more than a few are needed. The system is customarily supported on wide-flange or tubular columns and requires wall bracing for lateral stability.

3.5.2   Hot-rolled wide-flange beams

Wide-flange beam and girder system supporting steel roof deck should be familiar to most people involved in building construction. It is the simplest and most versatile of the framing systems, easily adaptable to any roof slope and accommodating suspended, concentrated, and axial loads with ease. The beams can be cantilevered and arranged in complex configurations for nonrectangular plans, altered or reinforced for localized loads. The flexibility has a price, of course. Unless some of these complications are actually present, steel beams are likely to be more expensive than bar joists. The beams become overly deep and heavy as the spans exceed about 40 ft.

Structural engineers have identified two ways to increase the efficiency of this system. The first one is a *continuous-beam* principle. Three simply supported beams (Fig. 3.13a) have higher maximum bending moments and larger vertical deflections than one continuous beam (Fig. 3.13b) covering the same three spans. Therefore, a three-span continuous beam is more efficient and requires less metal than three single-span beams.

Continuous framing has its limitations. Being statically indeterminate, it does not tolerate well any differential settlement of the supports, which can result in large secondary stresses threatening its integrity. Another problem is a possibility of large temperature stress buildup in a long single piece of metal. The design professional of record should carefully investigate a potential for problems caused by these two factors before specifying the continuous framing scheme.

A second way to increase the efficiency of the system lies in the *cantilevered-beam* scheme. Instead of one continuous beam, this framing consists of alternating cantilevered and simply supported beams (Fig. 3.14). The beam connections form *hinges*, designed not to transmit any bending moments. The length of cantilevers is selected to produce approximately equal negative

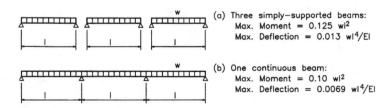

(a) Three simply—supported beams:
Max. Moment = 0.125 wl^2
Max. Deflection = 0.013 wl^4/EI

(b) One continuous beam:
Max. Moment = 0.10 wl^2
Max. Deflection = 0.0069 wl^4/EI

Figure 3.13 The efficiency of continuity.

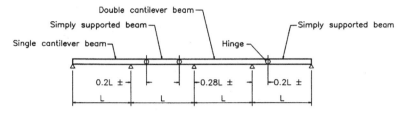

Figure 3.14 Cantilevered beam framing.

moments in the cantilevers and positive moments in the simply supported beams. This system is statically determinate and is less affected by differential settlement or by temperature stresses. The design success hinges, so to speak, on the proper joint design. Connections that are excessively rigid tend to convert this system into a continuous beam, with all its limitations.

Other avenues to efficiency could be taken by using the LRFD method (see Sec. 3.3.1) and by specifying high-strength steels.

3.5.3 Steel trusses

Long-span steel trusses have been used for many decades in both bridge and building structures. Prior to the advent of metal building systems, roof trusses were the framing of choice for industrial, warehouse, and commercial applications. Roof trusses have been, and still are, quite economical for clear spans ranging from about 40 to 140 ft. For example, a portal truss spanned some 132 ft over an existing building;[7] in some recent projects, trusslike open-web joists have been employed to span almost 200 ft. Steel trusses can be designed with a single or double slope of the top chord. The double slope results in a deeper section at midspan, structurally beneficial.

Historically, trusses supported steel wide-flange purlins, and truss spacing was limited by the length the purlins could span. Today's choices include not only those still popular sections but also open-web joists and even extra-deep steel roof deck. The optimum truss spacing, governed by the purlin capacity, is generally between 20 and 30 ft, the same as in pre-engineered buildings.

As in the previous two structural systems, lateral stability is provided by horizontal roof diaphragms made of either steel deck or diagonal bars, in combination with vertical wall bracing. In the past, trusses were commonly braced laterally by knee braces to the first panel point (Fig. 3.15a). This solution sacrificed some interior headroom and has lost its popularity in favor of a truss design with some depth at supports, incorporating the column into the trusswork (Fig. 3.15b). In the latter case, the column is erected and braced first, followed by relatively simple truss-to-column connections. The trusses are partially assembled in the shop to the maximum permissible shipping width, currently around 12 ft, and fully assembled in the field.

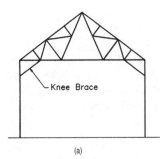
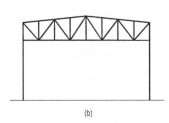

(a)   (b)

Figure 3.15   Roof trusses. (*a*) Fink truss; (*b*) Warren truss.

3.5.4 Hot-rolled steel rigid frames

A rigid frame, also known as a moment-resisting frame, consists of the column and beam sections rigidly joined together by moment-resisting connections. The resulting unified structure is stable and does not need bracing in its own plane. We have already mentioned rigid frame as the most popular structure for metal building systems, but the frames can be built by others, too.

In fact, low-rise buildings had traditionally utilized gable frames made of regular wide-flange members or tapered sections. The taper was achieved by cutting a wide-flange beam web at an angle, turning one section around and welding the webs together, or simply welding the frame from steel plates. With the benefits of pre-engineered construction becoming apparent, and with labor costs rising, custom fabrication of rigid frames fell into disrespect.

A recent report[8] states that an efficient steel fabricator in cooperation with an innovative structural engineer produced gable frames priced under the lowest bids from pre-engineered building manufacturers. Among other features, the engineer could accommodate column fixity in the sidewall direction into the foundation design, since both the foundation and the superstructure were designed in-house. This would have been difficult were the foundations and the frames designed by different parties (see discussion in Chap. 12). It remains to be seen whether this experiment can be successfully replicated.

Rigid frames offer many advantages over the other framing types, such as more effective use of steel than in simple beams, ease of maintenance and cleaning, as compared to trusses, and ability to support heavy concentrated loads. The disadvantages include relatively high material unit cost and susceptibility to differential settlement and temperature stresses. The frames produce horizontal reactions on the foundations, an additional design complication. And, as in any solid-web framing, pipes and conduits must be placed below the bottom flange unless expensive web openings are provided.

3.5.5 Other structural systems

There are many other types of structural framing which could, in some circumstances, be more appropriate for the project at hand than metal building systems. Lack of space precludes discussing all of them even in passing detail. Among the most popular systems we would mention the following:

- *Laminated wood arches.* These structures are similar in function to steel rigid frames and share the same design concepts with their steel brethren. The major advantage of wood arches lies in their charming and warm appearance, as opposed to a cold, utilitarian look of steel frames. Wood arches in combination with timber roof decking and masonry walls are used for churches, community buildings, and upscale residences. Laminated wood arches are cost-effective for spans from 30 to 70 ft. Roof slopes range from 3:12 to 14:12. This system offers a unique combination of beauty, strength, ease of installation, and cost efficiency. Among the disadvantages is the fact that wood, unlike steel, can rot and be infested by termites.

- *Precast concrete framing.* Precast concrete is heavier and usually more expensive than metal building systems, but in some circumstances factors like fire resistance and sound protection are more important. Concrete offers both. Precast, prestressed hollow-core planks are commonly available from 6 to 12 in deep and can span distances from 20 to 50 ft. Double-tee panels, 12 to 32 in in depth, are able to span distances from 12 to 100 ft. Precast roof panels are normally supported on precast concrete frames or masonry walls and rely on shear walls for lateral resistance. Building structures made completely of precast concrete offer some of the same advantages as metal building systems: speed of erection (some projects have been erected in $2\frac{1}{2}$ weeks—during the winter), single-source responsibility for structural work, and even flexibility of expansion.

- *Special construction.* Some truly extraordinary structural systems have been developed for applications requiring bold appearance, very long spans, and other unusual criteria. Suspension systems using exterior steel cables for roof support are more common for bridge applications but occasionally find their way into building construction as well. Air-supported fabric structures, such as the one used recently in Denver International Airport, offer a breathtaking way of covering massive amounts of space. The special structures can be used for clear spans in excess of 1000 ft. As the name suggests, specialists should be sought to help with this type of design.

3.6   Structural System Selection Criteria

Having discussed the topics of structural loads, design philosophies, and the available framing systems, we can at last consider some of the criteria for system selection.

3.6.1   Architectural requirements

The system of choice should satisfy both architectural and structural requirements; the relative importance of each needs to be established during the schematic design. It is wise to put in practice the timeless words of Louis Sullivan, "Form follows function." For most buildings in manufacturing and other "utilitarian" occupancies, such as factory, storage, and warehouse space, the harmony of function and structural form is obvious. For other uses such as churches, community, and commercial, architectural expression is probably of dominant importance and might override considerations of pure structural efficiency.

3.6.2   Fire resistance

Fire protection requirements specified in local building codes often dictate the choice of structure. Pre-engineered buildings of light-gage steel construction, trusses, and bar joists systems are difficult to cover with spray fireproofing; these should be specified with caution when the fireproofing is needed. Fortunately, this is rarely required for single-story buildings that conform to the "noncombustible, unprotected" classification.

3.6.3   Cost efficiency

Throughout the discussion in this chapter, we mentioned the optimum clear span ranges, advantages, and disadvantages of various systems. Where structural efficiency and cost are of paramount importance, these guidelines are intended to help an experienced practitioner to narrow down the system choices. However, the design team should choose a framing system that results in the lowest *overall* costs for the building, not just the least expensive structure. If a structural system penalizes other building systems, it may not be the bargain it appears to be.

3.6.4   Flexibility of use and expansion

The design team should carefully evaluate the owner's requirements for clear span, height, and building layout, and check this information against recently designed similar buildings. With the information and technology revolution in full swing, it is unlikely that a manufacturing plant being designed today will still contain the same production operations 20 years from now. On the other hand, a church layout may not change at all.

There is an obvious trade-off between cost efficiency and planning flexibility. For maximum flexibility, framing should be easy to remove, alter, or reinforce to accommodate future demands; all these are easiest to accomplish with simple-span framing. Of all the framing materials, hot-rolled structural steel beams are still the most adaptable. Unfortunately, the most economical building systems utilize continuity, multispan cantilevered beams, or prestressing. The owner and its team must decide whether it is wise to spend a little more now for a complete planning flexibility later.

3.6.5   Construction time

Frequently, the owner will put a premium on shortening duration of construction. This is understandable: Time is money. Several months shaved off the schedule may mean real savings on the construction financing, perhaps greater than the differences between the competing structural schemes. The framing with faster erection time (such as metal building systems) scores some extra points on this item.

3.6.6   Soil data

All too often, preliminary design and design development proceed without adequate geotechnical information; the engineers are expected to recommend a structural system without any soils data. This is quite unfortunate, because soil properties are crucial to the system selection. With good soil, economical spread footings are possible; with poor soil, expensive deep foundations might be called for. Much of the land still available near big cities probably has poor soils considered unsuitable for earlier development. A belated realization that expensive piles will be needed can kill a project with tight budgets. In such circumstances, the choice of a lightweight and flexible building system capable of tolerating some differential settlements can spell the difference between proceeding with the project or not.

3.6.7   Local practices

Prevailing local practices can weigh heavily on the system selection and should never be ignored. On the island of Guam, for example, most buildings are made of concrete; specifying laminated wood arches, however seemingly suitable for a new building, might raise many eyebrows. An abundance of local contractors skilled in a certain type of construction means that there will always be qualified people interested in submitting a bid with few contingencies. It probably also means that the needed materials are plentiful and inexpensive.

3.6.8   The choice

There are many other factors, often conflicting, that can influence a choice of structural system; most only remotely relate to structural issues. People skilled in making such decisions realize that selection of a structural system is more art than science. Various members of the design team may even initially disagree; in many cases, studies of alternative schemes accompanied by cost estimates are developed. Most importantly, the decision-making process should allow everyone involved to make their first and second choice of the systems that are later thoroughly analyzed and debated. The best solution is not always the most obvious.

Swensson and Robinson[9] tell about selection of a structural scheme for a large athletic facility. Four final schemes were considered: a basic gable metal building, a gable truss, a flat truss, and an arch. The designers eliminated the flat truss because it needed a much larger volume of air-conditioned space than others. The arches were ruled out as providing less workable finished space than the gabled frames. And finally, the gable truss system was chosen over the gable metal building despite its higher cost. Why? Because "the quality and flexibility of design provided by [this scheme] more than made up for the approximate 10% cost premium...."

With perceptions like this still widespread among engineers, it might be difficult to justify the selection of a pre-engineered building system for a high-visibility project. In the future, as the metal building industry continues to prove its mettle in nontraditional applications, and as its technical sophistication continues to increase, the quality of pre-engineered construction will likely rival that of stick-built structures. This book is but a small effort in this endeavor.

References

1. Cindi Crane, "Designing Buildings That Work," *Modern Steel Construction,* October 1994.
2. Minimum Design Loads for Buildings and Other Structures, ASCE Standard 7-95, American Society of Civil Engineers, New York, 1995.
3. *Low Rise Building Systems Manual,* 1986 ed. with 1990 Supplement, MBMA, Cleveland, Ohio.
4. Load and Resistance Factor Design Specification for Structural Steel Buildings, American Institute of Steel Construction, Inc., Chicago, Ill., 1986.
5. James R. Miller, "Performance of Pre-engineered Buildings in the CA Earthquake," *Metal Construction News,* July 1994.
6. Standard Specifications for Open Web Steel Joists, K-Series, Steel Joist Institute, Myrtle Beach, S.C., 1994.
7. Alexander Newman, "Boston Edison No. 514," *Modern Steel Construction,* no. 3, 1989.
8. Thomas Sputo, "Innovative Design of Gable Frame Buildings," *Modern Steel Construction,* May 1994.
9. Kurt D. Swensson and Douglas W. Robinson, "Field of Dreams," *Modern Steel Construction,* March 1995.

4

Primary Framing

4.1 Introduction

This chapter examines a palette of primary structural systems used in pre-engineered buildings. As discussed in the previous chapter, a complex process of choosing a framing system involves much more than structural considerations alone. Assuming that a metal building system is selected for the project at hand, the next milestone is choosing among the available types of pre-engineered primary framing. Proper selection of primary framing, the backbone of metal buildings, goes a long way toward a successful implementation of the design steps to follow. Some of the factors that influence the choice of main framing include:

- Dimensions of the building: width, length, and height
- Roof slope
- Required column-free clear spans
- Occupancy of the building and acceptability of exposed steel columns
- Proposed roof and wall materials

After all these factors are considered, the most suitable type of primary framing system frequently becomes obvious.

4.2 The Available Systems

Manufacturers call their framing systems many different names, often distilled into an alphabet soup of abbreviations. Still, only five basic types of metal building framing are currently on the market:

- Tapered beam
- Single-span rigid frame

- Multispan rigid frame
- Single-span and continuous trusses
- Lean-to

Each type can be supplied with either single or double roof slope. The most common primary frame systems are shown in Fig. 4.1. Primary framing is normally made either from high-strength steel conforming to ASTM A 572 with a minimum yield strength of 50,000 lb/in^2 or from common ASTM A 36 steel.

Each system has an optimum range of clear spans, as described below, but prior to that discussion we should first define the terms related to measurement of metal buildings. *Frame width* is measured between the outside surfaces of girts and eave struts, while the *clear span* is the distance between the inside faces of columns.* *Eave height* is measured between the bottom of the column base plate and the top of the eave strut; the *clear height* is the distance between the floor and the lowest point of the structure, usually the *rafter* (see Fig. 1.2).

How to dimension metal buildings on contract drawings? Manufacturers expect building width and length to be shown as the distances between the outside surfaces of wall girts (that plane is known as the *sidewall structural line*), not between the centerlines of exterior columns. Misunderstanding this convention leads to arguments between designers and manufacturers and to buildings being supplied in sizes slightly less than the designers had anticipated.

4.3 Tapered Beam

Tapered beam, also known as wedge beam or slant beam, is a logical extension of conventional post-and-beam construction into metal building systems. Indeed, what makes this system different from a built-up plate girder resting on two wide-flange columns are variability of the beam depth and rigidity of beam-to-column connections.

Most often, the beam is tapered by sloping the top flange for water runoff and keeping the bottom flange horizontal for ceiling applications (Fig. 4.1a). A less common version, reminiscent of a scissors truss, involves the beam with both flanges sloped. That configuration may be especially useful for the roof with a steep pitch used in combination with a low-slope cathedral ceiling. Usually the beams are spliced at midspan. Tapered-beam system is appropriate when:

- The frame width is between 30 and 60 ft, and eave height does not exceed 20 ft.
- Straight columns are desired (an important consideration for office and retail buildings with drywall interiors).
- The roofing material can tolerate a relatively low roof slope.

---

*The term "clear span" is occasionally misunderstood, despite its name, as some people measure it at the base and some at the widest point of the column such as the knee.

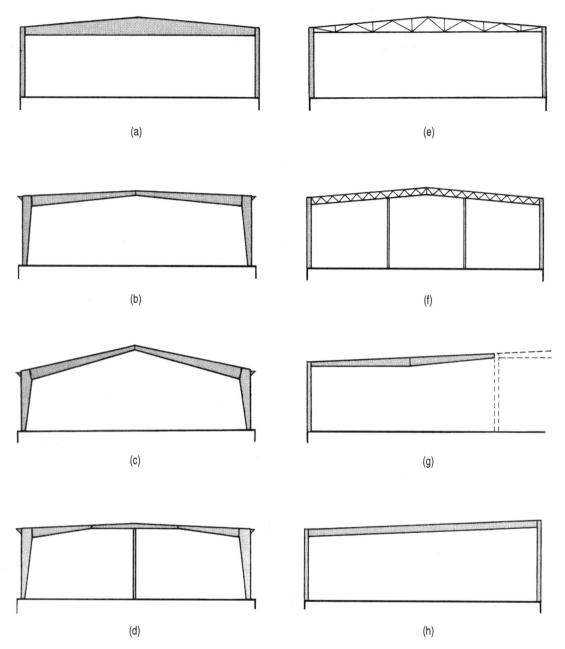

Figure 4.1 Types of primary frames. (*a*) Tapered beam; (*b*) single-span rigid frame, low profile; (*c*) single-span rigid frame, medium profile; (*d*) Multispan rigid frame; (*e*) single-span truss; (*f*) continuous truss; (*g*) lean-to; (*h*) Single-slope post-and-beam. (*Adapted from Star Building Systems design manual.*)

Tapered beams lose their attractiveness at spans exceeding 60 ft. Similarly, if the frame width is under 30 ft, standard hot-rolled framing might be less costly. Tapered-beam frames are typically specified for offices and small commercial and retail uses with moderate clear span requirements. Tapered beams are sometimes preferred for buildings with bridge cranes, because their bottom flanges, being horizontal, make for easy attachments and local reinforcing.

The system is shown in detail in Fig. 4.2. Typical frame dimensions for various spans and roof live loads are indicated in Fig. 4.3.

Design of tapered beams involves a frequently overlooked nuance. The manufacturers sometimes assume that beams in this system are connected to columns with "wind connections," rigid enough to resist lateral loads but flexible enough to behave under live loads in a single-span fashion. The question is, how realistic is this assumption? In structural steel design, as *AISC Manual of Steel Construction,* vol. II, *Connections*[1] points out, there are certain definitive criteria that these semirigid connections must satisfy. One such requirement is that "the connection material must have sufficient inelastic rotation capacity" to prevent it from failure under combined gravity and wind loads.

The manual's semirigid connection of choice is a pair of flexible clip angles that attach top and bottom beam flanges to the column (see Fig. 4.4). A flexible behavior of this connection has been experimentally demonstrated. On the

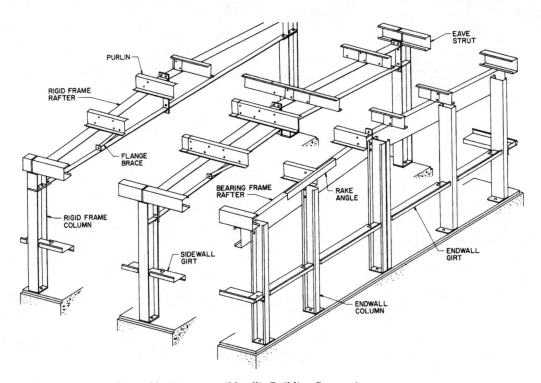

Figure 4.2 Details of tapered-beam system. (*Metallic Building Systems.*)

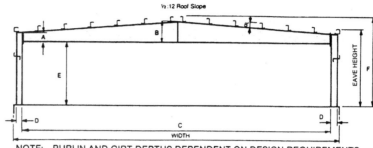

NOTE: PURLIN AND GIRT DEPTHS DEPENDENT ON DESIGN REQUIREMENTS.

WIDTH	EAVE HEIGHT (ACTUAL)	*20 PSF LL				30 PSF LL				40 PSF LL			
		B	C	D	E	B	C	D	E	B	C	D	E
30	9'-10"	1'-2"	28'-4"	8"	8'-7"	1'-9"	28'-3"	9"	8'-1"	1'-11"	28'-1"	10"	7'-10"
	11'-10"	1'-2"	28'-4"	8"	10'-7"	1'-9"	28'-3"	9"	10'-7"	1'-11"	28'-1"	10"	9'-10"
	13'-10"	1'-3"	28'-3"	8"	12'-6"	1'-9"	28'-3"	9"	12'-1"	1'-11"	28'-1"	10"	11'-10"
	15'-10"	1'-4"	28'-1"	9"	14'-6"	1'-9"	28'-3"	9"	14'-1"	1'-11"	28'-1"	10"	13'-10"
40	11'-10"	1'-7"	38'-3"	9"	10'-5"	2'-2"	38'-3"	9"	9'-10"	2'-6"	38'-1"	10"	9'-6"
	13'-10"	1'-7"	38'-1"	10"	12'-5"	2'-2"	38'-3"	9"	11'-10"	2'-6"	38'-1"	10"	11'-6"
	15'-10"	1'-7"	38'-3"	9"	14'-4"	2'-2"	38'-3"	9"	13'-10"	2'-6"	38'-1"	10"	13'-6"
	19'-10"	1'-9"	38'-1"	10"	18'-3"	2'-2"	38'-1"	10"	17'-10"	2'-6"	38'-1"	10"	17'-6"
50	11'-10"	2'-1"	48'-1"	10"	10'-2"	2'-6"	48'-1"	11"	9'-8"	2'-11"	47'-9"	1'-0"	9'-3"
	13'-10"	2'-1"	48'-1"	10"	12'-2"	2'-6"	48'-0"	11"	11'-8"	2'-11"	47'-9"	1'-0"	11'-3"
	15'-10"	2'-1"	48'-0"	10"	14'-2"	2'-6"	48'-1"	10"	13'-8"	2'-10"	47'-9"	1'-0"	13'-4"
	19'-10"	2'-1"	48'-1"	10"	18'-2"	2'-6"	47'-9"	1'-0"	17'-8"	2'-10"	47'-9"	1'-0"	17'-4"
60	13'-10"	2'-5"	58'-1"	10"	11'-11"	3'-2"	57'-8"	1'-1"	11'-3"	3'-4"	57'-8"	1'-1"	11'-1"
	15'-10"	2'-5"	58'-0"	10"	13'-11"	3'-0"	57'-9"	1'-1"	13'-5"	3'-4"	57'-8"	1'-1"	13'-1"
	19'-10"	2'-5"	58'-1"	10"	17'-11"	3'-0"	57'-8"	1'-0"	17'-5"	3'-2"	57'-8"	1'-1"	17'-3"
	23'-10"	2'-5"	57'-8"	1'-0"	21'-11"	3'-0"	57'-8"	1'-0"	21'-5"	3'-2"	57'-8"	1'-1"	21'-3"
70	13'-10"	2'-11"	67'-8"	1'-1"	11'-9"	3'-8"	67'-8"	1'-1"	10'-11"	4'-0"	67'-8"	1'-1"	10'-7"
	15'-10"	2'-11"	67'-8"	1'-1"	13'-9"	3'-8"	67'-8"	1'-1"	12'-11"	4'-0"	67'-8"	1'-1"	12'-7"
	19'-10"	2'-11"	67'-8"	1'-0"	17'-9"	3'-6"	67'-8"	1'-1"	17'-1"	4'-0"	67'-8"	1'-1"	16'-7"
	23'-10"	2'-11"	67'-8"	1'-0"	21'-8"	3'-7"	67'-9"	1'-1"	21'-1"	4'-0"	67'-8"	1'-1"	20'-7"

*12 PSF LL FRAME

Dimensions shown are for 25' bays. 20' and 30' bays also available.
Building components and dimensions shown are subject to change due to final design.

Figure 4.3 Typical dimensions of tapered-beam system. (*American Buildings Co.*)

other hand, in metal building systems members are normally connected to each other by through-bolted end plates as shown in Fig. 4.2. This type of joint is rather rigid and does not pass the flexibility muster for "wind connections."

If a joint lacks reserve rotational capacity, it is considered nearly rigid and thus capable of transmitting bending moments across the interface. The ends of the beam so connected to columns will have some bending moments partial-

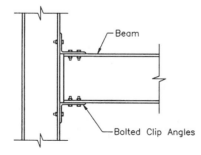

Figure 4.4 Conventional semi-rigid connection.

ly transmitted to the columns. If a simple-span assumption was made by the manufacturer, the columns were not designed for this bending moment. This dangerous oversimplification could conceivably result in the columns becoming overloaded. Whenever a tapered-beam system is proposed, it is wise to investigate the manufacturer's approach to this issue.

There are certain steps manufacturers can take to increase the connection flexibility. One solution might be to introduce compressible deflection pads shown in Fig. 4.11 (in a context of another system) that absorb some movement of the beam's ends.

The specifiers could of course sidestep the entire issue by simply adding a note to the contract documents stating that all beam-to-column connections in the tapered-beam system are to be considered rigid for the design purposes, but this approach might result in overly heavy column sections.

4.4 Single-Span Rigid Frame

If a tapered-beam system is a carryover from conventional construction, the single-span gabled rigid frame (Fig. 4.1*b* and *c*) is a quintessential pre-engineered product. Indeed, one reason for the success of the metal-building industry is the rigid frame. In contrast with the tapered-beam system, the single-span rigid frame is designed to take full advantage of connection rigidity: The frame members are tapered following the shape of the bending-moment diagram.

The deepest part of the frame is the *knee,* a joint between the beam and the column. For a two-hinge frame, the usual version of the system, the frame section is most shallow approximately midway between the knee and the ridge (Fig. 4.5); for a less common three-hinge frame, the most shallow section occurs at the ridge (Fig. 4.1*b*). The splices are made at the knee, at the ridge, and depending on the frame width, perhaps elsewhere in the rafter.

The main reason for the popularity of the gabled rigid-frame system lies in its cost efficiency—it requires less metal than most other structural systems of the same span and eave height. As McGuire[2] has demonstrated and as can be easily verified, a two-hinge gabled rigid frame spanning 60 ft, with the eave height of 14 ft plus the gable height of 10 ft, is 19 percent more efficient than a similar flat-roof rigid frame, and an incredible 53 percent more efficient than a statically determinate frame designed on the simple-span principle. This framing system is appropriate when:

■ Frame width is between 60 and 120 ft. Both smaller and larger spans are increasingly less economical.

■ Eave height is between 10 and 24 ft.

■ Tapered columns are acceptable.

■ Headroom at the exterior walls is not critical.

Single-span rigid frames can be classified as being high profile (slope 4:12), medium profile (slope 2:12), and low profile (slope from $\frac{1}{4}$:12 to 1:12). The

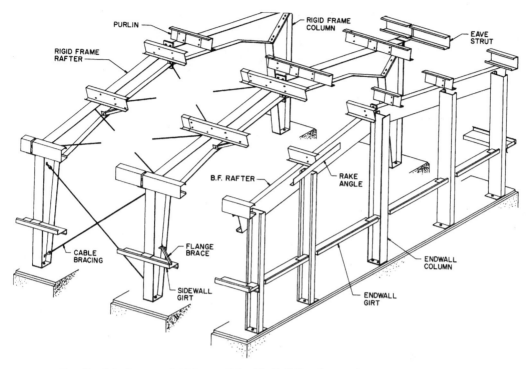

Figure 4.5 Details of single-span rigid frame. (*Metallic Building Systems.*)

frames of high profile are especially suitable for the roofing that requires substantial roof slope and for the applications demanding large clear heights near the midspan. The inward-tapered columns are the norm, but some other column configurations are possible for special conditions (Fig. 4.6).

The single-span rigid-frame system is extensively used anywhere an unobstructed working space is desired. It is suitable for such diverse applications as auditoriums, gymnasiums, aircraft hangars, showrooms, churches, recreational facilities, and industrial warehouses (Fig. 4.7).

While the frame width is best kept between 60 and 120 ft, single-span frames over 200 ft wide can be built for the cases where planning flexibility is paramount. The tables indicating typical dimensions of single-span rigid frames can be found in Figs. 4.8 and 4.9.

4.5 Multispan Rigid Frame

Multispan rigid frame, also known as continuous-beam, post-and-beam, or modular frame (Fig. 4.1d), utilizes the same design principles as single-span rigid frame. The multiplicity of spans allows for a theoretically unlimited building size, although in reality a buildup of thermal stresses requires that expansion joints be used for buildings wider than 300 ft.

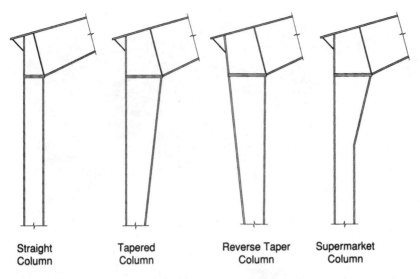

| Straight Column | Tapered Column | Reverse Taper Column | Supermarket Column |

Figure 4.6 Various column profiles. (*Varco-Pruden Buildings.*)

Figure 4.7 This open-front industrial facility is ideally suited to single-span rigid-frame construction. (*Photo: Maguire Group Inc.*)

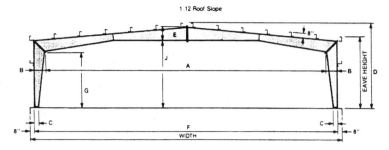

NOTE: PURLIN AND GIRT DEPTHS DEPENDENT ON DESIGN REQUIREMENTS.

WIDTH	EAVE HEIGHT (ACTUAL)	*20 PSF LL					30 PSF LL					40 PSF LL				
		A	D	E	G	J	A	D	E	G	J	A	D	E	G	J
30	10'-0"	27'-1"	11'-3"	9"	8'-8"	9'-10"	25'-11"	11'-3"	1'-0"	8'-1"	9'-7"	25'-7"	11'-3"	1'-2"	7'-11"	9'-5"
	12'-0"	27'-1"	13'-3"	9"	10'-8"	11'-10"	25'-11"	13'-3"	1'-0"	10'-1"	11'-7"	25'-7"	13'-3"	1'-2"	9'-11"	11'-5"
	14'-0"	27'-1"	15'-3"	9"	12'-8"	13'-10"	25'-11"	15'-3"	1'-0"	12'-1"	13'-7"	25'-7"	15'-3"	1'-2"	11'-11"	13'-5"
	16'-0'	27'-1"	17'-3"	9"	14'-8"	15'-10"	25'-11"	17'-3"	1'-0"	14'-1"	15'-7"	25'-7"	17'-3"	1'-2"	13'-11"	15'-5"
	20'-0"	26'-9"	21'-3"	9"	18'-6"	19'-10"	25'-11"	21'-3"	1'-0"	18'-1"	19'-7"	25'-7"	21'-3"	1'-2"	17'-11"	19'-5"
40	10'-0"	36'-1" *	11'-8"	9"	8'-2"	10'-3"	35'-1"	11'-8"	1'-2"	7'-8"	9'-10"	34'-7"	11'-8"	1'-5"	7'-6"	9'-7"
	12'-0"	36'-1"	13'-8"	9"	10'-2"	12'-3"	35'-1"	13'-8"	1'-2"	9'-8"	11'-10"	34'-7"	13'-8"	1'-5"	9'-6"	11'-7"
	14'-0"	36'-1"	15'-6"	9"	12'-2"	14'-3"	35'-1"	15'-8"	1'-2"	11'-8"	13'-10"	34'-7"	15'-8"	1'-5"	11'-6"	13'-7"
	16'-0"	36'-1"	17'-8"	9"	14'-2"	16'-3"	35'-1"	17'-8"	1'-2"	13'-8"	15'-10"	34'-7"	17'-8"	1'-5"	13'-6"	15'-7"
	20'-0"	36'-1"	21'-8"	9"	18'-2"	20'-3"	35'-1"	21'-8"	1'-2"	17'-8"	19'-10"	34'-7"	21'-8"	1'-5"	17'-6"	19'-7"
50	12'-0"	45'-3"	14'-1"	1'-2"	9'-9"	12'-3"	43'-11"	14'-1"	1'-11"	9'-2"	11'-6"	43'-10"	14'-1"	1'-11"	9'-2"	11'-6"
	14'-0"	45'-3"	16'-1"	1'-2"	11'-9"	14'-3"	43'-11"	16'-1"	1'-11"	11'-2"	13'-6"	43'-10"	16'-1"	1'-11"	11'-2"	13'-6"
	16'-0"	45'-3"	18'-1"	1'-2"	13'-9"	16'-3"	43'-11"	18'-1"	1'-11"	13'-2"	15'-6"	43'-10"	18'-1"	1'-11"	13'-2"	15'-6"
	20'-0"	45'-3"	22'-1"	1'-2"	17'-9"	20'-3"	43'-11"	22'-1"	1'-11"	17'-2"	19'-6"	43'-10"	22'-1"	1'-11"	17'-2"	19'-6"
	24'-0"	45'-3"	26'-1"	1'-2"	21'-9"	24'-3"	43'-11"	26'-1"	1'-11"	21'-2"	23'-6"	43'-10"	26'-1"	1'-11"	21'-2"	23'-6"
60	12'-0"	55'-2"	14'-6"	1'-5"	9'-9"	12'-5"	53'-4"	14'-6"	2'-1"	8'-11"	11'-9"	52'-8"	14'-6"	2'-1"	8'-7"	11'-9"
	14'-0"	55'-2"	16'-6"	1'-5"	11'-9"	14'-5"	53'-4"	16'-6"	2'-1"	10'-11"	13'-9"	52'-8"	16'-6"	2'-1"	10'-7"	13'-9"
	16'-0"	55'-2"	18'-6"	1'-5"	13'-9"	16'-5"	53'-4"	18'-6"	2'-1"	12'-11"	15'-9"	52'-8"	18'-6"	2'-1"	12'-7"	15'-9"
	20'-0"	55'-2"	22'-6"	1'-5"	17'-9"	20'-5"	53'-4"	22'-6"	2'-1"	16'-11"	19'-9"	52'-8"	22'-6"	2'-1"	16'-7"	19'-9"
	24'-0"	55'-2"	26'-6"	1'-5"	21'-9"	24'-5"	53'-4"	26'-6"	2'-1"	20'-11"	23'-9"	52'-8"	26'-6"	2'-1"	20'-7"	23'-8"
70	12'-0"	63'-11"	14'-11"	1'-9"	9'-2"	12'-7"	62'-6"	14'-11"	2'-5"	8'-7"	11'-10"	61'-10"	14'-11"	2'-9"	8'-3"	11'-6"
	14'-0"	63'-11"	16'-11"	1'-9"	11'-2"	14'-6"	62'-6"	16'-11"	2'-5"	10'-7"	13'-10"	61'-10"	16'-11"	2'-9"	10'-3"	13'-6"
	16'-0"	63'-11"	18'-11"	1'-9"	13'-2"	16'-6"	62'-6"	18'-11"	2'-5"	12'-6"	15'-10"	61'-10"	18'-11"	2'-9"	12'-3"	15'-6"
	20'-0"	63'-11"	22'-11"	1'-9"	17'-2"	20'-6"	62'-6"	22'-11"	2'-5"	16'-6"	19'-10"	61'-10"	22'-11"	2'-9"	16'-3"	19'-6"
	24'-0"	63'-11"	26'-11"	1'-9"	21'-2"	24'-6"	62'-6"	26'-11"	2'-5"	20'-6"	23'-10"	61'-10"	26'-11"	2'-9"	20'-3"	23'-5"
80	12'-0"	73'-7"	15'-4"	1'-11"	9'-0"	12'-9"	71'-10"	15'-4"	2'-7"	8'-3"	12'-1"	71'-2"	15'-4"	2'-10"	7'-11"	11'-10"
	14'-0"	73'-6"	17'-4"	1'-11"	11'-0"	14'-9"	71'-10"	17'-4"	2'-7"	10'-3"	14'-1"	71'-2"	17'-4"	2'-10"	9'-11"	13'-10"
	16'-0"	73'-6"	19'-4"	1'-11"	13'-0"	16'-9"	71'-10"	19'-4"	2'-7"	12'-3"	16'-1"	71'-2"	19'-4"	2'-10"	11'-11"	15'-10"
	20'-0"	73'-6"	23'-4"	1'-11"	17'-0"	20'-9"	71'-10"	23'-4"	2'-7"	16'-3"	20'-1"	71'-2"	23'-4"	2'-10"	15'-11"	19'-9"
	24'-0"	73'-6"	27'-4"	1'-11"	21'-0"	24'-9"	71'-10"	27'-4"	2'-7"	20'-3"	24'-1"	71'-2"	27'-4"	2'-10"	19'-11"	23'-9"
100	14'-0"	92'-2"	18'-2"	2'-4"	10'-5"	15'-1"	90'-10"	18'-2"	2'-10"	9'-9"	14'-8"	89'-10"	18'-2"	3'-5"	9'-4"	14'-1"
	16'-0"	92'-2"	20'-2"	2'-4"	12'-5"	17'-1"	90'-10"	20'-2"	2'-10"	11'-9"	16'-7"	89'-10"	20'-2"	3'-5"	11'-4"	16'-1"
	20'-0"	92'-2"	24'-2"	2'-4"	16'-5"	21'-1"	90'-10"	24'-2"	2'-10"	15'-9"	20'-7"	89'-10"	24'-2"	3'-5"	15'-3"	20'-0"
	24'-0"	92'-2"	28'-2"	2'-4"	20'-5"	25'-1"	90'-10"	28'-2"	2'-10"	19'-9"	24'-7"	89'-10"	28'-2"	3'-5"	19'-4"	24'-0"
120	14'-0"	110'-10"	19'-0"	2'-10"	9'-9"	15'-6"	109'-10"	19'-0"	3'-4"	9'-4"	15'-0"	108'-10"	19'-0"	4'-2"	8'-10"	14'-2"
	16'-0"	110'-10"	21'-0"	2'-10"	11'-9"	17'-6"	109'-10"	21'-0"	3'-4"	11'-4"	17'-0"	108'-10"	21'-0"	4'-2"	10'-10"	16'-2"
	20'-0"	110'-10"	25'-0"	2'-10"	15'-9"	21'-6"	109'-10"	25'-0"	3'-4"	15'-4"	21'-0"	108'-10"	25'-0"	4'-2"	14'-10"	20'-2"
	24'-0"	110'-10"	29'-0"	2'-10"	19'-9"	25'-6"	109'-10"	29'-0"	3'-4"	19'-4"	25'-0"	108'-10"	29'-0"	4'-2"	18'-10"	24'-2"
	30'-0"	110'-10"	35'-0"	2'-10"	25'-9"	31'-6"	109'-10"	35'-0"	3'-4"	25'-4"	31'-0"	108'-10"	35'-0"	4'-2"	24'-10"	30'-2"

*12 PSF LL FRAME

Dimensions shown are for 25' bays, 20' and 30' bays also available
Building components and dimensions shown are subject to change due to final design.

Figure 4.8 Typical dimensions of low-profile single-span rigid frame. (*Metallic Building Systems.*)

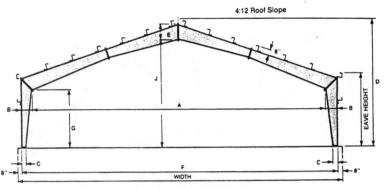

NOTE: PURLIN AND GIRT DEPTHS DEPENDENT ON DESIGN REQUIREMENTS.

WIDTH	EAVE HEIGHT (ACTUAL)	*20 PSF LL					30 PSF LL					40 PSF LL				
		A	D	E	G	J	A	D	E	G	J	A	D	E	G	J
30	9'-10"	26'-11"	14'-10"	7"	8'-8"	13'-6"	25'-11"	14'-10"	10"	8'-4"	13'-3"	25'-7"	14'-10"	1'-1"	8'-3"	13'-0"
	11'-10"	26'-11"	16'-10"	7"	10'-8"	15'-6"	25'-11"	16'-10"	10"	10'-4"	15'-3"	25'-7"	16'-10"	1'-1"	10'-3"	15'-0"
	13'-10"	26'-11"	18'-10"	7"	12'-8"	17'-6"	25'-11"	18'-10"	10"	12'-4"	17'-3"	25'-7"	18'-10"	1'-1"	12'-3"	17'-0"
40	9'-10"	36'-3"	16'-6"	9"	8'-5"	15'-0"	34'-7"	16'-6"	1'-3"	7'-10"	14'-6"	34'-1"	16'-6"	1'-6"	7'-8"	14'-3"
	11'-10"	36'-3"	18'-6"	9"	10'-5"	17'-0"	34'-7"	18'-6"	1'-3"	9'-10"	16'-6"	34'-1"	18'-6"	1'-6"	9'-8"	16'-3"
	13'-10"	36'-3"	20'-6"	9"	12'-5"	19'-0"	34'-7"	20'-6"	1'-3"	11'-10"	18'-6"	34'-1"	20'-6"	1'-6"	11'-8"	18'-3"
	15'-10"	36'-3"	22'-6"	9"	14'-5"	21'-0"	34'-7"	22'-6"	1'-3"	13'-10"	20'-6"	34'-1"	22'-6"	1'-6"	13'-8"	20'-3"
	19'-10"	36'-3"	26'-6"	9"	18'-5"	25'-0"	34'-7"	26'-6"	1'-3"	17'-10"	24'-6"	34'-1"	26'-6"	1'-6"	17'-8"	24'-3"
	23'-10"	36'-3"	30'-6"	9"	22'-5"	29'-0"	34'-7"	30'-6"	1'-3"	21'-10"	28'-6"	34'-1"	30'-6"	1'-6"	21'-8"	28'-3"
50	11'-10"	45'-5"	20'-2"	1'-1"	10'-2"	18'-4"	44'-0"	20'-2"	1'-5"	9'-8"	18'-0"	43'-6"	20'-2"	1'-6"	9'-6"	17'-11"
	13'-10"	45'-5"	22'-2"	1'-1"	12'-2"	20'-4"	44'-0"	22'-2"	1'-5"	11'-8"	20'-0"	43'-6"	22'-2"	1'-6"	11'-6"	19'-11"
	15'-10"	45'-5"	24'-2"	1'-1"	14'-2"	22'-4"	44'-0"	24'-2"	1'-5"	13'-8"	22'-0"	43'-6"	24'-2"	1'-6"	13'-6"	21'-11"
	19'-10"	45'-5"	28'-2"	1'-1"	18'-2"	26'-4"	44'-0"	28'-2"	1'-5"	17'-8"	26'-0"	43'-6"	28'-2"	1'-6"	17'-6"	25'-11"
	23'-10"	45'-5"	32'-2"	1'-1"	22'-2"	30'-4"	44'-0"	32'-2"	1'-5"	21'-8"	30'-0"	43'-6"	32'-2"	1'-6"	21'-6"	29'-10"
60	11'-10"	54'-7"	21'-10"	1'-1"	9'-10"	20'-0"	53'-6"	21'-10"	1'-7"	9'-6"	19'-6"	53'-4"	21'-10"	1'-7"	9'-5"	19'-6"
	13'-10"	54'-7"	23'-10"	1'-1"	11'-10"	22'-0"	53'-6"	23'-10"	1'-7"	11'-6"	21'-6"	53'-4"	23'-10"	1'-7"	11'-5"	21'-6"
	15'-10"	54'-7"	25'-10"	1'-1"	13'-10"	24'-0"	53'-6"	25'-10"	1'-7"	13'-6"	23'-6"	53'-4"	25'-10"	1'-7"	13'-5"	23'-6"
	19'-10"	54'-7"	29'-10"	1'-1"	17'-10"	28'-0"	53'-6"	29'-10"	1'-7"	17'-6"	27'-6"	53'-4"	29'-10"	1'-7"	17'-5"	27'-5"
	23'-10"	54'-7"	33'-10"	1'-1"	21'-10"	32'-0"	53'-6"	33'-10"	1'-7"	21'-6"	31'-6"	53'-4"	33'-10"	1'-7"	21'-5"	31'-5"
70	11'-10"	64'-3"	23'-6"	1'-7"	9'-9"	21'-2"	62'-10"	23'-6"	1'-10"	9'-3"	20'-10"	62'-6"	23'-6"	1'-9"	9'-1"	20'-11"
	13'-10"	64'-3"	25'-6"	1'-7"	11'-9"	23'-2"	62'-10"	25'-6"	1'-10"	11'-3"	22'-10"	62'-6"	25'-6"	1'-9"	11'-1"	22'-11"
	15'-10"	64'-3"	27'-6"	1'-7"	13'-9"	25'-2"	62'-10"	27'-6"	1'-10"	13'-3"	24'-10"	62'-6"	27'-6"	1'-9"	13'-1"	24'-11"
	19'-10"	64'-2"	31'-6"	1'-7"	17'-9"	29'-2"	62'-10"	31'-6"	1'-10"	17'-2"	28'-10"	62'-6"	31'-6"	1'-9"	17'-1"	28'-11"
	23'-10"	64'-2"	35'-6"	1'-7"	21'-9"	33'-2"	62'-10"	35'-6"	1'-11"	21'-2"	32'-9"	62'-6"	35'-6"	1'-9"	21'-1"	32'-11"
80	13'-10"	73'-7"	27'-2"	1'-4"	11'-6"	25'-1"	71'-6"	27'-2"	1'-11"	10'-9"	24'-5"	71'-0"	27'-2"	1'-11"	10'-7"	24'-5"
	15'-10"	73'-7"	29'-2"	1'-4"	13'-6"	27'-1"	71'-6"	29'-2"	1'-11"	12'-9"	26'-5"	71'-0"	29'-2"	1'-11"	12'-7"	26'-5"
	19'-10"	73'-6"	33'-2"	1'-4"	17'-6"	31'-1"	71'-6"	33'-2"	1'-11"	16'-9"	30'-5"	71'-0"	33'-2"	1'-11"	16'-7"	30'-5"
	23'-10"	73'-6"	37'-2"	1'-4"	21'-6"	35'-1"	71'-6"	37'-2"	1'-11"	20'-9"	34'-5"	71'-0"	37'-2"	1'-11"	20'-7"	34'-5"
100	13'-10"	92'-8"	30'-6"	1'-5"	11'-2"	28'-4"	91'-2"	30'-6"	1'-9"	10'-7"	27'-11"	90'-1"	30'-6"	2'-1"	10'-3"	27'-7"
	15'-10"	92'-8"	32'-6"	1'-5"	13'-2"	30'-4"	91'-2"	32'-6"	1'-9"	12'-7"	29'-11"	90'-1"	32'-6"	2'-1"	12'-3"	29'-7"
	19'-10"	92'-8"	36'-6"	1'-5"	17'-2"	34'-4"	91'-2"	36'-6"	1'-9"	16'-7"	33'-11"	90'-1"	36'-6"	2'-1"	16'-3"	33'-7"
	23'-10"	92'-8"	40'-6"	1'-5"	22'-2"	38'-4"	91'-2"	40'-6"	1'-9"	20'-8"	37'-11"	90'-1"	40'-6"	2'-1"	20'-3"	37'-7"
120	13'-10"	112'-6"	33'-10"	1'-6"	11'-1"	31'-6"	110'-4"	33'-10"	2'-1"	10'-4"	30'-11"	109'-6"	33'-10"	2'-2"	10'-0"	30'-10"
	15'-10"	112'-6"	35'-10"	1'-6"	13'-1"	33'-7"	110'-4"	35'-10"	2'-1"	12'-4"	32'-11"	109'-6"	35'-10"	2'-2"	12'-0"	32'-10"
	19'-10"	112'-6"	39'-10"	1'-6"	17'-1"	37'-6"	110'-4"	39'-10"	2'-1"	16'-4"	36'-11"	109'-6"	39'-10"	2'-2"	16'-0"	36'-10"
	23'-10"	112'-6"	43'-10"	1'-6"	21'-1"	41'-6"	110'-4"	43'-10"	2'-1"	20'-4"	40'-11"	109'-6"	43'-10"	2'-2"	20'-0"	40'-10"

*12 PSF LL FRAME

Dimensions shown are for 25' bays. 20' and 30' bays also available.
Building components and dimensions shown are subject to change due to final design.

Figure 4.9 Typical dimensions of high-profile single-span rigid frame. (*American Buildings Co.*)

Multispan rigid frames may have straight or tapered columns, the latter usually at the exterior. The rafters are normally tapered. The construction details are similar to single-span rigid frames save for the additional interior columns. Some typical framing dimensions are shown in Fig. 4.10.

Multispan rigid frames are often the only solution for the largest of buildings, such as warehouses, distribution centers, factories, and resource recovery facilities. Multispan rigid frames utilize continuous framing and are normally more economical than their single-span cousins. The disadvantages of continuous construction include susceptibility to differential settlement of supports, as noted in Chap. 3. Soil conditions at the site should be carefully evaluated before this system is specified. Also, the interior column locations are difficult to change in the future, should that be required because of a new equipment layout.

4.6   Single-Span and Continuous Trusses

Single-span (Fig. 4.1e) and continuous trusses (Fig. 4.1f) are similar in function to single-span and multispan rigid frames. The crucial difference between the frames and the trusses lies in the construction of the rafter's web—open for trusses and solid for frames. An open web allows for passage of pipes and ducts and thus permits the eave height in a truss building to be lower, which results in a smaller building volume to be heated or cooled and thus in lower energy costs. Therefore, trusses are most appropriate for the applications with a lot of piping and utilities, such as manufacturing facilities and distribution centers.

An example of simply supported truss framing is Butler Manufacturing Company's Landmark Structural System. Figure 4.11 illustrates the details of 3- and 4-ft-deep trusses common in that system. Note the deflection pad between the bottom chord of the truss and the column, intended to allow for some member rotation under gravity loads without inducing bending moments into the column. In effect, this is a good "wind connection," discussed above for a tapered-beam system, that provides lateral resistance in the plane of the truss. In Landmark, lateral resistance along the length of the building is provided by fixed-base endwall columns, an approach not without some pitfalls, as will be discussed later in this chapter.

4.7   Lean-to Framing

Lean-to framing, also known as wing unit (Fig. 4.1g), is not a true self-contained structural system but rather an add-on to the existing building. Tapered beams and straight columns are common in this type of construction (Fig. 4.12). For the optimum efficiency, the system is best specified for clear spans from 15 to 30 ft.

Lean-to framing is typically used for building additions, equipment rooms, storage, and a host of other minor attached structures. Structural details are similar to those of a tapered-beam system, except that a single slope is usually provided at the top surface and the beam taper precludes the bottom surface from being horizontal.

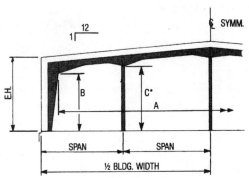

**REPRESENTATIVE
CLEARANCE DIMENSIONS
1:12 ROOF SLOPE**

BAY SPACING	–	25 ft.
ROOF SLOPE	–	1:12
LIVE LOAD	–	LL (psf)
WIND LOAD	–	WL (mph)

WIND LOAD APPLIED IN ACCORDANCE
WITH MBMA (1986)

*C = Minimum clearance other than at knee (dimension B). All points where rafter changes shape are checked and vertical dimension to lowest of these points is given.

		CLEARANCE DIMENSIONS (FEET-INCHES) **										
LL/WL→		20/80			20/100			25/90			30/90		
SPAN	E.H.	A	B	C	A	B	C	A	B	C	A	B	C
2 at 40	10	76-10	7- 5	10- 7	76- 8	7- 5	10- 7	75- 4	7- 6	10- 7	75- 0	7- 3	10- 5
	14	76- 8	11- 5	14- 7	76- 8	11- 5	14- 7	76- 2	11- 5	14- 7	76- 0	11- 3	14- 5
	16	76-10	13- 5	16- 7	76-10	13- 5	16- 7	76- 8	13- 3	16- 5	76- 0	13- 3	16- 5
	20	76-10	17- 5	20- 7	76- 2	17- 5	20- 7	76- 8	17- 3	20- 5	76- 8	17- 3	20- 5
2 at 60	10	114- 0	6- 8	8- 6	114- 0	6- 8	8- 6	113- 6	6- 8	8- 9	113- 2	6- 1	8- 9
	14	113- 8	11- 1	12- 9	113-10	10- 8	12- 9	113- 6	10-10	12- 9	113- 0	10- 8	12- 7
	16	113- 6	12- 8	14- 9	113- 6	12- 8	14- 9	112-10	12-10	14-10	112- 8	12- 6	14-10
	20	113- 8	16- 9	18- 9	113-10	16- 8	18- 9	113- 2	16- 9	18-10	112-10	16-10	18- 7
3 at 40	12	116- 8	9- 7	14- 6	116- 8	9- 7	14- 6	115-10	9- 5	14- 4	115- 2	9- 6	14- 4
	16	116-10	13- 7	18- 6	116-10	13- 7	18- 6	115- 8	13- 6	18- 4	115- 6	13- 3	18- 1
	20	116-10	17- 7	22- 6	116- 0	17- 7	22- 6	116- 8	17- 5	22- 4	116- 6	17- 2	22- 1
	24	116- 4	21- 7	26- 6	115- 4	21- 8	26- 6	116- 0	21- 5	26- 4	116- 0	21- 5	26- 3
3 at 60	12	174- 0	8- 8	10- 6	174- 0	8- 8	10- 6	173- 8	8- 8	10- 6	172-10	7-11	10-10
	16	173- 6	12- 8	14- 9	173- 6	12- 8	14- 9	172- 8	12-10	14- 7	172- 2	12- 9	14- 7
	20	173- 6	16- 8	18- 9	173- 6	16- 8	18- 9	172- 8	16- 9	18-10	172- 6	16- 9	18- 7
	24	173- 6	20- 8	22- 9	173- 6	20- 8	22- 9	172- 8	20- 9	22-10	172- 4	20- 9	22-10
4 at 40	12	156-10	9- 7	16- 1	156-10	9- 7	16- 1	156- 2	9- 5	15-11	155- 8	9- 3	15- 9
	16	156-10	13- 7	20- 1	156-10	13- 7	20- 1	156- 0	13- 5	19-11	155- 8	13- 3	19- 9
	20	156-10	17- 5	24- 0	156- 2	17- 7	24- 1	156- 8	17- 5	23-11	156- 8	17- 3	23- 9
	24	156- 8	21- 5	28- 0	155- 6	21- 6	28- 0	156- 2	21- 5	27-11	156- 2	21- 3	27- 9
4 at 60	16	233- 6	12- 8	14- 9	233- 6	12- 8	14- 9	232-10	12- 9	14- 7	232- 6	12- 9	14- 7
	20	233- 8	16- 8	18- 9	233- 8	16- 8	18- 9	233- 6	16- 8	18- 6	232- 8	16- 9	18- 7
	24	233- 6	20- 8	22- 9	233- 6	20- 8	22- 9	232-10	20- 9	22-10	232- 8	20- 8	22- 7
	30	233- 4	26- 9	28- 9	233- 8	26- 8	28- 9	232- 6	26- 9	28-10	232- 2	26- 9	28- 7
5 at 60	16	293- 6	13- 1	14- 9	293- 6	13- 1	14- 9	292-10	12- 9	14- 7	292- 6	12- 9	14- 7
	20	293-10	16- 8	18- 9	293- 8	16- 8	18- 9	292-10	16- 9	18-10	292- 6	16- 9	18- 7
	24	293- 6	20- 8	22- 9	293- 6	20- 9	22- 9	292-10	20- 9	22-10	292- 8	20- 8	22-10
	30	293- 6	26- 8	28- 9	293- 8	26- 8	28- 9	292- 6	26- 9	28-10	292- 2	26- 9	28- 7

**Clearances shown are approximate. Actual clearances may be somewhat different.

Figure 4.10 Typical dimensions of multispan rigid frames. (*Ceco Building Systems.*)

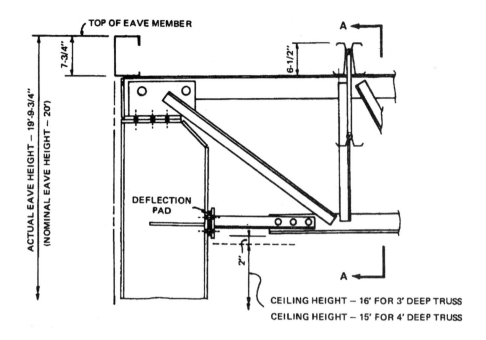

CEILING HEIGHT — 16' FOR 3' DEEP TRUSS
CEILING HEIGHT — 15' FOR 4' DEEP TRUSS

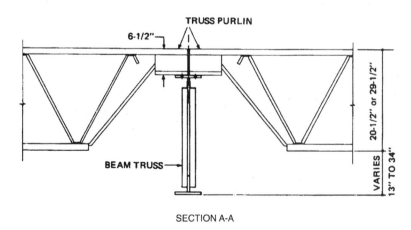

SECTION A-A

Figure 4.11   Details of Butler's Landmark Building. (*Butler Manufacturing Co.*)

4.8 Other Framing Systems

In addition to the framing described above, the marketplace contains several proprietary systems that are truly unique as well as those that only pretend to be different by adopting an unfamiliar name and some unusual details. Some of the "significant others" are mentioned below.

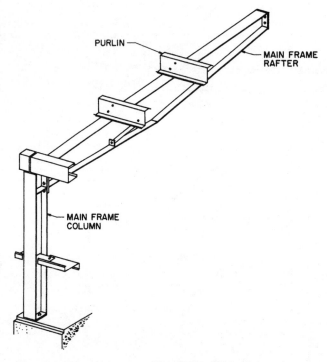

Figure 4.12 Lean-to framing. (*Metallic Building Systems.*)

4.8.1 Single-slope frames

Each of the five basic pre-engineered framing systems can be produced in a single-slope, rather than gable, configuration. The single-slope feature does not significantly affect structural behavior of the framing, its clear span capacity, or typical details. Confusing the terminology, some companies call their single-slope rigid-frame products (Fig. 4.1*h*) "lean-to rigid frames." In such cases, a picture is truly worth a thousand words.

Single-slope framing is frequently used for office complexes and strip shopping malls, where rainwater needs to be drained away from the parking areas or from the adjacent buildings.

4.8.2 Trussframes

Coronis Building Systems, Inc., started production of its proprietary framing line in 1956 and has since developed over 8000 variations of the framing. A typical trussframe* resembles a tapered beam, except for its web, which is made of truss-type members rather than being solid (Fig. 4.13). Other trussframe varieties include multispans, cantilevered pole-type shelters, canopies, and lean-tos.

---

*Trussframe is a trademark of Coronis Building Systems, Inc.

Figure 4.13 Trussframes. (*Coronis Building Systems.*)

A major advantage of this system is the absence of horizontal reactions at the columns under gravity loads, a natural property of tapered-beam framing (or any other simply supported straight beam, for that matter).

4.8.3 Delta Joist system

Delta Joists* are unlikely to be confused with any other structure. We would have placed this system in the next chapter, were these triangulated three-dimensional joists produced by Butler Manufacturing Co. not conceived and sold as a complete roof-support system rather than as mere roof purlins. The joists, which are available in 1-ft increments, have a constant depth of $25\frac{1}{4}$ in, regardless of loading. Their top and bottom chords are made of hot-rolled steel angles; the diagonals consist of round bars. The joists possess a very desirable characteristic—lateral stability—which makes them truly different, since it obviates the need for purlin bracing and for traditional horizontal roof diaphragms.

The Delta Joist system is best suited for buildings with load-bearing masonry or precast walls, where exterior columns and wall bracing are not needed; it

---

*Delta Joist is a trademark of Butler Manufacturing Co.

can be adapted for non-load-bearing endwalls if optional steel frames are used. The system normally provides a roof slope of $\frac{1}{4}$:12 and requires the building width to be a multiple of 4 ft. The joists can span up to 60 ft, a distance normally unattainable with the secondary members traditionally used in metal building systems. The Delta Joist system is intended to support Butler's proprietary standing-seam roof panels; top flanges of the joists come prepunched for attachment of the roof clips.

4.8.4   Flagpole-type systems

Occasionally, a manufacturer proposes a system that does not look much different from the competing ones but costs less. The savings can sometimes be explained by the fact that all the building columns, or the exterior columns only, are designed with fixed bases. In this design, rigid-frame action is partially or completely transferred from the frame knee to the column base. Each fixed column becomes a flagpole-type structure. This approach allows the manufacturer to eliminate not only the expensive rigid frames but the wall and roof bracing as well—and submit a lower price.

Why doesn't everybody design that way? The answer is simple: This kind of design, while allowing the manufacturer to realize some savings, penalizes the foundations by subjecting them to high bending moments that would be absent under a pinned-base scenario. Whether fixed-base design results in a higher or a lower overall building cost depends on the project specifics. *If anticipated from the beginning,* it may be a viable option in cases where the foundation capacity can be increased at a low cost. Havoc might result, however, if such fixed-base framing is proposed for a building where the foundations have already been designed—or worse, built—based on a pinned-base assumption, as we will see in Chap. 10.

Another problem with fixed-base columns: Base fixity is easy to assume, not easy to achieve. For example, a widespread industry habit of not providing grouted leveling plates under column bases often leaves the base plates bearing only on one edge and allows for some "play" between steel and concrete. If, in addition, the anchor bolts are not properly tightened, a fixed-base assumption is simply unrealistic.

4.9   A Role of Frame Bracing

Every structural system we have considered, except perhaps Butler's Delta Joist, requires lateral bracing of the rafter's compression flange for full structural efficiency. Under downward loads (dead, live, and snow), the top flange of primary members is mostly in compression. Fortunately, this flange carries roof purlins, which provide the necessary bracing. Under wind uplift, however, it is the bottom flange that is mostly in compression. Lacking any help from secondary members, the bottom flange needs to be stabilized against buckling by flange bracing, consisting usually of bolted angle sections (Fig. 4.14).

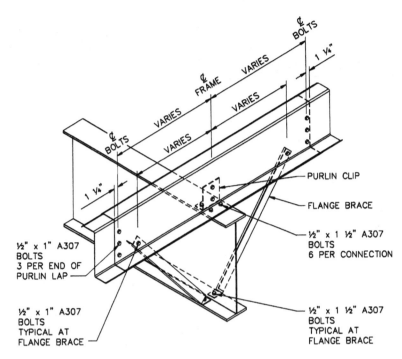

Figure 4.14 Typical flange brace at roof. (*Varco-Pruden Buildings.*)

Similar bracing is needed at interior flanges of rigid-frame columns that are normally in compression under downward loads. The bracing connects the interior flange to the wall girts (Fig. 4.15).

Locations of the flange bracing are determined by the metal building manufacturer and need not concern the specifiers. An absence of any flange bracing at all, however, warrants further inquiry.

4.10 Choices, Choices

Any large manufacturer can provide most of the primary framing systems discussed above. Which one to specify? Hopefully, the information provided in this chapter, as well as in Chap. 3, will be helpful. (The dimensions and details shown in the illustrations should be considered preliminary, as each manufacturer has its own peculiarities.) Also, Fig. 4.16, adapted from 1995 *Means Materials Cost Data*,[3] provides material and total cost per square foot for various types of pre-engineered framing.

In general, in a large facility that can tolerate interior columns and is unlikely to undergo drastic changes in layout, multispan rigid-frame construction should be tried first; it usually offers the lowest cost. If interior columns are objectionable, a clear-span system such as a single-span rigid frame should be considered. Smaller buildings can be economically framed with tapered beams

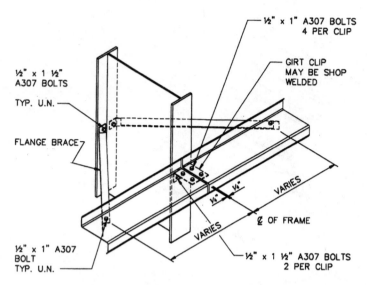

Figure 4.15 Typical column flange brace. (*Varco-Pruden Buildings.*)

or even lean-tos. Proprietary framing systems are difficult to specify for competitive bidding without severely restricting the competition and are most suitable for negotiated private work.

It helps to understand what is behind the client's clear-span requirements, as discussed in the previous chapter. Most clients would like a column-free plan that allows them unlimited flexibility; it's when the cost of that flexibility becomes clear that the budgets begin to vote. It is rather disheartening to see a building with huge clear spans—paid for dearly—promptly subdivided by the partitions erected by the owner. Whether intended as noise barriers or privacy screens separating various activities, each partition could have held a column and thus would have afforded some cost savings. Moreover, specifying buildings with unusually large clear spans restricts the list of bidders to the largest manufacturers able to produce heavy structures.

The architect has to decide whether straight, tapered, or other columns are appropriate for the project. While a utility or a manufacturing building can easily tolerate tapered columns of rigid frames, a plasterboard-clad library or a retail establishment probably would not. Trying to wrap a tapered column in sheetrock is usually not worth the effort and the cost; a system with straight columns could fit better even if slightly more expensive.

4.11 Endwall Framing

The foregoing discussion has dealt only with interior frames. What about the endwall framing? While each manufacturer has a slightly different approach and details of endwall framing, the basic design is essentially the same. The

R131-210 Pre-engineered Steel Buildings

These buildings are manufactured by many companies and normally erected by franchised dealers throughout the U.S. The four basic types are: Rigid Frames, Truss type, Post and Beam and the Sloped Beam type. Most popular roof slope is low pitch of 1″ in 12″. The minimum economical area of these buildings is about 3000 S.F. of floor area. Bay sizes are usually 20′ to 24′ but can go as high as 30′ with heavier girts and purlins. Eave heights are usually 12′ to 24′ with 18′ to 20′ most typical. Pre-engineered buildings become increasingly economical with higher eave heights.

Prices shown here are for the building shell only and do not include floors, foundations, interior finishes or utilities. Typical erection cost including both siding and roofing depends on the building shape and runs $1.35 to $2.60 for one in twelve roof slope and

$1.40 to $3.65 per S.F. of floor for four in twelve roof slope. Site, weather, labor source, shape and size of project will determine the erection cost of each job. Prices include erector's overhead and profit.

Table below is based on 30 psf roof load, 20 psf wind load and no unusual structural requirements. Costs assume at least three bays of 24′ each. Material costs include the structural frame, 26 ga. colored steel roofing, 26 ga. colored steel siding, fasteners, closures and flashing but no allowance for doors, windows, gutters or skylights. Very large projects would generally cost less than the prices listed below. Typical budget figures for above material delivered to the job runs $1250 to $1575 per ton. Fasteners and flashings (included below) run $.46 to $.68 per S.F.

Material Costs per S.F. of Floor Area Above the Foundations						
Type of Building	Total Width in Feet	Eave Height				
		10 Ft.	14 Ft.	16 Ft.	20 Ft.	24 Ft.
Rigid Frame	30-40	$3.90	$4.22	$4.55	$4.98	$5.63
Clear Span	50-100	3.95	3.95	4.27	4.48	4.98
	110	–	–	–	–	–
	120	–	–	–	4.55	–
	130	–	–	–	–	–
Tapered Beam	30	4.38	4.87	5.41	6.22	–
Clear Span	40	4.17	4.43	4.65	5.30	–
	50-80	4.55	4.38	4.55	4.98	–
Post & Beam	80	–	3.35	3.68	3.90	4.27
1 Post at Center	100	–	3.24	3.47	3.84	4.17
	120	–	3.19	3.30	3.52	3.95
Post & Beam	120	–	–	–	–	–
2 Posts	150	–	3.24	3.47	3.68	4.17
@ 1/3 Points	180	–	–	–	–	–
Post & Beam	160	–	3.19	3.30	3.68	4.00
3 Posts	200	–	3.24	3.47	3.73	4.05
@ 1/4 points	240	–	–	–	–	–

Typical accessory items are listed in the front of the book. All normal interior work, floors, foundations, utilities and site work should be figured the same as usual.

Costs in the table below include allowance for erection, normal doors, windows, gutters and erector's overhead and profit. Figures

do not include foundations, floors, interior finishes, electrical, mechanical or installed equipment.

Total Cost per S.F. Above the Foundations, 16′ Eave Height				
Project Size: Rigid Frame 30′ to 60′ Spans 1 in 12 Roof Slope	Basic Building Using 26 ga. Galvanized Roof & Siding S.F. Floor Area	Add to Basic Building Price		
		R13 Field Insulation S.F. Floor Area	Exterior Finish	S.F. of Skin
4,000 S.F.	$7.65	$1.44	Sandwich wall	$6.10
10,000 S.F.	6.78	1.23	Corrugated fiberglass	3.28
20,000 S.F.	6.27	1.17	Corr. fiberglass-insulated	4.09
			10 year paint	.23

Figure 4.16 Typical square-foot costs of pre-engineered buildings. (*From Means Building Construction Cost Data 1995. Copyright R.S. Means Co., Inc., Kingston, MA, 617-585-7880, all rights reserved.*)

function of endwall framing is to resist all loads applied to the building's end-walls and to support wall girts. In buildings with expandable endwalls, a regular interior frame is provided at the top of the walls. This frame resists all vertical loads as well as the lateral loads applied to the sidewalls, and the endwall framing is needed only to support wall girts. In buildings where endwalls are not expandable, the endwall framing also supports vertical loads and contains wall cross-bracing (or fixed-base wind posts).

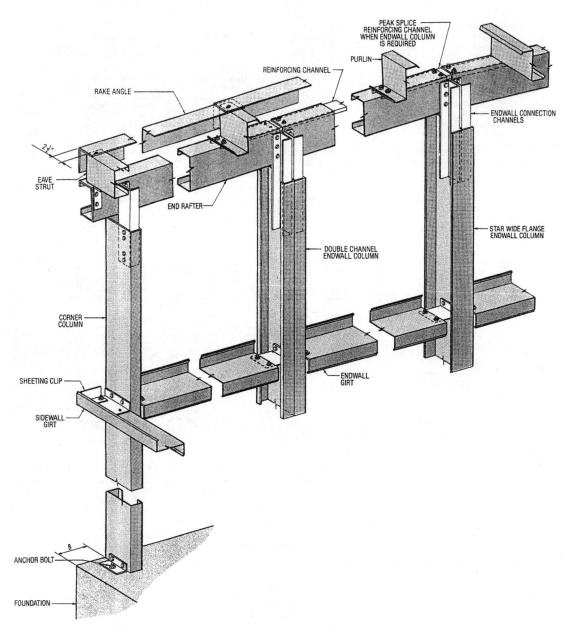

Figure 4.17 Endwall framing viewed from inside the building. (*Star Building Systems.*)

Endwall framing consists of columns (posts), roof beams, and corner posts, all with base plates and other accessories. Endwall columns are frequently made of either single or double cold-formed channels with a metal thickness of at least 14 gage, as shown in Figs. 4.2, 4.5, and 4.17. Alternatively, endwall columns may be made of hot-rolled or built-up wide-flange sections. End rafters are usually made of cold-formed channels unless a regular rigid frame is used for future expansion. The rafters are designed as simple-span members and spliced at each column; a reinforcing channel may be needed at each splice (Fig. 4.17).

Endwall columns are commonly spaced 20 ft on centers, a distance governed mostly by the girt spanning capacities. The column layout normally starts from a center column at the ridge line. Where no center column is provided, two endwall columns straddle the ridge line.

In nonexpandable endwalls, a connection between the endwall column and the rafter may consist of simply bolting the column flange to the rafter's web (Fig. 4.2) and connecting the purlin to the rafter—not to the column—with a clip angle (Fig. 4.18a). Or, the column may be attached to the rafter by small endwall connection channels (Fig. 4.17). In either case, a rake angle is needed at the top of the purlins to support the wall siding.

In expandable endwalls, the column-to-rafter connection requires an additional clip angle between the column and the frame rafter (Fig. 4.19a) and between the endwall girt and the frame column (Fig. 4.19b).

The endwall girts may have a flush or bypass inset. In the former case, the girts are designed as simple-span members framing into the column webs, except at the corners where sidewall girts are framed into the web of the corner column (Fig. 4.17 and 4.18b). In the latter case, a continuous girt design is possible, as described in the next chapter.

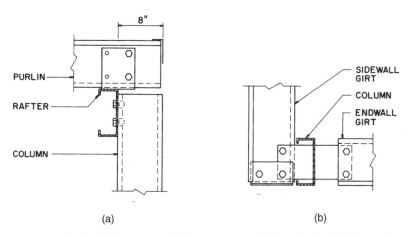

Figure 4.18 Endwall framing details for nonexpandable endwalls. (a) Connection between endwall column and purlin; (b) plan at corner. (*Metallic Building Systems.*)

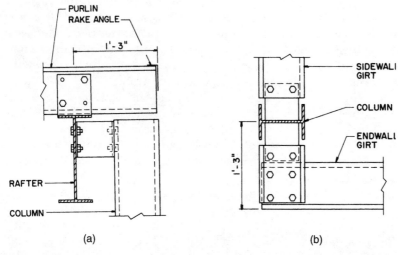

Figure 4.19 Endwall framing details for expandable endwalls. (*a*) Connection between endwall column and frame rafter; (*b*) plan at corner. (*Metallic Building Systems.*)

In some buildings with masonry, glass, or concrete walls, the curtain-wall structure can span directly from foundation to the roof. There, the endwall framing may consist only of a clear-spanning rigid frame, similar to the case of expandable endwalls but without any endwall girts and columns.

References

1. *Manual of Steel Construction,* vol. II, *Connections,* American Institute of Steel Construction, Chicago, Ill., 1992.
2. William McGuire, *Steel Structures,* Prentice-Hall, Englewood Cliffs, N.J., 1968.
3. *Means Materials Cost Data,* R.S. Means Co., Kingston, Mass., 1995.

5

Secondary Framing: Girts and Purlins

5.1 Introduction

Secondary framing plays a complex role in metal building systems that extends beyond supporting roof and wall covering and carrying exterior loads to main frames. Secondary structurals, as these members are sometimes called, may serve as flange bracing for primary framing and may function as a part of the building's lateral load–resisting system. Roof secondary members, known as *purlins,* often form an essential part of horizontal roof diaphragms; wall secondary members, known as *girts,* are frequently found in wall bracing assemblies.

A third type of secondary framing, known by the names of *eave purlin, eave strut,* or *eave girt,* acts as part purlin and part girt—its top flange supports roof panels, its web wall siding (Fig. 5.1). In addition to the traditional channel-like sections, eave purlins are available in some unique configurations such as the one offered by Varco-Pruden Buildings (Fig. 5.2). Eave purlins also participate

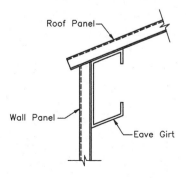

Roof Panel

Wall Panel

Eave Girt

Figure 5.1 Typical eave purlin.

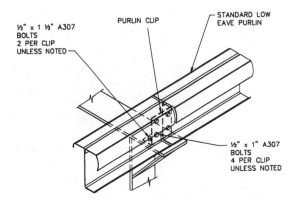

Figure 5.2 Proprietary eave purlin assembly. (*Varco-Pruden Buildings.*)

in wall cross-bracing assemblies and serve as ties between bracing bents along the same sidewall. And finally, the building's eave height is measured to the top of this member.

Since girts and purlins exhibit similar structural behavior, this chapter considers both types. And, since most secondary members normally encountered in metal building systems are made of cold-formed steel, our discussion starts with some relevant issues in design of cold-formed steel structures.

5.2 Design of Cold-Formed Framing

As mentioned in Chap. 2, the main design standard for cold-formed framing is Specification for the Design of Cold-Formed Steel Structural Members by American Iron and Steel Institute (AISI).[1] The Specification, Commentary, Design Examples, and other information constitute the *AISI Manual*. The first edition of the Specification appeared in 1946, with subsequent editions following in 1960, 1968, 1980, 1986, and 1989 (by Addendum). As this book went to print, the 1989 edition was recognized by all three model building codes as current; a 1995 edition was expected to be released soon. The LRFD-based Specification was first issued in 1991.[2]

The changes between various editions are substantial, a fact that reflects on the continuing research in this area of steel design. Since the Specification provisions are so fluid, framing manufacturers are challenged to comply with the latest requirements. Unfortunately, some have fallen behind, still using the previous editions.

Anyone who has ever attempted to design a light-gage member following the Specification provisions probably realized how tedious and complex the process was. This fact helps explain why cold-formed steel framing is rarely designed in most structural engineering offices. When such framing is needed, one of two things tends to happen to the engineers: They either uncritically rely on the suppliers' literature or simply avoid any cold-formed design at all by specifying hot-rolled steel members and hoping for a contractor to make the substitution and to submit the required calculations.

In this chapter, we limit our immersion into the actual Specification formulas that could easily have become obsolete by the time you read this book. Instead, we point out but a few salient concepts.

What makes cold-formed steel design so time-consuming?

First, materials suitable for bending are usually quite thin and thus susceptible to local buckling—distortion affecting only a part of the member (remember how easy it is to dent a tin can?). This mode of failure is of much less concern in the design of thicker hot-rolled members. Second, the flanges of light-gage sections cannot be assumed to be under a uniform stress distribution, as the flanges of an I beam might be (a shear lag phenomenon). To account for both the local buckling and the shear lag, the Specification utilizes a concept of "effective design width," in which only certain parts of the section are considered effective in resisting compressive stresses (Fig. 5.3). This concept is pivotal for stress analysis and deflection calculations performed for cold-formed members.

The trouble is, the effective design width depends on the stress in the member that, naturally, cannot be computed until some section properties are assumed first. Because of this "vicious circle," a few design iterations are needed. A common simplified yet conservative procedure for the effective width calculations assumes the level of stress to be the maximum allowable.

Another complication caused by a nonuniform stress distribution across thin, often nonsymmetrical, sections is their lack of torsional stability. Light-gage compression and flexural members can fail in torsional-flexural buckling mode by simultaneous twisting and bending, a failure that can occur at relatively low levels of stress. Torsional-flexural buckling can be prevented by keeping the compressive stresses very low or by plenty of bracing, as will be discussed later in this chapter.

The complexities of light-gage member design do not stop at bending and compression calculations. Tedious shear calculations are often accompanied by even more cumbersome web crippling checks. To be sure, web crippling failures occur in hot-rolled steel members too, but light-gage sections are incomparably more susceptible. Web crippling failures such as that shown in Fig. 5.4 are most likely to occur at supports, where shear stresses are at their maximum.

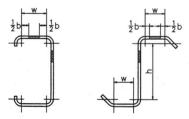

Figure 5.3 Effective width concept for C and Z sections (shaded areas are considered ineffective).

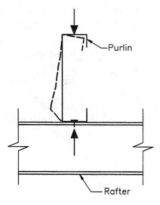

Figure 5.4 Web crippling.

Web crippling stresses are additive to bending stresses, and a combination of both needs to be investigated.

Whenever web crippling stresses are excessive, bearing stiffeners are required at supports, in which case it is common to assume that the total reaction force is transferred directly through the stiffener into the primary framing, neglecting any structural contribution of the member's web. A small gap might even be left under the flange of a girt or purlin. The stiffeners are usually made of clip angles, plates, or channel pieces. In Fig. 5.5, the load is transmitted from the web of a Z purlin via screws or bolts to the clip-angle stiffener and then from the stiffener to the rafter.

The Specification recognizes the fact that analytical methods of establishing load-carrying capacities of some cold-formed structural framing may not always be available or practical and allows determination of structural performance by load testing. The testing procedure is described in detail in the Specification section entitled "Tests for Special Cases." Summarizing, the member or assembly being tested should be able to carry twice the live load plus 1.5 times the dead load (the strength test) *and* not to distort excessively under 1.5 times the live load plus 1.0 dead load (the deflection test). The values

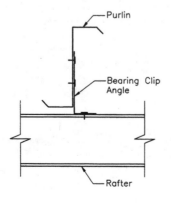

Figure 5.5 Bearing clip angle acting as web stiffener.

of the effective section modulus and moment of inertia are established based on the measurements of strains and deflections. The test results apply only to the specimen being tested. If the testing is intended to apply to the whole class of sections, as it usually is, the material properties such as yield strength are measured to verify that the tested specimen has the desired—and not higher—material strength. If the tested strength is too high, the test results are adjusted to be applicable to the minimum specified material properties.

5.3 Cold-Formed Steel Purlins

5.3.1 Available sizes and designs

Cold-formed C and Z purlins are the workhorses of the industry. Configurations of these members have originated at the bending press—they represent the two basic ways to bend a sheet of metal into a section with a web and two flanges. Light-gage purlins of 8 to 12 in in depth can span 25 to 30 ft, depending on the loading, material thickness, and deflection criteria. Purlin spacing is dictated by a load-carrying capacity of the roof panels; a 5-ft spacing is common. Appendix B includes section properties for purlin sizes offered by some manufacturers.

Cold-formed purlins are normally made of high-strength steel. Uncoated cold-formed members, still in the majority, usually conform to ASTM A 570 or A 607. Occasionally, galvanized purlins are provided. The old designation for galvanized members, ASTM A 446, has been replaced with a new ASTM Standard Specification A 653.[3] The new ASTM A 653 uses a grade of the material to define its yield point. The new standard includes the designations of zinc coating, G60 and G90, which used to be a part of a separate standard, ASTM A 525. (The latter has been replaced by ASTM A 924, which now covers all kinds of metal coatings applied by a hot-dip process.) For the products of structural quality (SQ), three grades—33, 40, and 80—are available, corresponding to the old grades A, C, and E of ASTM A 446. For example, ASTM A 653 SQ grade 40 with coating designation G60 takes the place of the old ASTM A 446 grade C with G60 coating.

A minimum yield strength for steel sections 16 gage and heavier is normally specified as 55,000 lb/in^2, although the Light Gage Structural Institute (LGSI) bases its load tables[4] on a minimum yield strength of 57,000 lb/in^2. Similarly, LGSI member companies have adopted slightly different section properties for their cold-formed sections than those of most metal building systems manufacturers (Fig. 5.6). LGSI products use less metal owing to the optimized flange and lip sizes: In the traditional C and Z sections, the flanges are too wide to be fully effective, while the flanges of LGSI sections are narrower but with slightly longer stiffening lips.

Cold-formed purlins can be designed as simple-span or continuous members. The beneficial effects, as well as the disadvantages, of continuous framing are explained in Chap. 3. Unlike the other types, cold-formed purlins can be made continuous by a simple overlapping and fastening. Light-gage C sections can be easily lapped back to back. Z sections can be nested one inside another,

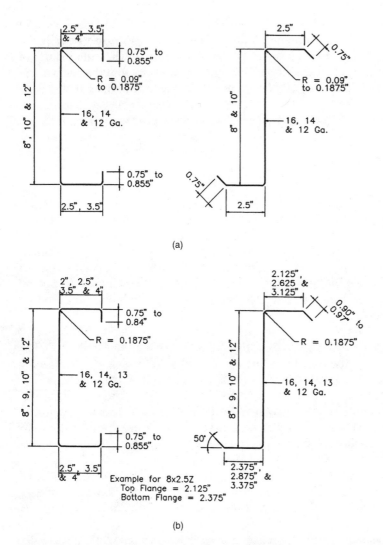

Figure 5.6 Typical C and Z girt and purlin sections. (*a*) Used by major metal building manufacturers; (*b*) offered by LGSI members.

although the traditional equal-flange Z sections of thicker gages might be difficult to nest. Zamecnik[5] observes in his investigation of a warehouse with the noticeably distorted Z purlins that it is "impossible to nest [the] two sections…without bending the web of the lower purlin away from the bottom flange," a situation that contributes to undesirable rotation of the purlins at the supports. Noteworthy, LGSI Z sections have flanges of slightly unequal width to facilitate splicing and provide better fit.

(A reminder to specifiers: LGSI sections should not be forced on the metal building systems manufacturers or specified indiscriminately, since the manufacturers have their production lines geared toward their own standard members. Please investigate the availability first.)

5.3.2 Design for continuity

To achieve some degree of continuity, cold-formed sections are lapped and bolted together for a distance of at least 2 ft; i.e., each member extends past the support by at least 1 ft (Fig. 5.7). The degree of continuity may be increased with a longer lap distance, albeit at a cost of the extra material used in the lap. Recent research[15] indicates that load capacity of Z purlins continues to increase until the length of the lap approaches one half of the span. The upper limit of such an increase is about 100 percent. Further increases in lap length do not seem to be cost-effective. A purlin can be attached to rafters in various ways, depending on the magnitude of crippling stress in the purlin's web. A simple bolting through the member flanges is acceptable if the stress is not critical; otherwise support clips acting as web stiffeners are needed.

Continuous framing, while offering significant material savings, requires careful consideration. The effects of potential problems caused by temperature changes and differential settlement have already been discussed. Further, continuous purlins are subjected to variable bending moments at different spans,

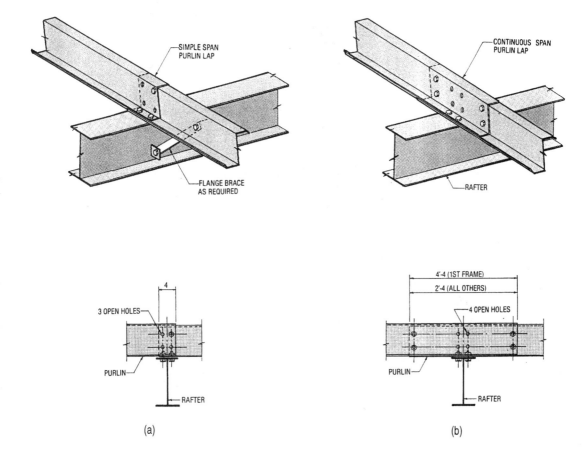

Figure 5.7 (*a*) Simple-span and (*b*) continuous-span purlin laps. (*Star Building Systems.*)

even from uniform loads: the most critical bending stresses in a continuous beam occur at the end spans. It follows that the end-bay purlins must have stronger sections than the interior ones. Alternatively, some manufacturers prefer to utilize the same purlins throughout the building and provide additional splice lengths for the end-bay purlins. Either approach is fine; a potential red flag might be raised only if the shop drawings indicate the same purlin sections and lap distances at all locations, although it could simply mean that some cost efficiency has been forgone.

5.3.3 A need for purlin and girt bracing

As structural engineers have long known, an unbraced compression flange of any single-web flexural member, even of a perfectly symmetrical one loaded through its web, has a tendency to buckle laterally under vertical loading. A singly symmetrical (C section) or a point-symmetrical (Z section) cold-formed purlin is even more susceptible to buckling because it has its shear center in a location quite different from the point of loading application, which is typically the middle of the top flange. Plus, the principal axes of a Z section are inclined to the web, and any downward load produces a lateral component. Because of these factors, the unbraced C and Z sections tend to twist and to become unstable even under gravity loading on a perfectly horizontal roof.

In sloped roofs, the purlin web is tilted from the vertical position, a fact that further complicates the problem of twisting. Gravity loading acting on a sloped purlin can be resolved into the components parallel and perpendicular to the roof, both of which tend to overturn a C or Z purlin, although in the different directions if the purlins are properly oriented as shown in Fig. 5.8. A computation based on the member geometry quickly finds that the two components equalize each other when the slope equals the ratio of the dimensions of the purlin's flange to its depth. For example, for an 8.5.in-deep Z purlin with a 2.5.in-wide flange, this slope is about 3.5:12.

At the roofs with appreciable slopes (over ½ to 12), proper purlin orientation is facing upslope, as shown in Fig. 5.9. For near-flat through-fastened roofs the purlins are frequently located in alternating positions (Fig. 5.10), a design that

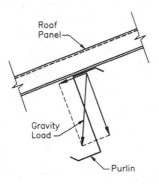

Figure 5.8 Gravity load applied at the top flange of Z purlin can be resolved into components parallel and perpendicular to the roof.

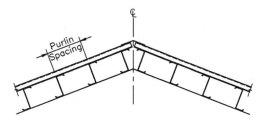

Figure 5.9 Purlin orientation at medium-sloped roofs.

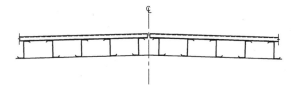

Figure 5.10 Purlin orientation at roofs with slopes less than 1:12.

relies on the roofing acting as a compression brace between the two purlins facing each other. This design is not applicable for standing-seam roofs because, as discussed below, standing-seam roofs may not qualify as lateral bracing for purlins.

The conclusion is clear: Lateral bracing is essential for stability of light-gage C and Z shapes.

5.3.4   The types of purlin and girt bracing

What kind of bracing can be used for secondary members?

First, the lateral bracing may be provided by some types of metal roofing, mainly of through-fastened variety. To qualify, the panels must be of proper thickness and configuration, with attachments that provide a continuous load path.

Standing-seam roofing (SSR) poses a quandary. SSR is intended to expand and contract with the temperature changes, as described in Chap. 6; it therefore allows some purlin rotation to take place before becoming effective as bracing. Many engineers consider SSR totally devoid of any bracing ability, while some recognize only the roofs with short and sturdy metal clips to be capable of the bracing function. Other clip types, especially a slender "articulating clip" (Fig. 6.6) which allows for up to 2 in of purlin movement, do not qualify as bracing. MBMA and AISI are presently sponsoring a test procedure to determine the bracing capability of various SSR types and are also sponsoring the development of a *Standing Seam Roof Design Guide* which will address this issue.

A second type of bracing can be provided by several lines of angles or flat bars running from eave to eave and attached to each purlin and eave girt. Flat strapping is connected to purlin flanges by screws; it is the easiest and cheapest to install. The strapping can function only in tension, however, and therefore is not as effective as angle bracing. Also, purlin bracing needs to be taut to perform properly, yet flat straps and round rods have a tendency to sag and are near useless in that condition.

Angle bracing ("sag angles") can be supplied in sections corresponding to the purlin spacing and be secured by fitting precoped tabs into the prepunched holes in purlin webs (Fig. 5.11). Alternatively, the angles can be supplied in longer sizes, field coped, and attached to the purlin flanges with screws (Fig. 5.12).

The spacing of bracing lines is determined by analysis. An additional brace is normally provided at each concentrated load.

Continuous purlins are sometimes considered braced at the points of lap splices, where the sections are doubled and sturdy bearing clips are often present. The unbraced length of such purlins can be counted either from a point of inflection or from an end of the lap. The latter location is better defined and is therefore recommended.

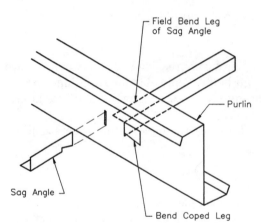

Figure 5.11 Purlin bracing by sag angles installed in prepunched holes.

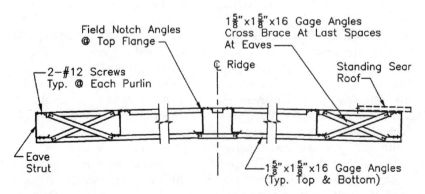

Figure 5.12 Purlin bracing for standing-seam roofs utilizing self-drilling screws.

5.3.5 Bracing for uplift conditions

Whenever purlins are stabilized by roofing or top-flange bracing, they are considered laterally supported only for *downward* loads, which produce mostly compressive stresses in the purlin's top flange. (Owing to continuity effects, some areas of the top flange located near supports will be in tension.) But what about the situations when wind produces upward forces and the *bottom* flange acts mostly in compression? According to one model,[6] the maximum compressive stress during uplift occurs at the intersection of the purlin's bottom flange and its web. There, the purlin is unbraced. As Tondelli[10] demonstrates, the bottom flange can also be in compression under downward loading in short interior spans of uniformly loaded continuous purlin runs.

Behavior of roof purlins braced only on the tension side is extremely complex and poorly understood. The AISI Specification has undergone many changes in this regard, the latest one with publication of a 1989 Addendum. In order to calculate the nominal moment strength M_n of the member with one flange attached to through-fastened roofing, the Addendum introduced a "reduction factor" method contained in Equation C3.1.3-1:

$$M_n = R \cdot S_e \cdot F_y$$

where M_n = nominal moment strength
S_e = *effective* elastic section modulus computed with the compression flange assumed to be stressed to yielding
F_y = design yield stress
R = reduction factor that can be taken as
0.4 for simple-span channels
0.5 for simple-span Z sections
0.6 for continuous-span channels
0.7 for continuous-span Z sections

The M_n in this equation is the ultimate moment strength of the section, not the moment caused by the maximum allowable service load, and the R factors are not supposed to be automatically transferred to the allowable load values. Nevertheless the R factors are widely used to convert the allowable uniform-load values for fully braced conditions found in the manufacturer's tables into the design values for purlins braced only on one side.

The "reduction factor" method of analysis is based on an extensive testing program[7] performed within some very specific limits. It does not apply to situations where *any* of the 14 conditions specifically mentioned in the Specification are not met. For those nonconforming cases, the user is directed to apply an unspecified "another rational analysis procedure" or to perform a load test mentioned above. The values of R factors might be changed in the upcoming Specification editions, since the latest research indicates that R factors for channel and Z sections might in fact be similar. The engineer who gets involved in these issues should obtain a copy of the latest Specification.

5.3.6   Bracing design: from conservative to ignorant

Whenever top flanges of purlins are stabilized by steel deck, qualifying through-fastened roofing, or bracing, conservative design approaches utilize the above-mentioned method of reduction factors to determine maximum spacing of bottom-flange bracing. If both purlin flanges are unbraced, the Specification Section D.3.2.2 *requires* the braces to be attached to the top and bottom flanges of C and Z sections "at the ends and at intervals not greater than one-quarter of the span length..."[1]; an additional bracing at the concentrated load locations is also required.

An unconservative method of unbraced purlin design that relied on "hat section analogy" was contained in the provisions of an obsolete Part III, Section 3 of the 1980 *AISI Manual.*[8] Under that method, two purlins connected by metal roofing were assumed to form a "hat section," which resulted in almost no capacity reduction at all due to unsupported conditions. This method is no longer recognized as valid by the 1986 edition of the *Manual,*[9] where it is pointed out that the old methodology was developed based on the tests of isolated members, not complete roof systems.

As a result of the different methods being used and, of course, of different skills of people who apply them, it is possible to see identical buildings being designed with either three rows of bracing in each bay or none at all.

Then, there are more complications. Structural behavior of purlins braced at large intervals is not yet completely understood. According to LGSI's consultant M. Loseke, such bracing may lead to a new mode of failure not described in the AISI Specification—a local buckling of the purlin's stiffening lip. If too few braces are provided, such as only one at midspan, a lip section near the brace could become crippled, ceasing its contribution to the section's strength and losing its stiffening ability. Further research is needed to better understand this phenomenon, but it appears that if bracing is provided at all, it should be spaced at relatively close intervals.

5.3.7   The recommended bracing

Ideally, the bracing locations should be established by proper analysis or realistic testing, such as a new testing procedure based on research sponsored by MBMA and AISI. In the absence of such testing, some guidelines for preliminary design are needed. One set of such guidelines is offered by Loseke, who recommends the following bracing intervals based not on the span length but on the thickness of the purlin steel:

- Less than 5 ft for all 16-gage sections
- From 5 to 6 ft for all 14-gage sections
- 6 ft for sections heavier than 14 gage

For purlin spans of 20 to 30 ft, we further recommend that the bracing be provided at least at one-quarter span points and at supports, as required by the

1989 AISI Specification for totally unbraced conditions. For unusually short spans, fewer rows of bracing might be needed, for longer spans, more.

The type of bracing we recommend is made of angle sections, such as L 1⅝ in by 1⅝ in by 16 gage, not the flat strapping favored by some manufacturers for being inexpensive and easy to use, as already mentioned. For standing-seam roofs, the bracing angles are to be provided for each flange, top and bottom; for through-fastened roofs, only for the bottom flange.

The angles should be cross-braced at the ends, as shown in Fig. 5.12, to provide stability for the whole assembly. Also, the bracing at the ridge is crucial for tying the two opposing lines of bracing together. Without ridge bracing, there is little to prevent all the purlins from buckling in the same direction in the dreaded progressive collapse scenario. We favor a ridge connection of Fig. 5.12 that utilizes self-drilling, self-tapping screws, over the common bent-tab details shown in Figs. 5.11 and 5.13, because unlike the bent tabs, the screws possess a definite load-carrying capacity.

Since the manufacturers' bracing practices vary so widely, we recommend that the bracing requirements be spelled out in the contract documents. That way, the playing field is level, and the manufacturer with an unconservative design approach will not gain an unfair advantage over a more respectable supplier.

5.3.8 Failures due to improper bracing

Why do we devote so much attention to purlin and girt bracing? The answer is simple: Improperly braced purlins and girts are likely to fail in a real hurricane

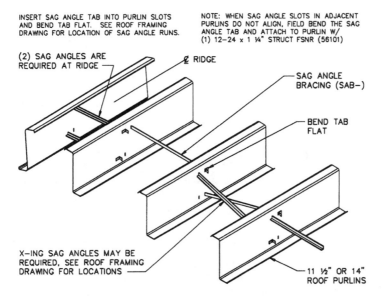

Figure 5.13 Typical sag angle details. (*Varco-Pruden Buildings.*)

or during a major snow accumulation. Tondelli[10] reports: "The most prevalent purlin and girt failure is due to wind suction at walls and at roofs, usually at end bays." Also, a study of damage caused by the 1970 Lubbock Storm and Hurricane Celia, published by the American Society of Civil Engineers, pointed out that wind uplift "caused buckling of purlins and girts in many metal building system structures. Buckling of the laterally unsupported inner flanges of purlins and girts was the initial type of damage incurred in many cases."[11]

By way of comparison, pre-engineered buildings designed in Canada have a very good loss experience, according to Canadian offices of Factory Mutual Insurance Co. Canadian National Building Code requires purlin bracing at quarter points.

Without proper bracing, a theoretical possibility of progressive collapse can become reality, with devastating consequences. As we discuss in Chap. 10, among some additional examples of building failures, one large metal building collapsed in 10 seconds from start to finish.

Curiously, many in the metal building industry regard some lateral flexibility of the purlins as a positive factor (and, logically, too much bracing as a vice). It is widely acknowledged that most metal building manufacturers rely upon purlin roll—rotation or horizontal displacement of the purlin's top flange—to reduce slotting of the through-fastened metal roofing.[12] The problem is, a little too much of the "roll," and the purlin might end up lying flat and fail.

Up to this day, some metal building manufacturers ignore the need for purlin bracing and will state if asked that none is required. Whenever the shop drawings indicate no purlin bracing at all, it is prudent to investigate the manufacturer's design approach to determine whether it is unconservative.

5.4   Other Types of Purlins for Metal Building Systems

5.4.1   Hot-rolled steel beams

Hot-rolled steel purlins predate modern metal building systems by decades. A multitude of industrial buildings constructed since the beginning of the century utilized hot-rolled channel and I-beam purlins spanning the distance between roof trusses, a then-dominant type of primary roof framing. The beams are still popular among many engineers for heavy-duty industrial applications and can be used in pre-engineered metal buildings as well. The main advantage of hot-rolled steel beams lies is their higher load-carrying capacities as compared with light-gage sections. The beams may be useful for spans longer than 30 ft, an upper limit for economical use of cold-formed framing. Also, hot-rolled purlins are quite appropriate for heavy suspended or concentrated loads. Among the disadvantages are their relatively high cost and a difficulty of obtaining continuity.

Hot-rolled shapes used as roof purlins still include channels and wide flanges. Both can either bear on top of primary-frame rafters or be framed flush. The top-bearing design is usually more economical, since it avoids expensive flange coping. Hot-rolled purlins are frequently used in combination with steel decking, which can span longer distances than through-fastened roofing and makes better bracing. Purlin spacing is governed by the deck's load-carrying capacity.

Hot-rolled purlins at sloped roofs do not escape the parallel-to-roof component of gravity loads (Fig. 5.14a). This component can be resisted either by a properly attached and continuous roof-deck diaphragm (Fig. 5.14b), or by sag rods (Fig. 5.14c), the spacing of which is determined by analysis. A typical sag rod bracing assembly is shown in Fig. 5.15.

The closer to the ridge, the greater the tension in the sag rods, since the upper rods collect the loads from all the purlins below them. In fact, the rod loaded the most is the tie rod over the ridge. Because of its critical function, the tie rod is often made of plates or structural shapes, rather than round bars.

If needed for bracing of the purlin's top flange, sag rods are located 2 or 3 in below the top flange, high enough to be effective and low enough for practicality of installation.

Unlike cold-formed C and Z framing, hot-rolled steel purlins can be readily designed for uplift, whether braced or unbraced between supports.

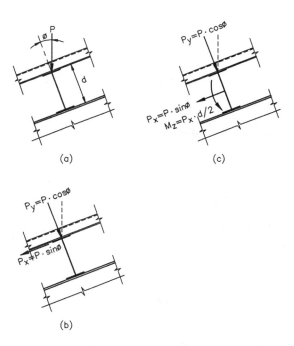

Figure 5.14 Forces acting on wide-flange purlins (channel purlins are subjected to an additional twisting component due to asymmetry). (a) Original force; (b) force resolution if roofing provides support for top flange, force P_x resisted by deck diaphragm; (c) force resolution if roofing provides no support, force P_x resisted by sag rods.

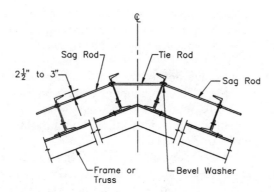

Figure 5.15 Typical sag rod details for hot-rolled purlins.

5.4.2 Open-web steel joists

Open-web steel joists, also known as bar joists, can span longer distances than both cold-formed and hot-rolled purlins. Open-web joists are discussed in Sec. 3.5.1, and the only point worthy of separate mention relates to our discussion about bracing for wind uplift. SJI's specification states that, where wind uplift forces are present, "a single line of *bottom chord* bridging must be provided near the first bottom chord panel point...."[13] In addition to this special uplift bracing, other top and bottom bridging is required, mainly for stability during erection.

5.5 Cold-Formed Steel Girts

Cold-formed C and Z girts are similar in most respects to cold-formed purlins, except that, of course, the girts are used in walls, not roofs. The discussion of available sections, basic design principles, continuity effects, and bracing needs for the purlins largely applies to cold-formed girts as well. One major exception is flange bracing, where girts often fare better than purlins: The exterior flanges of girts are typically braced by siding, while stability for the interior flanges can be provided by interior gypsum board or similar finishes. Some other differences worthy of note are summarized below.

5.5.1 Girt inset

Unlike cold-formed purlins which pass over the building frames to take advantage of continuity, light-gage girts can be positioned relative to columns in three different ways called insets. In *bypass* inset, the girts are located wholly outside the columns to take advantage of continuity (Fig. 5.16a). Bypass girts can be simply bolted to the outside column flange, if web crippling is not a problem, or be connected by bearing clips otherwise.

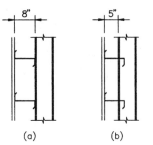

(a)   (b)   (c)

Figure 5.16 Girt insets. (a) Bypass; (b) semiflush; (c) flush.

Semiflush inset requires girt coping and allows some of the girt section to continue past the column (Fig. 5.16b). Bearing clips bolted to the outside column flange are typically required for attachment. (This design is not available from some manufacturers.)

Girts can also be positioned *flush* with the exterior of columns (Fig. 5.16c), being bolted to the column web with clip angles. Actually, the girts normally extend about 1 in past the column to allow for erection tolerances.

A close-up of the bypass girt assembly is shown in Fig. 5.17, and of the flush assembly in Fig. 5.18. Note the position of an eave strut in each illustration: It is simply bolted to the top of the column in the flush assembly but requires a special support bracket in the bypass configuration. At endwalls, incorporating any type of girt inset is easy, since there are no eave struts there; purlins simply cantilever over the end posts (Fig. 4.18a), and the rake angle spanning between the purlins picks up the top of the siding.

While the bypass inset allows for continuity, there may be a compelling reason to prefer the flush-type design whenever straight exterior columns are involved. Recall that wall panels are supported by fasteners at the exterior face of the girts and of the closure angle, or similar structure, attached to the foundation wall (Fig. 5.19). With bypass girts, the foundation must extend all the way to the inside face of the wall panel, and the resulting space between the inside surfaces of girts and columns is frequently unusable. A cost of this space could easily outweigh any savings derived from the girt continuity. Flush girts in combination with straight columns provide for a uniform and reasonable overall wall thickness.

5.5.2 Horizontal versus vertical girts

The most important function of girts is to transfer wind loads from wall materials to primary framing. Most commonly, girts are positioned horizontally to span between the frame columns. If the columns are spaced too widely, over 30 ft, intermediate *wind columns* are provided for girt support. Under this arrangement, metal siding—still the most popular wall material in metal building systems—is oriented vertically, being attached to each girt, to the base angle and the eave girt (Fig. 7.5). Girt spacing is governed by the load-resist-

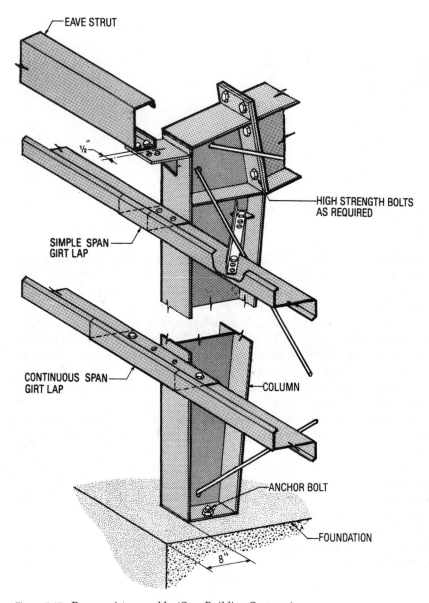

Figure 5.17 Bypass girt assembly. (*Star Building Systems.*)

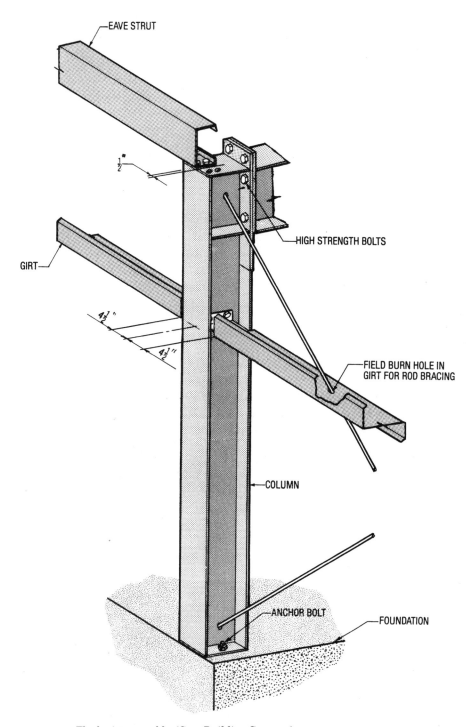

EAVE STRUT

$\frac{1}{2}$ "

HIGH STRENGTH BOLTS

GIRT

$4\frac{1}{2}$ "

$4\frac{1}{2}$ "

FIELD BURN HOLE IN
GIRT FOR ROD BRACING

COLUMN

ANCHOR BOLT

FOUNDATION

Figure 5.18   Flush girt assembly. (*Star Building Systems.*)

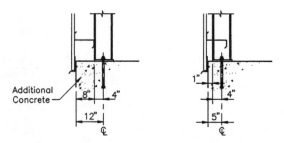

Figure 5.19 Some building space is lost with bypass girts.

ing properties of the wall panels; it is often between 6 and 8 ft for typical single-leaf siding. Figure 5.20 shows standard girt spacing for one manufacturer. The first girt is positioned to provide a clearance for doors.

At the base, the panels can be supported by a variety of means (see Fig. 5.21). With any of the details, base trim must be used to separate the panel from foundation concrete.

A less-known alternative is to run the girts vertically from foundation to the eave member and to utilize horizontally spanning extra-deep wall panels. This design solution can rise above the conventional by producing an interesting wall treatment with traditional materials. Vertical girts, akin to wall studs spaced at wide intervals, are framed into eave members, which act as beams spanning between the columns and resisting wind reactions from the girts. Standard cold-formed eave girts are in all probability not strong enough for that function, and hot-rolled beams are called for.

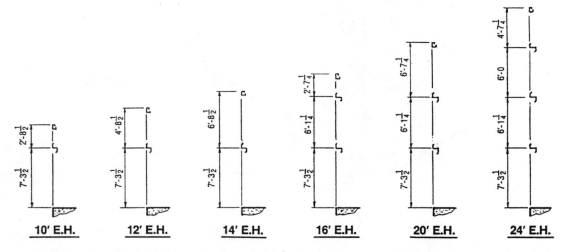

Figure 5.20 Typical horizontal girt spacing for various eave heights. (*Star Building Systems.*)

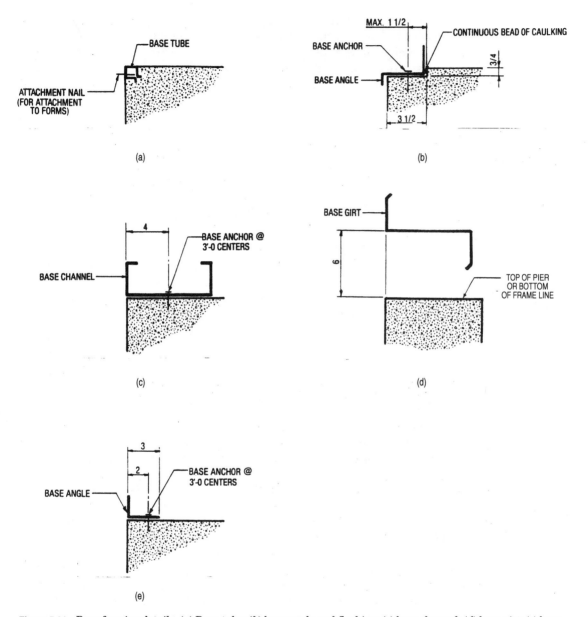

Figure 5.21   Base framing details. (*a*) Base tube; (*b*) base angle and flashing; (*c*) base channel; (*d*) base girt; (*e*) base angle. Note: A notch in concrete can be provided to align the exterior faces of metal panel and concrete. (*Star Building Systems.*)

Positioning girts vertically should be approached with caution when the building eave height exceeds 30 ft, a practical span limit for cold-formed framing, above which some intermediate horizontal framing is probably needed. Another issue to think about is how the column flange bracing will be handled since the traditional flange angles (Fig. 4.15) cannot be used with vertical girts. It is possible, of course, to design the main frames to be heavy enough not to need any flange bracing at all, in a sacrifice on the altar of appearance.

5.6   Hot-Rolled Steel Girts

Hot-rolled steel girts are specified for the same reasons hot-rolled purlins are—higher load-carrying capacity and designer's familiarity (sometimes bordering on distrust of cold-formed construction in general). Made of channel or wide-flange beam sections, these girts can be especially useful for spanning long distances and for custom framing around large windows and overhead doors. Since continuity is difficult to achieve with hot-rolled girts anyway, these sections are frequently designed with flush or semiflush insets.

Also, while the weight of cold-formed C and Z girts is insignificant, hot-rolled framing is rather heavy, tends to sag, and needs to be supported at regular intervals by the appropriately named sag rods. A channel girt is commonly analyzed as a simple-span beam for wind loading and as a continuously supported beam for gravity load, which consists of the girt's own weight and that of any supported wall materials. Sag rods are ultimately supported by the eave girt, also a hot-rolled member.

The issue of lateral bracing for hot-rolled girts is worthy of special note. With through-fastened metal siding, the girts can usually be considered braced at their exterior flanges. Room finishes such as drywall, if carried on steel studs or furring, can provide bracing for the interior flanges, which otherwise are deemed unbraced.

There are two ways to design a sag-rod supported girt with unbraced interior flange. The first approach simply assumes that the interior flange is unsupported from column to column and neglects any bracing contribution of the sag rods. The steel sections engineered under this assumption are so heavy that both clients and contractors tend to question their design.

The second approach recognizes a restraining action of the sag rods. The girts are considered laterally supported at each rod, and the sag rods are located as close together as needed for a full efficiency of the girt section. This seemingly unconservative approach has been used for decades and has withstood the test of time.

For the unconvinced, here is one rationalization, as advanced by the author earlier.[14] In order for the interior flange to buckle laterally—the most probable mode of failure—it must rotate and move vertically. This movement is prevented both at the exterior flange by the siding fasteners and at the sag rod locations. The interior and exterior flanges are, of course, tied together by the web, which acts as a cantilevered beam in restraining the unbraced flange (Fig. 5.22). It is commonly assumed that the compression flange of a flexural mem-

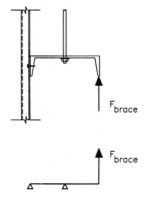

F_{brace}

F_{brace} **Figure 5.22** Web cross bending.

ber may be considered braced if the brace can resist some 2 percent of the compressive force in the flange. The bracing action of the web occurs, therefore, if the web is strong enough to resist this force by cross bending. The effective width of the web for this action is a matter of engineering judgment.

For girts deeper than 8 in, the web may be too thin for the cantilever action to work. In this case a few continuous lines of interior flange bracing, attached to the eave girt and to the foundations, may be needed.

References

1. Specification for the Design of Cold-Formed Steel Structural Members, 1986 with 1989 Addendum, American Iron and Steel Institute, Washington, D.C.
2. Design Standard for Load and Resistance Factor Design Specification for Cold-Formed Steel Structural Members, American Iron and Steel Institute, Washington, D.C., 1991.
3. ASTM A 653-94, Standard Specification for Steel Sheet, Zinc-Coated (Galvanized) or Zinc-Iron Alloy-Coated (Galvannealed) by the Hot-Dip Process, American Society for Testing and Materials (ASTM), Philadelphia, Pa., 1994.
4. *Light Gage Structural Steel Framing System Design Handbook,* Light Gage Structural Institute, Plano, Tex., 1995.
5. Frank Zamecnik, "Errors in Utilization of Cold-Formed Steel Members," *ASCE Journal of Structural Division,* vol. 106, no. ST12, December 1980.
6. Roger A. LaBoube, "Uplift Capacity of Z-Purlins," *ASCE Journal of Structural Engineering,* vol. 117, no. 4, April 1991.
7. Roger A. LaBoube, "Estimating Uplift Capacity of Light Steel Roof System," *ASCE Journal of Structural Engineering,* vol. 118, no. 3, April 1992.
8. *Cold-Formed Steel Design Manual,* American Iron and Steel Institute (AISI), Washington, D.C., 1980.
9. *Cold-Formed Steel Design Manual,* AISI, Washington, D.C., 1986.
10. Juan Tondelli, "Purlin and Girt Design," *Metal Architecture,* February 1992, p. 26.
11. Joseph Minor et al., "Failures of Structures Due to Extreme Winds," *ASCE Journal of the Structural Division,* vol. 98, ST11, November 1972.
12. "Serviceability Design Considerations for Low-Rise Buildings," *Steel Design Guide Series* no. 3, AISC, Chicago, Ill., 1990.
13. Standard Specifications for Open Web Steel Joist, K-Series, Steel Joist Institute, 1985.
14. Alexander Newman, "The Answer in 'Steel Interchange,'" *Modern Steel Construction,* AISC, Chicago, Ill., September 1994, p. 10.
15. Ahmad A. Ghosn and Ralph R. Sinno, "Load Capacity of Nested, Laterally Braced, Cold-Formed Steel Z-Section Beams," *ASCE Journal of Structural Engineering,* vol. 122, no. 8, August 1996.

6

Metal Roofing

6.1 Introduction

Roof's function goes far beyond protecting the interior from the weather. Architecturally, roof complements and accentuates the color and texture of the building and plays a major role in establishing its character and appearance. Structurally, roof covering may resist wind and live loads and may serve as bracing for roof purlins.

Metal roofing is among the most attractive features pre-engineered buildings have to offer, having contributed mightily to the growing popularity of metal building systems. This chapter examines the available metal roofing materials and discusses specifying metal roofing for new construction. Some special challenges of reroofing applications are outlined in Chap. 14.

Metal roofing has been used in Europe, and later in this country, for centuries. Traditionally, it was formed into pans by hand for subsequent manual crimping or seaming. The glorious golden dome covering the Massachusetts State House was fabricated and installed by none other than Paul Revere.[1] Designed by Charles Bulfinch and completed in 1797, this edifice is still considered to be one of the five best buildings in Boston.

Some popular turn-of-the-century roofing materials included terne roofs, made of steel coated with an alloy of 4 parts lead and 1 part tin. Unfortunately, those early terne roofs eventually rusted and had to be painted over.

Contemporary metal roofing is a far cry from its predecessors. Today's products offer long, largely maintenance-free service lives, reflected in 20-year warranties; the best may last for half a century with some periodic maintenance and spot repairs.

Occasionally, pre-engineered buildings are covered with nonmetal roofs— built-up or membrane. This fact could reflect many reasons ranging from the architect's desire to fit into the surrounding environment, where metal roofing might be out of character, to the owner's prosaic preference to avoid hearing the rain noise amplified by metal. Built-up or membrane roofs are easier to incorporate into the buildings which already deviate from the one-trade concept, such as those framed with bar joists or hot-rolled purlins and steel roof deck. Design guidelines for nonmetal roofing are widely available and are not repeated here.

6.2   Main Types of Metal Roofs

6.2.1   "Waterproof" vs. "water-shedding" roofing

Fundamentally, metal roofing can be classified by the way it resists water intrusion. *Water-shedding* roofing is functionally similar to roof shingles—it relies on steep slope to rapidly shed rainwater. As with shingles, the minimum slope required by this type of roofing is 4:12, although a 3:12 pitch is often considered acceptable. Water-shedding roofing is normally installed on top of underlayment, such as 30-lb roofing felt, and is sometimes separated from it by a friction-reducing paper slip sheet.

In contrast, *water barrier,* or "waterproof," roofs are intended to function under occasional standing water. The system is not designed to be completely leak-free under long-term water immersion, however, and still requires some minimum roof slope for best performance. Tobiasson and Buska[25] recommend a minimum slope of 1:12 (1 in/ft) for "waterproof" standing-seam roofs and state that the slopes larger than the minimum perform better, especially in cold regions. Others consider the slopes as low as $\frac{1}{4}$:12 to be adequate. In any case, as Ref. 26 points out, water-barrier metal roofs are "usually not watertight at their valleys, eaves, ridges, rakes and penetrations."

6.2.2   Architectural vs. structural roofing

The terms "architectural" and "structural" are somewhat misleading, as either type of metal roofing serves architectural purposes and is available in a variety of finishes and profiles. The main difference between the two types is this: Architectural roofing relies on structural support to be provided by decking or by closely spaced furring channels, while structural roofing can span the distance between roof purlins unassisted.

In practice, architectural roofing is akin to water-resisting cladding or, more specifically, to water-shedding roofing discussed above. True to the name, architectural roofing may be used to create dramatic visual effects not possible with other types of roofing (Fig. 6.1).

Structural roofing is often implied to be of "waterproof" design, although these two terms refer to different concepts. Structural roofing may be used on shallow slopes, perhaps as low as $\frac{1}{4}$:12, even though a larger slope is preferable, as was just pointed out. Most roofing types can be installed on very steep slopes, including vertical, although good sealants and sturdy structural supports become critical for steep-slope installations.

Structural roofing may be considered a form of roof decking and, as such, is required to meet certain wind uplift criteria as well as to support a worker's weight (250 lb). Architectural roofing does not have to meet these require-

Figure 6.1 Architectural roofing provides a bold visual effect.

ments. Both kinds of metal roofing are extremely lightweight, with weights of only 1 to 2 lb/ft^2.

As discussed in Chap. 3, wind loading is not uniform from one part of the roof to another. The loading is much higher along the roof's perimeter, for example, and sometimes along the ridge. Instead of using structural roofing panels of heavier gages in the areas of high localized loads, it is better to space the purlins closer. A common design involves supports for structural roofing at 5 ft in the field of the roof, but only half that within 10 ft or so from the roof edges.

The contract documents should indicate the maximum deflection criteria for structural roofing. A limit of $L/180$ is reasonable, although there could be circumstances when a more stringent limit is justified. Structural design of sheet-metal roofing follows the AISI Specification mentioned in Chap. 5. Many engineers, especially those following U.S. Army Corps of Engineers Guide Specifications,[10] specify minimum section properties of the roofing—the moment of inertia and the section modulus—right on the contract drawings.

6.2.3 Classification by method of attachment and by direction of run

Metal roofing can also be classified by method of attachment to supports. *Through-fastened roofs* are attached directly to purlins, usually by screws or rivets. *Standing-seam* roofing, on the other hand, is connected indirectly by concealed clips formed into the seams.

Metal roofing comes in ribbed panels with seams normally located along the slope. One exception to this rule is *Bermuda roofing,* which runs horizontally. The panels of this original roofing are through-fastened to supports with concealed clips and resemble clapboards with reveals of 9.5 to 11.5 in. The inherent design weakness posed by a horizontal seam orientation limits the use of Bermuda roofing to locales without snow and ice accumulation.

6.2.4 Metal roofing market

Metal roofs have been steadily gaining in popularity, and today cover about two-thirds of new low-rise nonresidential buildings. According to a survey by Metal Construction Association, 2 billion ft^2 of metal roofing was installed in 1995, a one-third increase over 1994. One-quarter of that amount has been used in conventional (non–pre-engineered) buildings.[2] A survey of metal roofing specifiers found that almost 64 percent had used architectural metal roofing and over 59 percent, structural. A so-called specialty metal roofing embossed to resemble tile, shakes, or shingles was specified by about 11 percent of the survey participants.[3]

Most panels are manufactured, or "preformed," from precoated coils of light-gage metal at the factory, where the coils are handled with great care to preserve the finish during forming. Attempts to apply coatings after forming tend to result in some quality problems with color, thickness uniformity, and durability.[4]

Recently, portable roll-forming machines have become available, and many types of panels can now be formed on-site. The survey[3] indicates that almost 16 percent of the respondents had specified a site-formed panel system. Factory forming, however, is still likely to provide a better-quality finish.

6.3   Various Seam Configurations

Contemporary roofing panels come in the following seam configurations:

1. *Lapped seam,* normally found in through-fastened roofs, offers the simplest and most economical design (Fig. 6.2*a*). The edges of corrugated roofing panels are simply overlapped, receive a bead of sealant, and are fastened to roof purlins. Despite the economy, the fasteners of lapped panels are exposed to weather—and sight. This system lacks a certain sophistication and is reserved for relatively basic, functional structures.

2. *Flat seam* is formed by bending the sides of two adjacent roofing sheets 180° and hooking them together (Fig. 6.2*b*). It is relatively rare.

3. *Batten seam* originated in the times of hand-forming, when the sides of two adjacent panels were bent up, separated by a wood batten strip, and covered with a snap-on cap (Fig. 6.2*c*). Most modern batten seam systems dispense with wood but preserve the metal batten design (Fig. 6.3).

4. *Standing-seam* roofing has the seams elevated 2 to 3 in above the flat panel part that carries water (Fig. 6.2*d*). The picture shows a so-called Pittsburgh double lock, a 360° roll-formed seam resembling the seam of a food

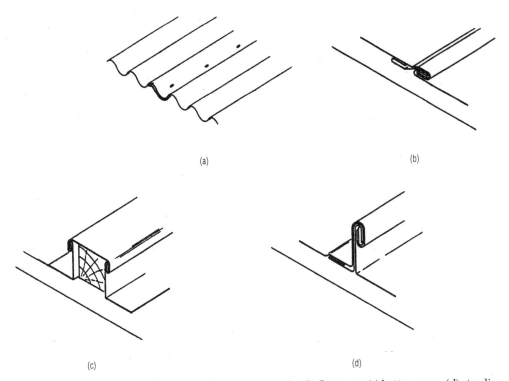

(a)

(b)

(c)

(d)

Figure 6.2 Various seam configurations. (*a*) Lapped panels; (*b*) flat seam; (*c*) batten seam; (*d*) standing seam. (*Adapted from Means Square Foot Cost Data 1995. Copyright R.S. Means Co., Inc., Kingston, MA, 617-585-7880, all rights reserved.*)

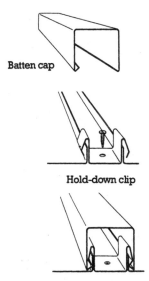

Batten cap

Hold-down clip

Clips are located
at a maximum
of 2′ on center.

Figure 6.3 Batten seam details.
(*Carlisle Engineered Metals.*)

can. Other types of standing seams include those that are simply snapped together, usually with a sealant in between. The majority of modern metal roof systems, about 60 percent by one count, use a standing-seam design.

6.4 Through-Fastened Roofing

Through-fastened lapped-seam roofing represents the oldest design approach to metal roofs. Still a popular choice for industrial and warehouse applications, it is inexpensive, straightforward, and easy to erect. It has two major disadvantages, however.

The first concerns roofing penetration by fasteners and a corresponding potential for leaks. Success of a typical cure for this problem, self-tapping screws with rubber or neoprene washers, depends heavily on the installer's skills. The second problem is that the roofing is prevented by the fasteners from thermal expansion and contraction. As discussed in Chap. 3, a long piece of metal can buckle if temperature stresses become excessive. Even if it does not, repeated expansion and contraction will tear the metal around the connecting screws and lead to leaks.[5] Or the screws may become so loosened by the continuing sidesway motions that the roofing might be blown off during a hurricane. To limit buildup of thermal stresses, the width of buildings with through-fastened roofs should not exceed approximately 60 ft. In such smaller buildings, through-fastened roofs may have a better chance of survival.

Another potential problem with screw-fastened roofing is metal fatigue. Xu[9] and others have shown that this type of roofing may fail locally, by cracking around the fasteners, when subjected to strong fluctuating wind loading. Lynn and Stathopoulos[28] have concluded that wind-induced fatigue was the only possible cause of several failures of through-fastened roofs in Australia during cyclone Tracy in 1974. Whether other kinds of metal roofing are equally vulnerable is not clear at this time.

Through-fastened roofing is usually 1 to 2 in deep and is made of steel 26 to 24 gage thick. We recommend the 24-gage material for better dimensional stability and impact resistance. The manufacturers offer load tables for various panel configurations, such as those of Figs. 6.4 and 6.5, to facilitate roofing selection given the roof live or snow loads, purlin spacing, and wind uplift rating. Another selection criterion concerns insurance requirements, which may actually control the design. Factory Mutual, for example, often requires panels of larger depth or gage than needed for strength alone.[6]

While many engineers would not want to impose any fastening requirements on manufacturers, some wish to offer such guidance. The following is recommended by U.S. Army Corps of Engineers[10] as the maximum fastener spacing:

- 8 in at end laps and at connections to intermediate supports
- 12 in at side laps

In order for gasketed washers to function properly, the fasteners should be driven to proper depth perpendicular to the roofing. All roofing joints should be sealed, and the sidelaps laid away from the prevailing winds.

"R" PANEL

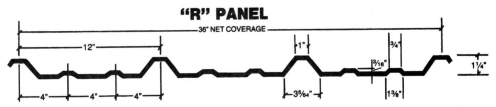

SECTION PROPERTIES									
			TOP FLAT IN COMPRESSION			BOTTOM FLAT IN COMPRESSION			
PANEL GAGE	F_y (KSI)	WEIGHT (PSF)	I_x (in.⁴/ft.)	S_x (in.³/ft.)	M_a (in.-Kips)	I_x (in.⁴/ft.)	S_x (in.³/ft.)	M_a (in.-Kips)	
29	80.0	0.75	0.0286	0.0255	0.9161	0.0276	0.0333	1.1973	
26	80.0	0.94	0.0423	0.0388	1.3928	0.0390	0.0437	1.5698	
24	50.0	1.14	0.0542	0.0518	1.5500	0.0517	0.0544	1.6296	
22	50.0	1.44	0.0696	0.0704	2.1084	0.0454	0.0610	1.8253	

NOTES
1. All calculations for the properties of panels are calculated in accordance with the 1986 edition of *Specifications for the Design of Light Gauge Cold Formed Steel Structural Members* - published by the American Iron and Steel Institute (A.I.S.I.).
2. I_x is for deflection determination.
3. S_x is for bending.
4. M_a is allowable bending moment.
5. All values are for one foot of panel width.

ALLOWABLE UNIFORM LIVE LOADS
IN POUNDS PER SQUARE FOOT

		29 Gage (F_y = 80 KSI)					26 Gage (F_y = 80 KSI)				
SPAN TYPE	LOAD TYPE	SPAN IN FEET					SPAN IN FEET				
		4.0	5.0	6.0	7.0	8.0	4.0	5.0	6.0	7.0	8.0
2-SPAN	POSITIVE WIND LOAD	51	33	22	17	13	78	50	34	26	19
	LIVE LOAD/DEFLECTION	43	22	13	8	6	63	32	18	12	8
3 OR MORE	POSITIVE WIND LOAD	64	41	28	21	16	97	62	43	32	24
	LIVE LOAD/DEFLECTION	54	28	16	10	7	79	40	23	15	10

		24 Gage (F_y = 50 KSI)					22 Gage (F_y = 50 KSI)				
SPAN TYPE	LOAD TYPE	SPAN IN FEET					SPAN IN FEET				
		4.0	5.0	6.0	7.0	8.0	4.0	5.0	6.0	7.0	8.0
2-SPAN	POSITIVE WIND LOAD	86	55	38	28	22	117	75	52	38	30
	LIVE LOAD/DEFLECTION	68	42	24	15	10	76	46	26	17	11
3 OR MORE	POSITIVE WIND LOAD	108	69	48	35	27	146	94	65	48	37
	LIVE LOAD/DEFLECTION	85	52	30	19	13	95	57	33	21	14

NOTES
1. Allowable loads are based on uniform span lengths and F_y of 80 KSI for 29 and 26 gage and F_y of 50 KSI for 24 and 22 gage.
2. Live load is allowable live load.
3. Wind load is allowable wind load and has been increased by 33⅓%.
4. Deflection loads are limited by a maximum deflection ratio of L/240 of span or maximum bending stress from live load.
5. Weight of the panel has not been deducted from allowable loads.
6. Load table values do not include web crippling requirements.
7. Minimum bearing length of 1½" required.

Figure 6.4 R panel by MBCI. (*MBCI.*)

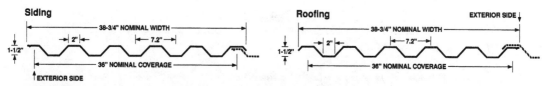

Siding **Roofing**

<table>
<tr><th colspan="13">SMITH STEELITE STYLE-RIB SIDING AND ROOFING MAXIMUM SPANS</th></tr>
<tr><td>Live Load</td><td rowspan="2">Defl.</td><td rowspan="2">Span</td><td colspan="2">20 PSF (98 kg/m²)</td><td colspan="2">30 PSF (146 kg/m²)</td><td colspan="2">40 PSF (195 kg/m²)</td></tr>
<tr><td>Gage/Weight</td><td>Wall</td><td>Roof</td><td>Wall</td><td>Roof</td><td>Wall</td><td>Roof</td></tr>
<tr><td rowspan="6">GALV. STEEL
18 Gage (0.047")
2.49 lbs./ft²</td><td rowspan="3">L/120</td><td>SS</td><td>11'-4" (3.45 m)</td><td>11'-3" (3.43 m)</td><td>9'-10" (3.00 m)</td><td>9'-10" (3.00 m)</td><td>9'-0" (2.74 m)</td><td>8'-11" (2.71 m)</td></tr>
<tr><td>DS</td><td>15-2 (4.62)</td><td>14-0 (4.27)</td><td>13-1 (3.98)</td><td>11-5 (3.48)</td><td>11-4 (3.45)</td><td>9-11 (3.02)</td></tr>
<tr><td>TS</td><td>13-11 (4.24)</td><td>13-11 (4.24)</td><td>12-2 (3.71)</td><td>12-2 (3.71)</td><td>11-1 (3.38)</td><td>11-0 (3.35)</td></tr>
<tr><td rowspan="3">L/180</td><td>SS</td><td>9-10 (3.00)</td><td>9-10 (3.00)</td><td>8-7 (2.62)</td><td>8-7 (2.62)</td><td>7-10 (2.39)</td><td>7-10 (2.39)</td></tr>
<tr><td>DS</td><td>13-3 (4.04)</td><td>13-2 (4.01)</td><td>11-7 (3.53)</td><td>11-5 (3.48)</td><td>10-6 (3.20)</td><td>9-11 (3.02)</td></tr>
<tr><td>TS</td><td>12-2 (3.71)</td><td>12-2 (3.71)</td><td>10-8 (3.25)</td><td>10-7 (3.22)</td><td>9-8 (2.94)</td><td>7-10 (2.39)</td></tr>
<tr><td rowspan="6">GALV. STEEL
20 Gage (0.036")
1.87 lbs./ft²</td><td rowspan="3">L/120</td><td>SS</td><td>10-3 (3.12)</td><td>10-2 (3.10)</td><td>9-0 (2.74)</td><td>8-11 (2.72)</td><td>8-2 (2.49)</td><td>8-1 (2.46)</td></tr>
<tr><td>DS</td><td>13-5 (4.09)</td><td>11-7 (3.53)</td><td>10-11 (3.33)</td><td>9-6 (2.90)</td><td>9-6 (2.90)</td><td>8-2 (2.49)</td></tr>
<tr><td>TS</td><td>12-8 (3.86)</td><td>12-7 (3.83)</td><td>11-1 (3.38)</td><td>10-7 (3.22)</td><td>10-1 (3.07)</td><td>9-2 (2.79)</td></tr>
<tr><td rowspan="3">L/180</td><td>SS</td><td>9-0 (2.74)</td><td>8-11 (2.72)</td><td>7-10 (2.39)</td><td>7-9 (2.36)</td><td>7-1 (2.16)</td><td>7-1 (2.16)</td></tr>
<tr><td>DS</td><td>12-0 (3.66)</td><td>11-7 (3.53)</td><td>10-6 (3.20)</td><td>9-6 (2.90)</td><td>9-6 (2.90)</td><td>8-2 (2.49)</td></tr>
<tr><td>TS</td><td>11-1 (3.38)</td><td>11-0 (3.35)</td><td>9-8 (2.95)</td><td>9-7 (2.92)</td><td>8-9 (2.67)</td><td>8-9 (2.67)</td></tr>
<tr><td rowspan="6">GALV. STEEL
22 Gage (0.030")
1.56 lbs./ft²</td><td rowspan="3">L/120</td><td>SS</td><td>9-7 (2.92)</td><td>9-7 (2.92)</td><td>8-5 (2.56)</td><td>8-4 (2.54)</td><td>7-7 (2.31)</td><td>7-5 (2.26)</td></tr>
<tr><td>DS</td><td>11-10 (3.61)</td><td>10-3 (3.12)</td><td>9-8 (2.94)</td><td>8-5 (2.56)</td><td>8-4 (2.54)</td><td>7-3 (2.21)</td></tr>
<tr><td>TS</td><td>11-10 (3.61)</td><td>11-6 (3.51)</td><td>10-4 (3.15)</td><td>9-5 (2.87)</td><td>9-4 (2.84)</td><td>8-1 (2.46)</td></tr>
<tr><td rowspan="3">L/180</td><td>SS</td><td>8-5 (2.56)</td><td>8-4 (2.54)</td><td>7-4 (2.23)</td><td>7-3 (2.21)</td><td>6-8 (2.03)</td><td>6-7 (2.00)</td></tr>
<tr><td>DS</td><td>11-3 (3.43)</td><td>10-3 (3.12)</td><td>9-8 (2.94)</td><td>8-5 (2.56)</td><td>8-4 (2.54)</td><td>7-3 (2.21)</td></tr>
<tr><td>TS</td><td>10-4 (3.15)</td><td>10-4 (3.15)</td><td>9-1 (2.77)</td><td>9-0 (2.74)</td><td>8-3 (2.51)</td><td>8-1 (2.46)</td></tr>
<tr><td rowspan="6">GALV. STEEL
24 Gage (0.024")
1.24 lbs./ft²</td><td rowspan="3">L/120</td><td>SS</td><td>8-10 (2.69)</td><td>8-9 (2.67)</td><td>7-8 (2.33)</td><td>7-4 (2.23)</td><td>7-0 (2.13)</td><td>6-4 (1.93)</td></tr>
<tr><td>DS</td><td>10-2 (3.10)</td><td>8-10 (2.69)</td><td>8-4 (2.54)</td><td>7-2 (2.18)</td><td>7-2 (2.18)</td><td>6-3 (1.91)</td></tr>
<tr><td>TS</td><td>10-10 (3.30)</td><td>9-10 (3.00)</td><td>9-4 (2.84)</td><td>8-1 (2.46)</td><td>8-1 (2.46)</td><td>7-0 (2.13)</td></tr>
<tr><td rowspan="3">L/180</td><td>SS</td><td>7-8 (2.33)</td><td>7-8 (2.33)</td><td>6-9 (2.06)</td><td>6-8 (2.03)</td><td>6-1 (1.85)</td><td>6-1 (1.85)</td></tr>
<tr><td>DS</td><td>10-2 (3.10)</td><td>8-10 (2.69)</td><td>8-4 (2.54)</td><td>7-2 (2.18)</td><td>7-2 (2.18)</td><td>6-3 (1.90)</td></tr>
<tr><td>TS</td><td>9-6 (2.89)</td><td>9-6 (2.90)</td><td>8-3 (2.51)</td><td>8-1 (2.46)</td><td>7-6 (2.29)</td><td>7-0 (2.13)</td></tr>
</table>

All above weights are per net square foot.
Loads and spans for carbon steel are based on AISI Cold-Formed Steel Design Manual.
The above span tables are in accordance with the 1986 Light Steel Code with material having a yield strength of 33,000 psi (2320 kg/cm²), one-third extra strength for wind loads only.
Roof spans are for positive loading. Wall spans are for positive or negative loading.

Minimum sheet length: 2'-0" (.61 m). Maximum sheet length: 40'-0" (12.19 m).
Consult Smith Steelite for sheet lengths less than 2'-0" (.61 m) or greater than 40'-0" (12.19 m).
Length tolerance: maximum variation ±1/2" (12.7 mm).
Roof spans do not include dead weight of panels.

Figure 6.5 Style-Rib siding and roofing by Smith Steelite. (*Smith Steelite.*)

Some manufacturers attempt to overcome the second weakness of through-fastened roofing by providing slotted holes in the panels. For example, Butlerib II roof system by Butler Manufacturing Company utilizes prepunched slotted holes in the bottom sheet of the end joint and regular holes in the top sheet (Fig. 6.6a). A typical panel is shown in Fig. 6.6b. The manufacturer points out that it has taken extraordinary steps to perfect this system by providing a long return leg, which increases dimensional stability under roof traffic, by using constant-grip lock rivets instead of sheet-metal screws and by incorporating a special sealant groove in the seam (Fig. 6.6c). These steps have resulted in a premium through-fastened system with a unique 10-year weathertightness warranty.

A few other manufacturers attempt to justify lack of slotted holes in their panels by relying on purlin roll—a slight rotation under thermal loading. The purlin roll, while quite real, exists mostly in roofs with cold-formed C or Z sec-

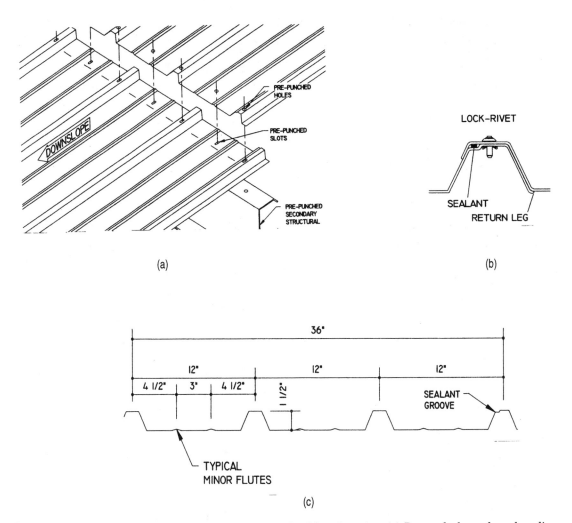

Figure 6.6 Exposed-fastener panels: Butlerib II by Butler Manufacturing. (*a*) Prepunched panels and purlins assure correct alignment; (*b*) seam details with lock rivet; (*c*) panel cross section. (*Butler Manufacturing Co.*)

tions without top-flange bracing. As discussed in Chap. 5, the practice of relying on purlin roll raises some serious questions.

Exposed fasteners without a durable corrosion-resisting coating invite trouble. The better ones are made of stainless steel or aluminum; galvanized or cadmium-plated screws are best left for interior applications. To reduce complaints about fastener visibility, exposed fasteners may have a color-coordinated head finish or be fitted with colored plastic caps.

Many problems associated with exposed fasteners, such as visibility and corrosion, can be solved with concealed-fastener systems. Since these systems are most commonly selected as wall materials, the discussion about them is deferred until the next chapter.

6.5 Standing-Seam Roof (SSR)

6.5.1 System components

The standing-seam roof is a major improvement over the through-fastened variety. Instead of being simply lapped together, the adjacent seams of SSR panels are formed in the field by a portable seaming machine or, in a step down, by hand tools. To accommodate expansion and contraction, the roofing sheets are attached to purlins with concealed clips allowing for movement. (Curiously, the first standing-seam roofing introduced by Armco Buildings in 1934 had exposed fasteners. The concealed-clip design was introduced only in 1969 by Butler Manufacturing Company.) Some of the most common types of commonly available clips are shown in Fig. 6.7.

Despite visual differences, all the clips consist of two pieces—the rigid base attached to the purlin and the movable insert rolled into the seam. The best clip designs are self-centering, i.e., preset to allow an equal amount of movement up and down the slope. The latest invention is a so-called articulating clip, intended to compensate for misaligned roof purlins.[7] The clip, first introduced by Elco Industries and now offered by MBCI, is shown in Fig. 6.7.

Some other features found in high-end systems include stainless steel, rather than galvanized, and movable inserts as well as prepunched holes in both panels and purlins, which reduces panel misalignment. Such prepunching, if available, is a good enough reason to buy a complete metal building system from one manufacturer instead of mixing and matching components from various suppliers.

Standing-seam roofing can be encountered in both architectural and structural incarnations. Joint sealant is normally required to improve seam watertightness in structural panels on low-rise roofs. The clips shown in Fig. 6.7 are reserved for structural roofs. Architectural roofing panels, owing to their shorter length and span, do not need such elaborate designs and instead are connected to supports with fixed clips or cleats (Fig. 6.8). The clips, which are spaced 2 to 3 ft apart, allow the panels to simply slide back and forth without binding. Architectural roofing normally requires no sealants in the seams.

The most common SSR seam configurations are shown in Fig. 6.9. As it indicates, there are two distinct groups of leg design—vertical and trapezoidal. Both types have their adherents among various manufacturers. The trapezoidal leg seems to be more popular, partly because it allows for an easy concealment of the clip, and partly because it is better able to accommodate thermal expansion and contraction of the roofing perpendicular to the slope (Fig. 6.10). Spacing of the corrugations and width of the panels vary among the manufacturers. Some panels with widely spaced seams are provided with cross flutes 6 in or so on centers for better rigidity and walkability, as well as for reduction of wind vibrations and noise. An example of the properties and configuration of standing-seam panels may be found in Fig. 6.11.

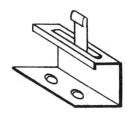

TAB MOVEMENT IN A SLOT AT THE TOP

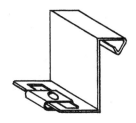

THE CLIP MOVES WITH THE PANELS THROUGH A SLOT IN THE BASE. THE "C" CLAMP ATTACHES AND ALLOWS THE CLIP TO MOVE

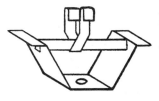

TAB MOVEMENT ON A BAR AT THE TOP

TAB MOVEMENT IN A SLOT IN THE MIDDLE WHILE BASE ARTICULATES TO ACCOUNT FOR SUBSTRUCTURE MISALIGNMENT

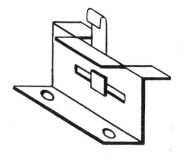

TAB MOVEMENT IN A SLOT IN THE MIDDLE

Figure 6.7 Common designs of standing-seam roof clips. (*From Nimtz,[7] courtesy of Metal Architecture.*)

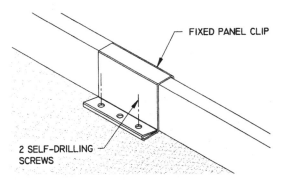

FIXED PANEL CLIP

2 SELF-DRILLING SCREWS

Figure 6.8 Fixed clips may be used for short architectural panels. (*Butler Manufacturing Co.*)

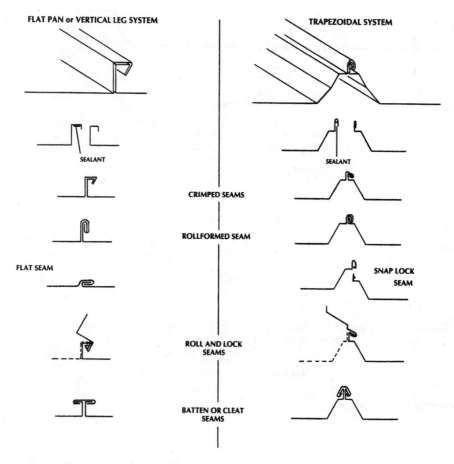

Figure 6.9 Common panel seams for standing-seam roofs. (*From Nimtz,*[7] *Courtesy of Metal Architecture.*)

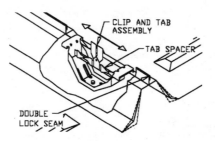

Figure 6.10 Trapezoidal seam allows for an easy concealment of the clip and accommodates transverse panel movement. (*Butler Manufacturing Co.*)

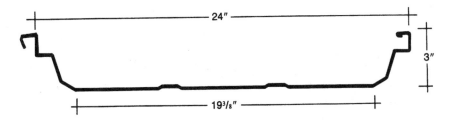

SECTION PROPERTIES								
			TOP FLAT IN COMPRESSION			BOTTOM FLAT IN COMPRESSION		
PANEL GAGE	F_y (KSI)	WEIGHT (PSF)	I_x (in.⁴/ft.)	S_x (in.³/ft.)	M_a (Kip in.)	I_x (in.⁴/ft.)	S_x (in.³/ft.)	M_a (Kip in.)
24	50.0	1.23	0.1282	0.0666	1.9945	0.0528	0.0485	1.4528
22	50.0	1.56	0.1905	0.1031	3.0861	0.0720	0.0680	2.0350

NOTES
1. All calculations for the properties of double-lok ℠. panels are calculated in accordance with the 1986 edition of *Specifications for the Design of Light Gauge Cold Formed Steel Structural Members* - published by the American Iron and Steel Institute (A.I.S.I.).
2. I_x is for deflection determination.
3. S_x is for bending.
4. M_a is allowable bending moment.
5. All values are for one foot of panel width.

ALLOWABLE UNIFORM LIVE LOADS
IN POUNDS PER SQUARE FOOT

24" double-lok ℠.

24 Gage (F_y = 50 KSI)

SPAN TYPE	LOAD TYPE	SPAN IN FEET						
		3.0	3.5	4.0	4.5	5.0	5.5	6.0
SINGLE	POSITIVE WIND LOAD	143	105	81	64	52	43	36
	LIVE LOAD/DEFLECTION	148	109	83	66	53	44	37
2-SPAN	POSITIVE WIND LOAD	197	145	111	88	71	59	49
	LIVE LOAD/DEFLECTION	108	79	61	48	39	32	27
3 OR MORE	POSITIVE WIND LOAD	224	165	126	100	81	67	56
	LIVE LOAD/DEFLECTION	135	99	76	60	48	40	34

22 Gage (F_y = 50 KSI)

SPAN TYPE	LOAD TYPE	SPAN IN FEET						
		3.0	3.5	4.0	4.5	5.0	5.5	6.0
SINGLE	POSITIVE WIND LOAD	201	148	113	89	72	60	50
	LIVE LOAD/DEFLECTION	229	168	129	102	82	68	57
2-SPAN	POSITIVE WIND LOAD	305	224	171	135	110	91	76
	LIVE LOAD/DEFLECTION	151	111	85	67	54	45	38
3 OR MORE	POSITIVE WIND LOAD	314	231	177	140	113	93	79
	LIVE LOAD/DEFLECTION	188	138	106	84	68	56	47

NOTES
1. Allowable loads are based on uniform span lengths and F_y of 50 KSI.
2. Live load is allowable live load.
3. Wind load is allowable wind load and has been increased by 33⅓%.
4. Deflection loads are limited by a maximum deflection ratio of L/240 of span or maximum bending stress from live load.
5. Weight of the panel has not been deducted from allowable loads.
6. Load table values do not include web crippling requirements.

Figure 6.11 Sample standing-seam roof panel properties. (*MBCI.*)

6.5.2   Guidelines and details for use of SSR

After corrugated sheets are seamed and engaged by clips, the individual sheets become parts of a metal roof membrane that moves as a unit with temperature changes. The movement capacity of the clips and the expansion details restricts the maximum uninterrupted roof width to about 200 ft, beyond which stepped expansion joints are needed. A typical detail of the stepped roof expansion joint is shown in Fig. 6.12.

Standard length of panels varies among manufacturers. The length is best kept under 40 ft, a practical limit imposed by shipping and handling constraints, although panels up to 60 ft can be shipped by rail. For wider roofs, the panels have to be spliced. The end splices, which are normally staggered from panel to panel, can be aided by special clamping plates and prepunched holes (Fig. 6.13). This detail avoids a direct panel-to-support connection which can restrict movement. The endlap details deserve close attention, since endlaps, along with roof penetrations, account for a lion's share of problems with metal roofs.

Through-the-roof fasteners are not totally eliminated in standing-seam roofs —after all, roofing must be positively attached to supports *somewhere*—but their number is reduced by about 80 percent. Through-fastening of panels usu-

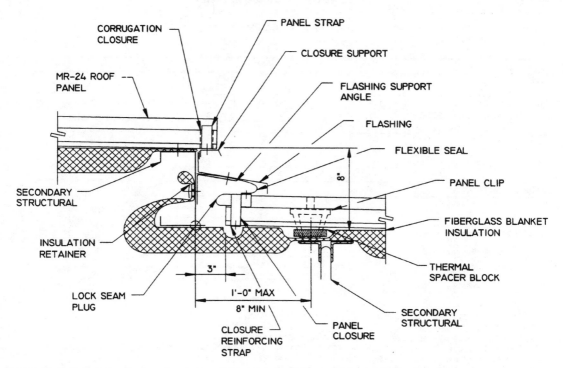

Figure 6.12   Stepped roof expansion joint with thermal spacer block. (*Butler Manufacturing Co.*)

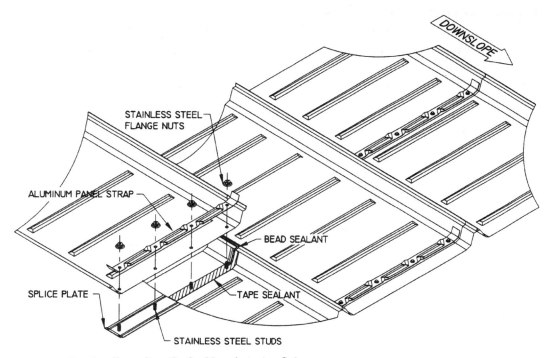

DOWNSLOPE

STAINLESS STEEL
FLANGE NUTS

ALUMINUM PANEL STRAP

BEAD SEALANT

SPLICE PLATE

TAPE SEALANT

STAINLESS STEEL STUDS

Figure 6.13   Panel endlap splice. (*Butler Manufacturing Co.*)

ally occurs at the eave strut, allowing the panels to expand toward the ridge covered with a flexible flashing cap. This design provides bracing for the eave strut and helps resist high wind-uplift forces that tend to occur near the eaves. (It is estimated that some 75 percent of all roofing damage due to wind occurs at roof edges.) Elimination of panel movement at the eaves holds an added benefit for the metal roofs located in cold regions, where the eaves are vulnerable to ice dam formation and snow sliding.

Exterior gutters, whether exposed or concealed, are common to all metal roofing, since two-directional slope toward internal drains is obviously impractical with this type of construction. Typical eave, ridge, and rake details for standing-seam roofs are illustrated in Figs. 6.14 to 6.16. Figure 6.17 shows a good method of sealing moisture out of roof-to-wall transitions.

6.5.3   Disadvantages and limitations of SSR

The biggest disadvantage of the standing-seam roof system can be traced to its biggest advantage—the movement ability. Lacking a direct attachment to supports, the roofing provides little or no lateral bracing to the purlins (see discussion in Sec. 5.3.4); it also offers no diaphragm action. Wherever SSR is used, a separate system of purlin bracing and a separate horizontal diaphragm structure are needed. For architectural roofing, which needs a supporting surface, both these functions can be served by a metal deck substrate. The metal deck

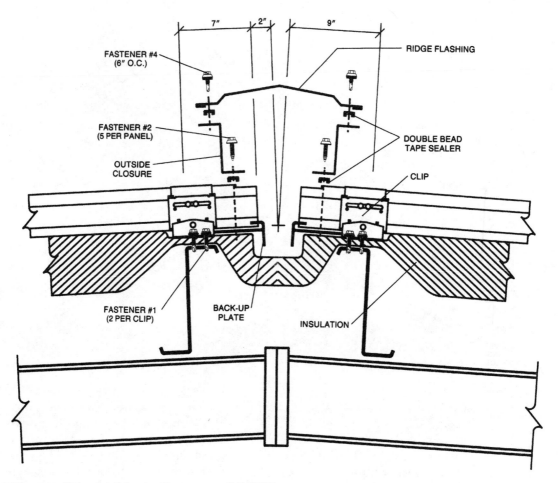

Figure 6.14 Ridge detail for standing-seam roof. (*MBCI.*)

can also be used under structural roofing, but most manufacturers prefer bracing. Alternatively, some manufacturers attempt to resolve this issue by offering separate liner panels which provide some limited bracing and diaphragm
action. To be truly useful, however, the liner panels must be as rigid as the steel
decking, an uncommon occurrence. Other suppliers simply claim that *their*
standing-seam roof is different and capable of serving as a diaphragm—without offering any supporting test data.

Another major disadvantage of SSR: It is best suited for rectangular buildings without too many design complications. For example, a large number of
roof openings requires too many through-fastened connections at the edges
that may defeat the roof's ability to "float." Similarly, complex plans tend to create situations for which standard design details, geared to a simple fixed
end–expansion end assumption, are ill suited. Nonrectangular roofing panels

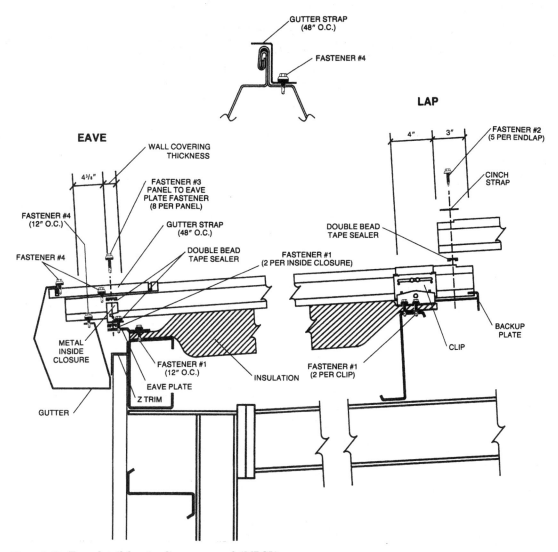

Figure 6.15 Eave detail for standing-seam roof. (*MBCI.*)

not only require expensive field cutting and fitting but also could compromise the available closure, sealant, and finishing details. Even a rectangular layout may require a lot of design ingenuity to allow for roof movement at the corners and other critical locations. With some systems, a simple hip roof may present enough complications to negate the economy of the standing-seam design.[8]

Standing-seam roofs may present some appearance problems, too. As Stephenson[8] pointed out, the often-utilitarian appearance of SSR closure and edge details may not be appropriate for aesthetically demanding applications, where conventional batten-type roofing may serve better. With SSR, it might

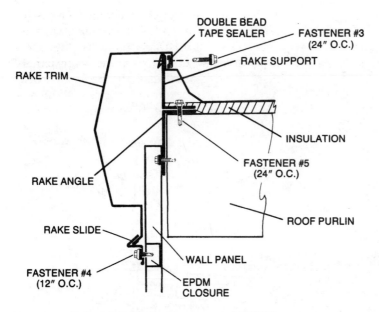

Figure 6.16 Rake detail for standing-seam roof. (*MBCI.*)

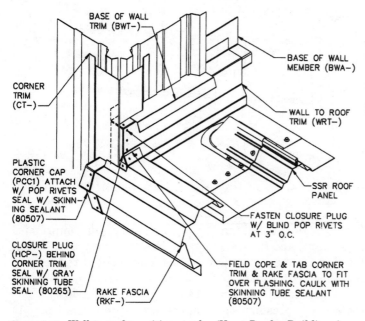

Figure 6.17 Wall-to-roof transition at rake. (*Varco-Pruden Buildings.*)

be difficult to achieve good-looking solutions for roof slope changes or fascia-soffit transitions. In such situations, architectural roofing systems, such as Butler Manufacturing's VSR* Roof System, or MBCI's Battenlock† (Fig. 6.18) are more appropriate. These are architectural vertical standing-seam panels that are recommended for roof-to-fascia transitions (Fig. 6.19). Be forewarned, however, that the real-life transition line may not come out as crisp as the pictures show (Fig. 6.20). The manufacturers caution that "oil-canning," or slight waviness, may occur in panels of this type.

A complex, yet common, detail of ridge-to-hip transition for VSR roofing is shown in Fig. 6.21.

Standing-seam roofs constructed in cold regions are especially prone to leakage caused by ice dam formation. After extensive studies at the Cold Regions Research and Engineering Laboratory (CRREL), Tobiasson and Buska[25] have concluded that the risk of leakage can be mitigated by several factors. These include using "waterproof" instead of "water-shedding" panel design and increasing the roof slope. (In addition to ice damming, the roofs with slopes less than 2 to 12 are difficult to ventilate and are vulnerable to condensation.) Reducing the roof overhang, adding insulation, and introducing ventilation also help achieve a "cold roof" where local snow melting is largely prevented.

Sliding SSR panels may produce a specific "metal" noise objectionable to some people. The noise can be masked by roof insulation.

On balance, all these limitations are far outweighed by the SSR's advantages. Standing-seam roofs remain a premium choice for metal building systems. Superior performance of standing-seam roofs is reflected in the warranties longer than those for lap-seam roofs. Indeed, the popularity of these roofs often attracts the prospective users to metal building systems. Standing-seam roofs, while more expensive initially, often prove very economical in long-term comparisons which take into account life-cycle costs.

6.6 Panel Finishes

Metal roofs can survive only if protected from corrosion. Nothing can detract more from the appearance of a metal building system than a sight of rusted corrugated roofing. Modern metal finishes not only offer good looks but also protect the roofing from moisture, its number one enemy, and from pollution. Durable is the finish that does not peel, crack, or discolor for a reasonable period of time. A good fading resistance is especially important for roofs in sunny locales, where ultraviolet radiation often destroys darker colors, such as reds and blues.

---

*VSR is a registered trademark of Butler Manufacturing Co.
†Battenlock is a registered trademark of MBCI.

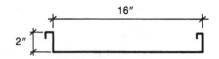

SECTION PROPERTIES								
			TOP FLAT IN COMPRESSION			BOTTOM FLAT IN COMPRESSION		
PANEL GAGE	F_y (KSI)	WEIGHT (PSF)	I_x (in.⁴/ft.)	S_x (in.³/ft.)	M_a (Kip in.)	I_x (in.⁴/ft.)	S_x (in.³/ft.)	M_a (Kip in.)
24	50.0	1.29	0.1005	0.0544	1.6300	0.0557	0.0489	1.4650
22	50.0	1.65	0.1413	0.0791	2.3700	0.0788	0.0652	1.9520

NOTES
1. All calculations for the properties of BattenLok panels are calculated in accordance with the 1986 edition of *Specifications for the Design of Light Gauge Cold Formed Steel Structural Members* - published by the American Iron and Steel Institute (A.I.S.I.).
2. I_x is for deflection determination.
3. S_x is for bending.
4. M_a is allowable bending moment.
5. All values are for one foot of panel width.

ALLOWABLE UNIFORM LIVE LOADS
IN POUNDS PER SQUARE FOOT

24 Gage (F_y = 50 KSI)

SPAN TYPE	LOAD TYPE	SPAN IN FEET							
		2.5	3.0	3.5	4.0	4.5	5.0	5.5	6.0
2-SPAN	POSITIVE WIND LOAD	232	161	118	91	72	58	48	40
	LIVE LOAD/DEFLECTION	156	109	80	61	48	39	32	27
3 OR MORE	POSITIVE WIND LOAD	290	201	148	113	89	72	60	50
	LIVE LOAD/DEFLECTION	195	136	100	76	60	49	40	34

22 Gage (F_y = 50 KSI)

SPAN TYPE	LOAD TYPE	SPAN IN FEET							
		2.5	3.0	3.5	4.0	4.5	5.0	5.5	6.0
2-SPAN	POSITIVE WIND LOAD	337	234	172	132	104	84	70	59
	LIVE LOAD/DEFLECTION	208	145	106	81	64	52	43	36
3 OR MORE	POSITIVE WIND LOAD	421	293	215	165	130	105	87	73
	LIVE LOAD/DEFLECTION	260	181	133	102	80	65	54	45

NOTES
1. Allowable loads are based on uniform span lengths and F_y of 50 KSI.
2. Live load is allowable live load.
3. Wind load is allowable wind load and has been increased by 33⅓%.
4. Deflection loads are limited by a maximum deflection ratio of L/240 of span or maximum bending stress from live load.
5. Weight of the panel has not been deducted from allowable loads.
6. Load table values do not include web crippling requirements.

Figure 6.18 Sample architectural panel data (Battenlock by MBCI). (*MBCI.*)

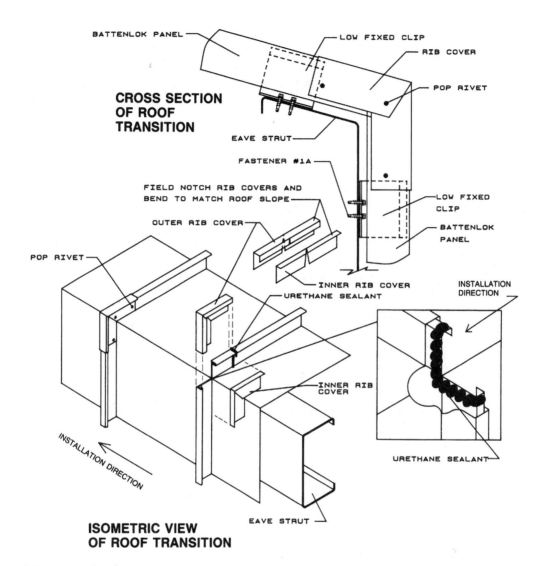

CROSS SECTION OF ROOF TRANSITION

BATTENLOK PANEL
LOW FIXED CLIP
RIB COVER
POP RIVET
EAVE STRUT
FASTENER #1A
FIELD NOTCH RIB COVERS AND BEND TO MATCH ROOF SLOPE
OUTER RIB COVER
POP RIVET
LOW FIXED CLIP
BATTENLOK PANEL
INNER RIB COVER
URETHANE SEALANT
INSTALLATION DIRECTION
INNER RIB COVER
INSTALLATION DIRECTION
URETHANE SEALANT

ISOMETRIC VIEW OF ROOF TRANSITION

EAVE STRUT

NOTES
1. Field cut legs of panels and bend to required angle.
2. Apply urethane sealant to both the roof portion and fascia portion of the male leg of the panel before the next panel is installed.
3. Field notch and bend inner and outer rib covers to match the roof transition.
4. Field apply a bead of urethane sealant over rib before installing rib covers.
5. Pop rivet inner and outer rib covers to rib of panel.
6. Using vise grip duckbills, crimp the outer rib cover to match the roof and fascia seams.

Figure 6.19 Roof-to-fascia transition detail with Battenlok. (*MBCI.*)

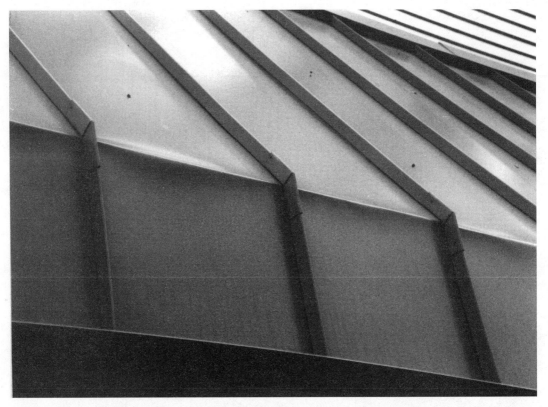

Figure 6.20 The line of roof-to-fascia transition occasionally comes out less than crisp.

6.6.1 Anticorrosive coatings

The most popular anticorrosive coatings for steel roofing are based on metallurgically bonded zinc, aluminum, or a combination of the two. The new ASTM specification A 924-94 covers both zinc and aluminum applied by the hot-dip process.

Zinc coating, found in the familiar galvanized steel, relies primarily on a sacrificial chemical action of zinc, slowly melting away while protecting the underlying metal. Obviously, the thicker the layer of zinc, the longer the protection; a coating conforming to new ASTM A 653[30] with a G60 or G90 designation is adequate for most applications. In addition to its sacrificial protection, galvanizing provides a barrier against the elements, although this action is secondary. The barrier action is helped by a white film formed by the products of zinc oxidation. Mill-galvanized steel has a familiar shiny spangled finish, while the appearance of hot-dip galvanized finish is rough and dull. According to some estimates, hot-dip galvanized panels may lose about $\frac{1}{2}$ mil of the coating thickness every 5 years.

Aluminum coating, on the other hand, acts primarily as a physical barrier formed by a transparent chemical-resistant residue of aluminum oxide, a prod-

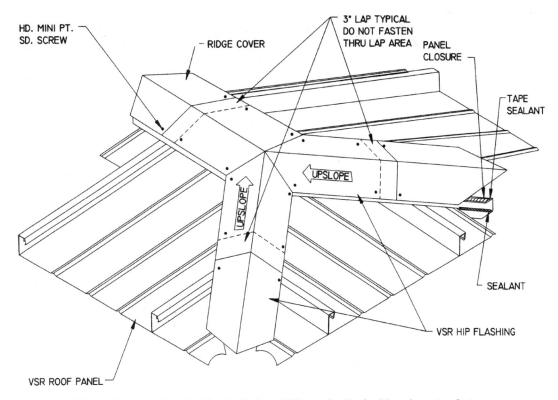

Figure 6.21   Ridge-to-hip transition detail with Butler's VSR panels. (*Butler Manufacturing Co.*)

uct of aluminum oxidation. ASTM A 463[31] type 2 specifies the minimum weight of aluminum coating for roofing as 0.65 oz/ft^2 on both sides (the coating designation T2-65). Aluminized steel has an even matte finish.

Aluminum-zinc coatings combine the sacrificial action of zinc and the barrier protection of aluminum. Two compositions are common—Galvalume and Galfan. Galvalume,* coating developed by Bethlehem Steel Corp., is made of 55 percent aluminum, 43.5 percent zinc, and 1.5 percent silicon and is described in ASTM A 792.[32] Galvalume sheets are available in commercial, lock-forming, and structural grades, and each grade can have one of three coating weights—AZ50, AZ55, and AZ60. A popular AZ55 coating weighs 0.55 ft^2; it is normally used for unpainted structural SSR.[29] Galfan, made of 95 percent zinc and 5 percent aluminum, is specified in ASTM A 875. Galvalume-coated steel looks like a cross between galvanized and aluminized steels, while Galfan finish is difficult to distinguish from pure galvanizing.

Aluminum-zinc coatings should offer excellent metal protection for decades. A recent survey of 82 Galvalume-coated metal roofs in the eastern United

---

*Galvalume is a registered trademark of BIEC International Inc.

States by the Galvalume Sheet Producers of North America found the roofs to be in excellent condition. The organization projects that these roofs can easily last 30 years in most regions before needing major maintenance; it estimates that Galvalume coating should last two to four times longer than G90 galvanizing in marine, industrial, and rural environments.

Galfan-coated steel is especially suitable for applications involving field bending and forming of panels, since it is virtually unaffected by cracking or flaking common in bent hot-dip galvanized members.

It is worth keeping in mind that, regardless of how great the coating might be, most roofing corrosion occurs at the field-cut edges. Despite the fact that zinc can extend its healing properties to cut edges, it is still best to require that all roofing be factory-trimmed and finished. The use of factory-supplied touch-up compound could greatly improve the roof's resistance to corrosion in the areas of cuts and handling damage. One also needs to remember that roofing with aluminum and zinc coatings should not be in direct contact with unprotected steel, to avoid galvanic action. A contact of such roofing with chemically treated wood decking can also be damaging; the two materials must be separated by a properly installed underlayment layer.

6.6.2 Paint coatings

Rarely are anticorrosive coatings left exposed: Both aesthetics and desire for another layer of protection call for a durable paint finish. (Unpainted Galvalume panels are likely to be found in structural SSR roofing.[29]) Paints for metal roofs are predominantly based on organic (carbon) compounds, such as polyester, acrylic, and fluorocarbon.

Acrylic- and *polyester-based* paints, common in residential use, are specified in the American Architectural Manufacturers Association (AAMA) standard 603. These synthetic polymers have a durable, abrasion-resistant surface. A slightly different, and better, product is *siliconized polyester* that offers a good chalk (gradual erosion of the film) resistance and gloss retention. Acrylic- and polyester-based paints normally offer 3- to 5-year guarantees, while siliconized polyesters may come with 10-year or even prorated 20-year warrantees.

Fluorocarbon-based paints are made with polyvinylidene fluoride (PVDF) resin, introduced around 1970. PVDF resin is an exceptionally stable compound that offers extraordinary durability, color stability, and resistance to ultraviolet radiation, heat, and chalking. The finished surface is dense, smooth, and stain-resistant. According to its manufacturers, PVDF does not absorb ultraviolet rays that account for common metal-roof color deterioration. In one study, it was the only paint type still virtually unaffected by Florida sun and salty air after 12 years of exposure. PVDF-based coatings can easily outlast their 20-year performance guarantees.

The PVDF resin is supplied by Elf Atochem North America, Inc., under the trademark of Kynar 500 and by Ausimont USA, Inc., under the trademark of Hylar 5000. These two companies license their products to some American paint companies; the complexities of the production process restrict the list to only a few. To reflect the dual origin of this finish, it is often referred to as Kynar 500/Hylar 5000 fluoropolymer resin. The actual coating is commonly made with 70 percent resin and 30 percent pigments and solvents.

The AAMA standard 605.2 specifies criteria for PVDF-based paints such as the acceptable limits for gloss retention, color change, and coating erosion. In addition, it prescribes testing of some other properties such as salt spray resistance, durability under heat, humidity resistance, adhesion, and chemical resistance.

PVDF finish can come in various thicknesses. The standard coating is 1 mil (0.025 mm) thick, often stated as 0.9 mil. Perfect for mildly corrosive environments, it may be insufficient for moderately corrosive situations, where a premium 2-mil (0.05-mm) finish may be called for. (This premium finish may require a special setup by the coil coater and will probably need a longer lead time.) For exceptionally aggressive or abrasive environments, a special 4-mil (0.1-mm) finish may be considered. However, the expense and difficulty of obtaining this thickness, coupled with the fact that any field-cut panel edges, holes, and scratches will reduce the effectiveness of the "superfinish," point to some other metals, such as stainless steel or aluminum, for such applications.[10]

The warranties offered on panel finishes are often prorated. Upon completion of the job, the manufacturer files a chip of the material which could be used for comparison if and when the fade-out claim is made.

6.6.3 Stainless steel, copper, and aluminum

For those special applications that demand a cut above the coated steel, or for the purists who believe that any paint will ultimately fail, stainless steel, pure copper, or aluminum may be the materials of choice.

Stainless steel has an excellent resistance to corrosion, although it is not corrosion-proof, as many people believe. Stainless steels of types 302 and 304 are commonly used; type 316 offers superior resistance to corrosion at a premium cost.

To further increase resistance to corrosion, stainless steel can be coated with a terne alloy, already mentioned in Sec. 6.1 as a popular material at the turn of the century. While the early carbon-steel panels did not last very long, modern terne-coated stainless steel is the most durable roofing material available. Clearly not intended for routine applications because of its cost, terne-coated stainless steel is found on some of the best-known corporate headquarters in America including IBM, Procter & Gamble, and Coca-Cola.

Working with stainless-steel roofing requires some know-how on the part of designers and installers alike. To avoid galvanic action, stainless steel should be physically separated from any mild steel objects, such as purlins, fasteners, and clip angles. This can be accomplished by installation of a moisture barrier, such as 30-lb roofing felt, use of stainless-steel fasteners, and following good industry practices when welding or soldering the material. Some useful information on working with stainless steel can be found in Ref. 24.

Copper roofing, an old and respected material, is still used to reproduce a rich and beautiful look of the years past. Apart from renovation of historic structures, copper has its place in new construction of all kinds. The panels, as in Paul Revere's times, are formed on site. Some of copper's disadvantages, apart from the cost, include problems with the water runoff staining the materials below. Copper in contact with aluminum, stainless steel, or galvanized or plain steel may initiate a galvanic action.

Aluminum is the most common of nonferrous metal roofing materials. It offers excellent resistance to corrosion and thus may be appropriate for buildings located near the ocean and in saline environments. Being relatively soft, aluminum is easy to bend and extrude but also easy to damage and dent. (Reports abound of aluminum-clad buildings dented by gravel thrown off during lawn mowing.) Another disadvantage of aluminum is its high coefficient of expansion; aluminum roofing will expand approximately twice as much as steel roof. Again, the question of joining dissimilar metals needs to be carefully addressed.

Aluminum panels are usually anodized by dipping into a tank with electrolyte. The panel length is limited by the available size of the electrolyte tank. Electric current passing through the tank deposits a layer of aluminum oxide coating that forms a layer of chemically resistant, hard, and durable finish. The panels can be left in a natural color, or a pigment could be added during anodizing to produce a choice of chemically bonded colors, such as bronze and black. The anodized finish retains colors well but is difficult to repair if scratched; it is susceptible to damage by pollutants.

Structural design of aluminum is covered by Aluminum Association (AA) standards[11] and specifications.[12] For stress analysis, structural section properties are computed using the actual dimensions of the cross section. For deflection check, the "effective width" concept is employed. The aluminum alloys used for panels normally conform to ASTM B 209. The panels should be at least 0.032 in (0.8 mm) thick.

6.7   Site-Formed Metal Panels

Despite the already mentioned—and obvious—quality advantages of shop-fabricated metal panels, there are circumstances when roll forming on-site is performed. The panels formed at a job site need not be constrained by the shipping limits and could extend from ridge to eave, thus eliminating the trouble-prone endlaps. Also, transportation charges are saved, although expensive field labor costs are incurred instead. On the balance, site-formed panels are normally less expensive than those supplied by leading metal building manufacturers.

Job-site roll forming was introduced in the 1970s and has been steadily expanding since, paralleling the improved quality of portable roll forming equipment. One of the leaders in the development and utilization of such equipment is Knudson Manufacturing. The company touts its state-of-the-art roll formers with rubberized drive rollers that are said to handle steel, aluminum, and copper coil for damage-free forming of standing-seam roofing. Knudson can reportedly produce continuous panels up to 150 ft long and can form some C, Z, and hat-channel sections on-site. Prefinished curved panels can be site-rolled by Berridge Manufacturing Co. of Houston and by some others.

Despite increasing product quality, job-site roll forming should be approached with some caution, since many site-produced panels still do not come out as good as shop-fabricated panels of major manufacturers, and their installers are not necessarily as experienced. Also, there is still a chance that the coil finish could be damaged during forming; for this reason, galvanized and aluminized steel, as well as anodized aluminum, are not recommended for field forming. Incidentally, metal coil formed during cold weather should be preheated prior to forming, a fact often forgotten at the job site.

6.8   Wind Uplift Ratings of Metal Roofs

Pictures of blown-off and damaged roofs often accompany media reports on hurricanes, tornadoes, and tropical storms. Damage to metal roofs from strong winds, which generate high suction forces, can manifest itself as panel buckling, fastener breakage or pullout, seam deformation or opening, and standing-seam clip failure. To assure specifiers of their products, metal building manufacturers seek to obtain a wind uplift designation from one of the leading testing bodies: Underwriters Laboratories (UL), Factory Mutual (FM), or U.S. Corps of Engineers (Corps). The American Society for Testing and Materials (ASTM) is also active in the quest to develop a perfect testing procedure. Unfortunately, as of this writing, no test is able to accurately predict roofing behavior during a "real-world" disaster. A brief explanation of the available procedures will help put roofing salespeople's claims in a proper perspective.

6.8.1   UL 580 standard for wind uplift testing

This classic test[23] has been used since 1973. It involves a 10- by 10-ft sample of roofing constructed on a testing platform in accordance with the manufacturer's typical specifications. The edges of the sample are sealed and clamped at the perimeter. The specimen is then subjected to alternating wind pressure and suction. Having safely resisted a 100-mi/h wind for 1 h and 20 min, the specimen earns a class 30 designation. To pass to the next rating level, class 60, the same specimen must withstand a pressure equivalent to a 140-mi/h wind for another 80 min. The highest designation, class 90, can be earned by testing the same specimen a third time for yet another 1 h and 20 min under pressure produced by a wind speed of 170 mi/h.

The panels with a UL 580 class 90 designation have generally performed well—until subjected to a real hurricane. Partial roofing blow-offs and seam separations have been reported to occur at wind speeds producing only about one-fifth of the roof uplift capacity that could be expected from a UL 580 class 90 rating.[13] How could this happen?

The experts point out that the test was developed for evaluations of the adhesive strength of built-up roofs and not for mechanically attached metal roofing. Also, this "static" test, conducted at constant pressures, does not account for real-world wind gusts and shifting pressure patterns.[14] Further, the limited specimen size (10 by 10 ft) and the continuous perimeter attachment do not accurately represent the real behavior of metal roofs, especially of the standing-seam type. For all these reasons, the results of UL 580 tests cannot be directly translated into the allowable wind uplift pressures, despite some roofing salespeople's claims to the contrary. A UL 90 Wind Uplift Classification does not mean the roofing can safely support a 90 lb/ft^2 uplift, although the specimen must resist a combined 105 lb/ft^2 upward load for 5 min to qualify. Essentially, the test results can only be used for "indexing"—comparing the tested panel with other similar products under a stringent set of conditions.

6.8.2 ASTM testing procedures

Once the serious limitations of the UL 580 test became understood, many architects and construction specifiers began to seek alternative testing methods. One such procedure, called ASTM E 330 Modified, has become quite popular. The original ASTM E 330 test[18] had been developed for curtain walls, not roofs. The two behave differently under wind loading, and the procedure suitable for a rather stiff wall assembly is not readily transferable to flexible metal roofing. Despite its widespread use, the "modified" test procedure is not approved by ASTM for testing metal roofs.

Recognizing the need to develop a proper standard for testing metal roofs, ASTM formed its Subcommittee E06.21.04 to sort out the complexities of the issue. The subcommittee's brainchild, ASTM E 1592-94, Structural Performance of Sheet Metal Roofing and Siding by Uniform Static Air Pressure Difference, covers both metal panels and their anchors. It essentially retains the basic approach of the E 330 test method, slightly changing it to allow for the roofing's flexibility. The new test specifies loading to be applied in a manner that facilitates detection of slowly developing failures such as seam separations.[19] Instead of being a pass-fail test, ASTM E 1592 provides a standardized procedure to evaluate or confirm structural performance of roofs under uniform static loading.

It took the subcommittee over 5 years to overcome the initial deadlock[15] and reach a consensus. Once published, however, the document was quickly endorsed by a major industry group, Metal Roofing Systems Association (MRSA), that was formed in January 1994. MRSA has voted to recommend both the ASTM E 1592 and UL 580 as the preferred test methods for standing-seam roofing[20] and produced a technical bulletin explaining the standards to the specifiers. MRSA recognizes the fact that through-fastened roofing behaves differently from SSR under wind loading and, in fact, can be rationally analyzed for uplift, unlike SSR. Accordingly, the Association does not feel that the ASTM E 1592 test is required for through-fastened roofing,[21] a reasonable argument supported by MBMA and MCA.

6.8.3 Factory Mutual standard 4470

Factory Mutual standard 4470 is another widely used test for wind uplift. The standard incorporates many other unrelated tests such as durability, and resistance to corrosion, fire, and leakage. Assemblies that satisfy all the tests are designated as class I or class II roof covers.

For wind uplift testing, a roofing assembly is mounted on a 5- by 9-ft platform, sealed, and clamped at the perimeter. The assembly is then subjected to increasing wind pressure from the underside until it either fails or is able to sustain a certain pressure for a minute. The highest FM rating, I-90, means that the roofing assembly could resist a 90 lb/ft^2 load for 1 min. Applying a safety factor of 2 yields a maximum allowable design load of 45 lb/ft^2.[16]

The FM 4470 testing procedure has been criticized along the same lines as UL 580. Both are considered "static" tests that do not account for wind gusts, and both operate with a limited specimen size continuously supported along the perimeter. These arguments are even stronger for the FM 4470 setup, since its platform is even smaller than the UL's, and distortion from perimeter clamping is even more pronounced.

6.8.4 U.S. Corps of Engineers testing procedures

Alarmed by a growing realization that the available testing procedures listed above fail to accurately measure wind uplift resistance of standing-seam roofing, the Corps has created its own testing methodology. Its Guide Specification for Military Construction Section 07416 (CEGS 07416) prescribes a new procedure for the testing believed to better reflect the effects of the actual field conditions. For this test, the edge details must correspond to the actual field construction; the perimeter clamping is out. The test must be performed under the supervision of an independent professional engineer using the procedures approved by the Corps. The first company that followed this methodology at its own research facility was Butler Manufacturing.[13]

Some consider the Corps-developed procedure a cross between the methods of "modified" ASTM E 330 and ASTM E 1592, plus the new supervision requirements. It attracts the same criticisms for being "static"—not accounting for nonuniform roof pressure distribution—requiring only a small number of loading cycles, and not being designed to determine an allowable load capacity of the roof.[14] Still, it is probably the most realistic testing procedure for metal roofing available today.

6.8.5 Which test to specify and why

Why test metal roofing at all, instead of, say, full-scale testing of pre-engineered buildings? The answer is, structural behavior of primary building frames subjected to wind uplift can be reasonably predicted by calculations, but that of standing-seam metal roofing cannot be.

Metal panels deflect and distort so much under uplift loading that the analysis based on a flat section is as relevant to the actual behavior as the beam theory is to arches. Moreover, many panel failures occur because of unlocking of the panel sidelaps at the clip locations and excessive bending stresses introduced into the "hook" portion of the clips.[14]

So, if the rational design is not available and testing procedures are imperfect, what is a specifier to do? A sensible course of action for critical applications might be to require the standing-seam roofing to be tested in accordance with the U.S. Corps of Engineers procedure. It is also wise to carefully investigate any alternative testing methods proposed by the supplier, since some manufacturers already conduct dynamic testing of their products, arguably superior to the Corps' procedure. The manufacturer's track record should provide some assurance as well.

Meanwhile, a quest for the perfect test continues. Among other studies, MBMA and AISI are sponsoring research conducted at Mississippi State University, Starkville. The new 32- by 14-ft air pressure chamber at that facility can reportedly simulate "real-world" wind gusts by electromagnetically providing a nonuniform pressure distribution that can be changed almost instantaneously. It is hoped that a better understanding of the ways the real wind acts on the roofs will lead to more reliable testing methods for standing-seam roofing.

6.9   Some Tips on Roofing Selection

Metal roofing can provide long and largely maintenance-free service life if designed and installed correctly. The designer's contribution is to select a proper roofing product and details for the building. It is hoped that the information provided in this chapter will help determine whether architectural or structural, standing-seam, or through-fastened roofing is appropriate.

Do not compromise the advantages of metal roofing by cluttering it with a forest of pipes, ducts, openings, and rooftop-mounted equipment. Every roof penetration or opening results in field-cut panels that can restrict temperature expansion, expose metal to corrosion, and invite future leaks. Where impossible to avoid, the penetrations and openings should be carefully detailed to allow for panel movement and to resist water intrusion. One of the product selection criteria should be clarity of the manufacturer's details for difficult conditions.

For architectural panels with felt underlayment, any penetrations through the felt should be detailed at least as carefully as through the metal. Also, underlayment must be allowed to drain at the eaves, which is impossible if the eave trim blocks it. To drain properly, the felt should run on top of the trim and into the gutter, not behind it.[27]

When standing-seam roofing is called for, carefully review the seam details offered by various manufacturers. As Stephenson[17] observed: "Some so-called weather-sealed designs are more weather-resistant than others, and some may prove to be not weather-sealed at all." Clearly, the Pittsburgh-style seam (Fig.

6.2*d*) is superior to the snap-on types of Fig. 6.9. Structural roofing recommended for low slopes ($\frac{1}{4}$:12) is likely to be more water-resistant than most architectural products designed for a 3:12 or steeper slope. From the standpoint of hurricane resistance, all seam designs are somewhat vulnerable, but the snap-on types seem to perform the worst.

Architectural roofing, even when used on the steep slope, does not work very well with snow guards. The snow and ice trapped by the guards may melt and seep through the joints, resulting in ice damming. When the snow guards are required, it is best to specify high-quality Pittsburgh-style seams.

Even the best-designed system will fail if installed incorrectly. This is why warranties and the manufacturer's and installer's reputation are important. A wealth of information about various metal roofing manufacturers can be found in NRCA's *Commercial Low-Slope Roofing Materials Guide.*[22] For projects with any nontypical features, details for which are not included in the manufacturer's standard assortment, shop drawing submittal should be required. The shop drawings should clearly indicate the installation details for all trim, fasteners, and sealants.

References

1. Paul D. Nimtz, "Metal Roofing: Past, Present and Future," *Metal Architecture,* November 1992.
2. "Market Trends/1996," *Metal Construction News,* January 1996.
3. "Specification Survey Results Indicate Metal Roofing Continuing Market Advances," *Metal Architecture,* May 1994.
4. "Masterspec-Evaluations," Sec. 07411, American Institute of Architects, Washington, D.C., 1990.
5. R. White and C. Salmon (eds.), *Building Structural Design Handbook,* Wiley, New York, 1987, pp. 1048–1050.
6. *Metal Building Systems,* 2d ed., Building Systems Institute, Inc., Cleveland, Ohio, 1990.
7. Paul D. Nimtz, "Metal Roof Systems," *Metal Architecture,* January 1993.
8. Fred Stephenson, "Specifying Standing Seam Roofing Systems for Conventional Buildings," *Roof Design* (now *Exteriors Magazine*), March 1984, as reprinted in *The Construction Specifier,* November 1986.
9. Y. L. Xu, "Fatigue Performance of Screw-Fastened Light-Gauge-Steel Roofing Sheets," *ASCE Journal of Structural Engineering,* vol. 121, no. 3, March 1995.
10. U.S. Army Corps of Engineers, Guide Specification for Military Construction, CEGS-07413 and CEGS-07416, Huntsville, Ala., 1994.
11. *Aluminum Standards and Data,* Aluminum Association, 1990.
12. *Aluminum Construction Manual Series,* Sec. 1, *Specifications for Aluminum Structures,* Aluminum Association, New York, 1986.
13. Charles H. Gutberlet, Jr., "Designing against Wind," *The Military Engineer,* September–October 1992.
14. Harold Simpson, "Wind Uplift Tests," *Metal Architecture,* January 1993.
15. John Gregerson, "Bracing for Wind Uplift," *Building Design & Construction,* September 1990.
16. Richard Coursey, "Keeping Wind from Raising the Roof," *Architecture,* August 1988.
17. Fred Stephenson, "Design Considerations for Architectural Metal Roofing," *Handbook of Commercial Roofing Systems,* Edgell Communications, Inc., Cleveland, Ohio, 1989.
18. ASTM E 330, Test Method for Structural Performance of Exterior Windows, Curtain Walls and Doors by Uniform Static Air Pressure Difference, American Society for Testing and Materials, Philadelphia, Pa., 1990.
19. "ASTM's Metal Roofing Standard Intended to Improve Performance," *Metal Architecture,* March 1995.
20. "MRSA Recommends Wind Uplift Tests for Standing-Seam Roofing," *Metal Architecture,* March 1995.

21. "Technical Bulletin on Wind Uplift Testing…," *Metal Architecture,* May 1995.
22. *Commercial Low-Slope Roofing Materials Guide,* NRCA, Rosemont, Ill., 1993.
23. UL 580, Standard for Tests for Uplift Resistance of Roof Assemblies, Underwriters Laboratories, Inc., Northbrook, Ill., 1989.
24. Standard Practices for Stainless Steel Roofing, Flashing, Copings, Stainless Steel Industry of North America, Washington, D.C.
25. Wayne Tobiasson and James Buska, "Standing-Seam Metal Roofing Systems in Cold Regions," CRREL Misc. Paper 3233, Hanover, N.H., 1993.
26. Wayne Tobiasson, "General Considerations for Roofs," *CRREL Misc. Paper* 3443, Hanover, N.H., 1994.
27. Richard Schroter, "Leak Prevention for Metal Roofs," *Architectural Specifier,* November/December 1995.
28. Brian A. Lynn and Theodore Stathopoulos, "Wind-Induced Fatigue on Low Metal Buildings," *ASCE Journal of Structural Engineering,* vol. 111, no. 4, April 1985, pp. 826–839.
29. Spec-Data for Section 07600, Standing Seam Metal Roofing, Galvalume Sheet Producers of North America, Bethlehem, Pa., 1995.
30. ASTM A 653-94, Steel Sheet, Zinc-Coated (Galvanized) or Zinc-Iron Alloy-Coated (Galvannealed) by the Hot-Dip Process, ASTM, Philadelphia, Pa., 1994.
31. ASTM A 463-94, Steel Sheet, Aluminum-Coated by the Hot-Dip Process, ASTM, Philadelphia, Pa., 1994.
32. ASTM A 792-93, Steel Sheet, 55% Aluminum-Zinc Alloy-Coated by the Hot-Dip Process, ASTM, Philadelphia, Pa., 1993.

Chapter

7

Wall Materials

7.1 Introduction

Gone are the days when metal buildings were uniformly clad in corrugated galvanized panels. Were Rip van Winkle to go to his nap in the 1940s and to awake today, he might have trouble recognizing those buildings. Now, the choices of exterior wall materials for metal building systems are as numerous as for conventional construction; it is difficult to tell which structure is hidden behind a sleek contemporary facade.

This chapter's discussion is focused on common wall materials for metal building systems: metal panels, masonry walls, concrete, and some modern lightweight finishes. A few possible combinations of these materials to serve various functional or aesthetic needs are explored. Ours is not an exhaustive study of all the available choices; we can only afford a briefest of discussions, omitting such familiar materials as glass, wood, and stone.

7.2 Metal Panels

First mass-produced pre-engineered buildings were clad in unpainted galvanized steel panels. Color was introduced in the late 1950s; paint was applied by spraying and baking on, as in refrigerators and car fenders. In contrast, modern metal panels are formed from factory-coated coils and come in many durable finishes.

Wall panels of metal buildings are normally supported on cold-formed C or Z girts. Most panels are made of 24-, 26-, or 28-gage galvanized steel with additional coatings discussed in Chap. 6. Metal roof and wall panels are similar in many ways, and some products can be used for both applications. Wall panels are normally shorter than their roof brethren and thus do not expand as much with temperature changes. Therefore, the standing-seam panel design, so popular in roofs, is not needed for walls. Metal wall panels can be of the shop- or field-assembled kind, and with either exposed or concealed fasteners.

7.2.1 Field-assembled panels

Field-assembled panels consist of exterior wall siding, fiberglass blanket insulation, and in some cases, liner sheets. The liners not only provide finished interiors but also readily accommodate acoustical surface treatments.

A panel can be assembled by either one of two methods. In the first method, insulation blankets are fastened to the eave girts and allowed to hang down, being held in place with retaining strips; then exterior sheets are attached to the girts through the insulation, and finally the liners, if needed, are installed (Fig. 7.1).

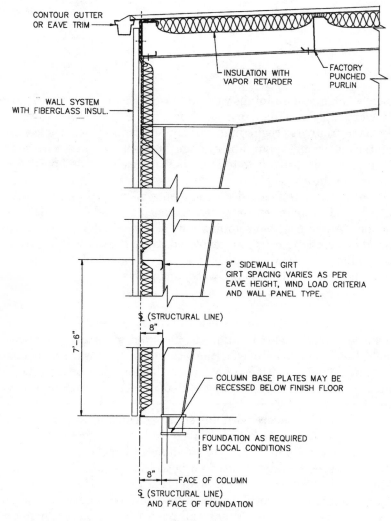

CONTOUR GUTTER OR EAVE TRIM

INSULATION WITH VAPOR RETARDER

FACTORY PUNCHED PURLIN

WALL SYSTEM WITH FIBERGLASS INSUL.

8" SIDEWALL GIRT
GIRT SPACING VARIES AS PER EAVE HEIGHT, WIND LOAD CRITERIA AND WALL PANEL TYPE.

₵ (STRUCTURAL LINE)

8"

7'-6"

COLUMN BASE PLATES MAY BE RECESSED BELOW FINISH FLOOR

FOUNDATION AS REQUIRED BY LOCAL CONDITIONS

8"

FACE OF COLUMN

₵ (STRUCTURAL LINE) AND FACE OF FOUNDATION

Figure 7.1 Field-assembled metal panels with exposed fasteners. (*Butler Manufacturing Co.*)

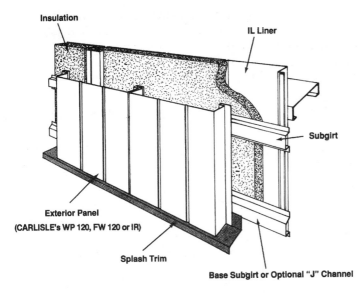

Figure 7.2 Field-assembled insulated panel. (*Carlisle Engineered Metals.*)

Another method uses liner panels with special edge ribs. Liner panels are first attached to the wall girts; then hat-shaped subgirts are fastened to the panel ribs, insulation placed, and finally, exterior sheets are fastened to the subgirts (Fig. 7.2).

Field-assembled panels require bottom closure strips made of rubber, metal, or foam to keep insects, dust, and vermin from getting inside the wall. Closure strips are attached to exterior sheets prior to erection.

The main advantages of field-assembled panels are rapid installation, low cost, and easy replacement of damaged pieces.[1] The panels are lightweight and normally do not require cranes for erection. Window openings are easy to make in the field and are easy to trim with wall girts or framing channels (Fig. 7.3). A fire-rated wall assembly can be readily made by installing fire-rated gypsum boards between the liner and face panels.

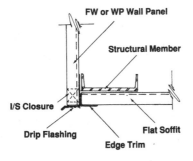

Figure 7.3 Detail at soffit. (*Carlisle Engineered Metals.*)

The disadvantages include loss of thermal performance at the points of panel attachment, where insulation is squeezed. Some manufacturers solve this problem by using thermal blocks, similar to the ones found in roofs, or by inserting a special caulking tape between the metal pieces in contact. It is also possible, of course, to place a layer of rigid insulation between the girt and the liner and to dispense with fiberglass, but this solution is uncommon.

Field-assembled panels may cost from $4 to $6 per square foot in place, not including girts.

7.2.2 Shop-assembled panels

In shop-assembled panels, the same three system components—exterior siding, insulation, and interior liner—are delivered as a unit. In addition to better fit, shop assembly saves field labor. Rigid insulation, mostly urethane foam or polystyrene, provides better R values than fiberglass does. Depending on the amount of insulation, panel thickness can vary from $2\frac{1}{2}$ to 6 in; the width can range from 24 to 42 in.

Factory-insulated panels are filled with foam insulation and have interlocking joints to reduce thermal bridging. A typical panel is attached to supports at its top and bottom and at intermediate girts with concealed fasteners or with expandable fasteners installed from the inside (Fig. 7.4).

Composite panels consist of steel or aluminum exterior sheets laminated over rigid insulation or corrugated paper. Joined by high-performance epoxy adhesives, the assembly is lightweight, strong, durable, and energy-efficient. However, as Hartsock and Fleeman[2] have demonstrated, recent growth in popularity of these panels was not accompanied by rising engineering knowledge.

Foam insulation plays a structural role in composite design of foam-filled sandwich panels, enabling them to span longer distances than is possible with a face sheet alone. Composite design, however, may be governed not by the limits on deflection and bending stress under load but by such esoteric criteria as thermal warp and skin buckling. Indeed, thermal warp from unequal temperature expansion or contraction of the two faces is a major cause of composite panel failure.

Thermal warp, or bowing, matters less with simply supported composite panels than with multiple-span panels. Likewise, the choice of lighter colors can reduce surface temperature and thus the warping. Another potentially critical item is panel anchorage: The fastener design capacity may control the panel's anchorage. Design of composite panels involves a certain amount of finesse and is best left to experienced and sophisticated manufacturers.

With shop-assembled panels, any field changes are difficult to make, and the locations of all wall openings must be established before the panels are made. Some installers attempt to "fix" any mislocated openings by using shears and nibblers to trim the metal and saws to cut the insulation, but the results are rarely perfect. Similarly, sloppy erection will make for a poor fit, defeating the advantages of this system. Installation of shop-assembled panels should be performed only by experienced erectors.

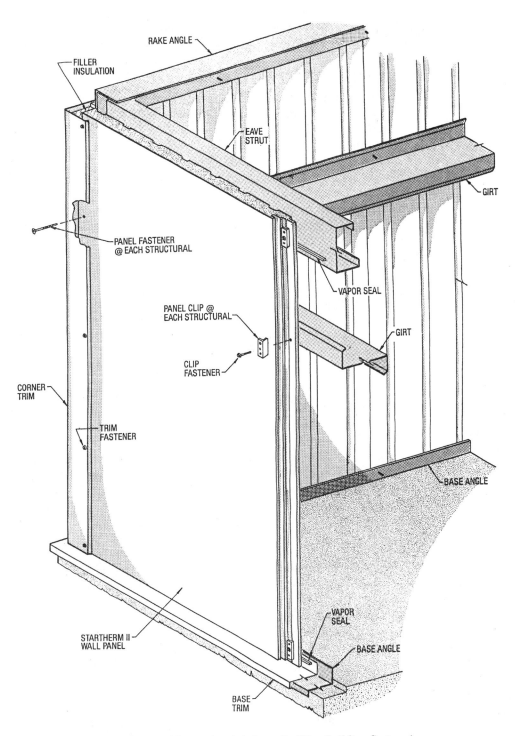

Figure 7.4 Typical installation of factory-insulated panels. (*Star Building Systems.*)

Shop-assembled metal panels may cost from $7 to $12 per square foot, installed.

7.2.3 Exposed-fastener panels

These panels originated from the basic corrugated sheets that came with the first pre-engineered buildings. Today's choices go well beyond the basic and are quite varied (Figs. 6.4, 6.5, and 7.6). Wall panels with exposed fasteners behave similarly to through-fastened roofing; the products shown in Figs. 6.4 and 6.5 may be used as both roofing and siding. Galvanized-steel siding is produced in thicknesses from 29 to 18 gage, with the midrange gages being the most popular. Aluminum sections are also available. Panel width can range from 2 to 4 ft; panel length is limited by shipping and handling constraints of about 40 ft.

A typical exposed-fastener panel is attached to supports with self-drilling screws or similar fasteners (Fig. 7.5). Sidelap fasteners might be spaced 24 in apart.

Exposed-fastener siding is still the most economical exterior wall material available; it is widely used for buildings ranging from factories to schools. For added versatility, the panels can run horizontally. Deep-rib panels (3 to 4 in in depth) may be especially appropriate for this purpose, if able to span the distance between the frame columns (Fig. 7.7).

While the exposed-fastener system is easier to install and is more "forgiving" to field errors than composite panels, the installer's experience is still important. As simple a mistake as overtightening the fasteners may dimple or damage the panels and invite water penetration.

7.2.4 Concealed-fastener panels

In this system, the fasteners connecting panels to supports are hidden from view by interlocking edge joints (Fig. 7.8). In addition to a pleasing appearance, concealed-fastener panels generally provide better protection from water infiltration than exposed-fastener siding. Physical protection is enhanced, too, since these panels are difficult to remove. The panels are usually 1 to $1\frac{1}{2}$ in in depth and are made of 18- to 24-gage steel; the length of 30 ft and longer can be procured.

Concealed-fastener design is often found in field-assembled insulated panels discussed in Sec. 7.2.1. The face and liner panels can be interconnected via hat-shaped subgirts (Fig. 7.2), or directly, if both the liner and the face sheet have identically spaced outstanding legs.

Concealed-fastener panels come in a variety of reveals, striations, and textures preferred by many architects seeking to avoid "oil-canning" (minor surface waviness), which tends to affect smooth light-gage metal panels.

7.2.5 The rain screen principle

Metal wall panels—as metal roofs—leak water mostly through the joints. The only waterproofing protection the joints typically offer is a bead of sealant applied between the edges of the adjacent sheets. When—not if—the sealant

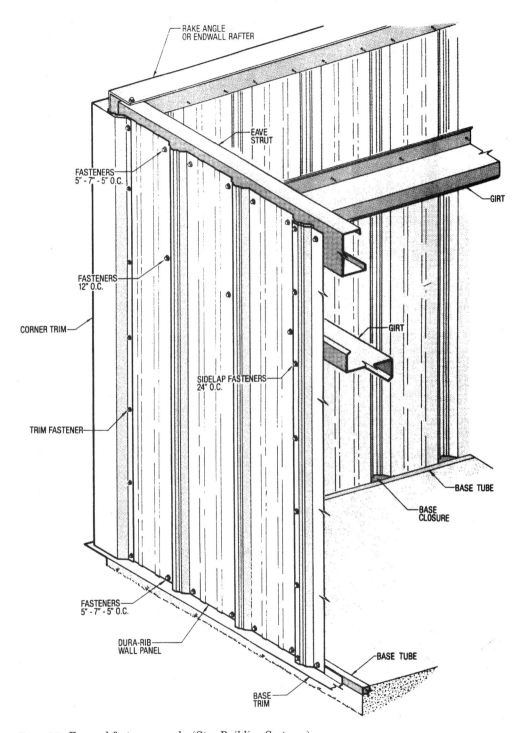

Figure 7.5 Exposed-fastener panels. (*Star Building Systems.*)

"A" PANEL

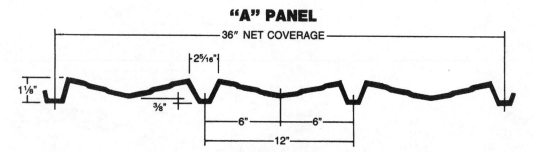

PANEL GAGE	Fy (KSI)	WEIGHT (PSF)	SECTION PROPERTIES					
			TOP FLAT IN COMPRESSION			BOTTOM FLAT IN COMPRESSION		
			I_x (in.⁴/ft.)	S_x (in.³/ft.)	M_a (in.-Kips)	I_x (in.⁴/ft.)	S_x (in.³/ft.)	M_a (in.-Kips)
29	80.0	0.74	0.0133	0.0243	0.8725	0.0183	0.0256	0.9182
26	80.0	0.94	0.0180	0.0318	1.1411	0.0244	0.0342	1.2278
24	50.0	1.14	0.0239	0.0402	1.2027	0.0314	0.0442	1.3228
22	50.0	1.44	0.0336	0.0590	1.7670	0.0403	0.0579	1.7334

NOTES

1. All calculations for the properties of panels are calculated in accordance with the 1986 edition of *Specifications for the Design of Light Gauge Cold Formed Steel Structural Members* - published by the American Iron and Steel Institute (A.I.S.I.).
2. I_x is for deflection determination.
3. S_x is for bending.
4. M_a is allowable bending moment.
5. All values are for one foot of panel width.

ALLOWABLE UNIFORM WIND LOADS IN POUNDS PER SQUARE FOOT

SPAN TYPE	LOAD TYPE	29 Gage (Fy = 80 KSI)					26 Gage (Fy = 80 KSI)				
		SPAN IN FEET					SPAN IN FEET				
		4.0	5.0	6.0	7.0	8.0	4.0	5.0	6.0	7.0	8.0
2-SPAN	POSITIVE WIND LOAD	44	22	13	9	6	58	30	17	10	7
3 OR MORE	POSITIVE WIND LOAD	55	28	16	11	7	73	37	21	13	9

SPAN TYPE	LOAD TYPE	24 Gage (Fy = 50 KSI)					22 Gage (Fy = 50 KSI)				
		SPAN IN FEET					SPAN IN FEET				
		4.0	5.0	6.0	7.0	8.0	4.0	5.0	6.0	7.0	8.0
2-SPAN	POSITIVE WIND LOAD	74	38	22	14	10	96	52	30	19	13
3 OR	POSITIVE WIND LOAD	92	48	28	17	12	120	65	37	24	16

NOTES

1. Allowable wind loads are based on uniform span lengths and F_y of 80 KSI for 29 and 26 gage and F_y of 50 KSI for 24 and 22 gage.
2. Allowable wind load has been increased by 33⅓%.
3. Minimum bearing length of 1½" required.
4. "A" panel is to be used as a wall panel only.

Figure 7.6 Exposed-fastener panel. (*A panel by MBCI.*)

Siding

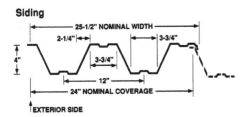

Roofing

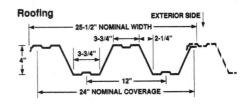

SMITH STEELITE SUPER-RIB SIDING AND ROOFING MAXIMUM SPANS									
Live Load				20 PSF (98 kg/m²)		30 PSF (146 kg/m²)		40 PSF (195 kg/m²)	
Gage/Weight		Span		Wall	Roof	Wall	Roof	Wall	Roof
GALV. STEEL 18 Gage (0.047") 3.04 lbs./ft²	L/120		SS	23'-11" (7.29 m)	22'-9" (6.93 m)	20'-11" (6.38 m)	19'-10" (6.05 m)	19'-0" (5.79 m)	17'-5" (5.31 m)
			DS	29-8 (9.04)	23-9 (7.24)	24-2 (7.37)	19-11(6.07)	20-11 (6.38)	17-5 (5.31)
			TS	29-7 (9.02)	26-7 (8.10)	25-10 (7.87)	22-3 (6.78)	23-5 (7.14)	19-6 (5.94)
	L/180		SS	20-11 (6.38)	20-0 (6.10)	18-3 (5.56)	17-9 (5.41)	16-7 (5.05)	16-3 (4.95)
			DS	28-1 (8.56)	23-9 (7.24)	24-2 (7.37)	19-11 (6.07)	20-11 (6.38)	17-5 (5.31)
			TS	25-10 (7.87)	24-6 (7.47)	22-7 (6.88)	21-9 (6.63)	20-6 (6.25)	19-6 (5.94)
GALV. STEEL 20 Gage (0.036") 2.28 lbs./ft²	L/120		SS	21-9 (6.63)	20-11 (6.38)	19-0 (5.79)	17-5 (5.31)	17-3 (5.26)	15-3 (4.65)
			DS	25-8 (7.82)	21-0 (6.40)	21-0 (6.40)	17-6 (5.33)	18-2 (5.54)	15-3 (4.65)
			TS	26-10 (8.18)	23-5 (7.14)	23-5 (7.14)	19-6 (5.94)	20-4 (6.20)	17-1 (5.21)
	L/180		SS	19-0 (5.79)	18-3 (5.56)	16-7 (5.05)	16-2 (4.93)	15-1 (4.60)	14-9 (4.50)
			DS	25-6 (7.77)	21-0 (6.40)	21-0 (6.40)	17-6 (5.34)	18-2 (5.54)	15-3 (4.65)
			TS	23-6 (7.16)	22-7 (6.88)	20-6 (6.25)	19-6 (5.94)	18-8 (5.69)	17-1 (5.21)
GALV. STEEL 22 Gage (0.030") 1.91 lbs./ft²	L/120		SS	20-6 (6.25)	19-3 (5.87)	17-11 (5.46)	16-0 (4.88)	16-3 (4.95)	14-0 (4.27)
			DS	23-5 (7.14)	19-4 (5.89)	19-2 (5.84)	16-0 (4.88)	16-7 (5.05)	14-0 (4.27)
			TS	25-3 (7.70)	21-7 (6.58)	21-5 (6.53)	17-11 (5.46)	18-6 (5.64)	15-8 (4.78)
	L/180		SS	17-11 (5.46)	17-3 (5.26)	15-8 (4.77)	15-3 (4.65)	14-2 (4.32)	13-11 (4.24)
			DS	23-5 (7.14)	19-4 (5.89)	19-2 (5.84)	16-0 (4.88)	16-7 (5.05)	14-0 (4.27)
			TS	22-1 (6.73)	21-4 (6.50)	19-4 (5.89)	17-11 (5.46)	17-6 (5.33)	15-8 (4.78)

All above weights are per net square foot.
Loads and spans are based on AISI Cold-Formed Steel Design Manual.
The above span tables are in accordance with the 1986 Light Steel Code with material having a yield strength of 33,000 psi (2320 kg/cm²), one-third extra strength for wind loads only.
Roof spans are for positive loading. Wall spans are for positive or negative loading.

Roof spans include dead weight of panel.
Minimum sheet length: 2'-0" (.61m). Maximum sheet length: 40'-0" (12.19 m).
Consult Smith Steelite for sheet lengths less than 2'-0" (.61 m) or greater than 40'-0" (12.19 m).
Length tolerance: maximum variation ±1/2" (12.7 mm).
Roof spans include dead weight of panel.

Figure 7.7 Deep-rib exposed fastener siding. (*Super-Rib by Smith Steelite.*)

fails, the joint starts to leak. The situation is exacerbated by a common installation technique of applying the sealant between two panel sheets and driving fasteners through the interface, squeezing the sealant in the process. The flattened sealant won't last long, and a failure is invited.

A radically different method of preventing water intrusion is based on the *rain screen–cavity wall* principle that dispenses with the idea of a single water barrier. The rain screen principle was formulated in 1963 by Kirby Garden, a Canadian researcher who recognized the impossibility of sealing every little opening that can develop in an exterior wall. Instead, he argued, the seals should be moved to an *interior* wall, where protection from the elements and solar radiation is easier to achieve.

A rain screen wall consists of three parts: exterior layer which resists most water intrusion, interior waterproofing, and air space in between. The interior waterproofing membrane is completely water-impermeable, with sealed joints and flashing. It is partly protected from the elements by the exterior sheet

Load Span Table—WP 120

NOTE: Numbers shown reflect wind loads for corresponding profiles. Deflections are less than L/180.

Material	Span	20 PSF Wall	30 PSF Wall	40 PSF Wall	50 PSF Wall
20 Gage	1	8-6	7-5	6-9	6-3
	2	11-5	9-11	9-0	8-5
	3	10-6	9-2	8-4	7-9
22 Gage	1	7-9	6-9	6-2	5-9
	2	10-5	9-1	8-3	7-8
	3	9-7	8-5	7-7	7-1
24 Gage	1	6-11	6-1	5-1	5-1
	2	9-4	8-2	7-5	6-9
	3	8-7	7-6	6-10	6-4
.032 Alum.	1	5-9	4-8	4-1	3-7
	2	6-11	5-8	4-11	4-5
	3	7-9	6-4	5-6	4-11
.04 Alum.	1	7-0	5-8	4-11	4-5
	2	8-2	6-8	5-8	5-2
	3	9-2	7-6	6-6	5-9

Standard Maximum Length - 32'.
Contact Carlisle for lengths over 32'.

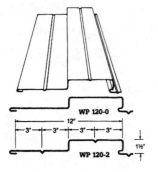

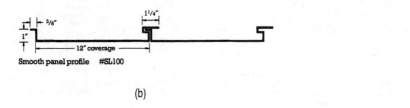

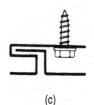

(a)

(b)

(c)

Figure 7.8 Concealed-fastener wall panel. (*a*) Exterior wall panel (WP by Carlisle); (*b*) interior liner (no. SL100 by Carlisle); (*c*) detail of interlocking connection and concealed fastener. (*Carlisle Engineered Metals.*)

("veneer") that sheds most water. The air space physically separates the two layers and facilitates drainage of accumulated condensate.

Exterior panels of rain screen walls have no exposed sealants. Instead, some specific design steps are taken to block the various pathways of water intrusion into the cavity that are described below for a horizontal panel joint:[3]

1. Kinetic energy of nearly horizontal wind-driven rain, the most common way of water penetration. The best defense is to build a ridge in the lower panel and make an internal baffle.

2. Surface tension: Water clings to and flows along the underside of the top panel. Remedied by a drip edge.

3. Gravity force: Water simply follows the panel's exterior surface downward. Can be overcome by sloping the joint surfaces upward.

4. Capillary action: Water seeps into a thin joint as through a wick. Disappears when the joint is at least $\frac{1}{2}$ in wide.

5. Air pressure drop and air currents: Water is sucked into the cavity by a pressure differential between the cavity and the outside. In a cavity wall, water intrusion can only be resisted by the backup waterproofing–air barrier.

Formawall, a typical wall panel that conforms to the rain screen–cavity wall principle is shown in Fig. 7.9.* To some degree, this principle is reflected in any cavity wall with interior waterproofing, flashing, and weep holes.

The design steps listed above will greatly reduce but not totally eliminate water intrusion into the cavity, especially due to an air-pressure differential. The interior waterproofing layer, flashing, and adequate cavity ventilation are therefore critical for success and should be designed and constructed with care.

Another type of rain screen is the *pressure-equalized wall*. The principle of pressure equalization assumes that if an air-pressure differential between the cavity and the exterior is eliminated, a problem of water leakage which it causes, the Achilles tendon of cavity-wall rain screens, can be solved. To achieve pressure equalization, sufficient openings are left in exterior veneer to make the cavity essentially a part of the outside world. The cavity itself is divided into relatively small compartments which restrict air movement within it and allow for rapid changes in air pressure.

Herein lies a problem. If the interior wall layer which carries an air-impermeable waterproofing is truly subjected to the full force of the wind, that layer, and not just the exterior veneer, must be structurally designed for the full wind loading. This is common practice for brick veneer walls but not for two-leaf metal panels.

Some panel manufacturers now reflect the pressure-equalization principle in joint design of their composite panels. The "cavity" in Fig. 7.9 is confined to the

---

*Formawall is a registered trademark of H.H. Robertson Company.

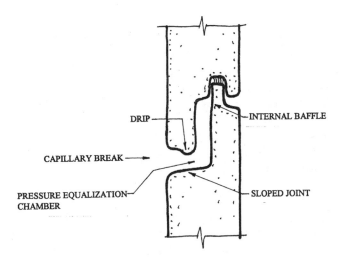

Figure 7.9 Panel joints based on rain screen principle.

joint area and ends with a concealed sealant between the panels. The best way to determine the effectiveness of this and other joint construction methods in containing water and air leaks is by full-scale testing, following the procedure of ASTM E 331.[27] This test is performed by spraying water at a specific rate, while maintaining an air-pressure differential between the inside and outside wall surfaces.

The promised benefits of pressure-equalized rain screen walls are yet to be fully proved. It is not even totally clear whether perfect pressure equalization is possible to achieve at all. Meanwhile, a basic cavity-wall rain screen, when properly designed and constructed, provides a reliable, simple, and cost-effective method of protection against the elements and water leakage.

7.3   Masonry

Masonry exteriors offer beauty, stability, security (remember the Third Little Pig?), and lateral-load resistance and require little maintenance. Long used in conventional construction, exterior masonry walls are now increasingly found in pre-engineered buildings.

Two basic kinds of masonry walls are single-wythe concrete masonry units (CMU) and two-wythe cavity walls. The latter may consist of brick veneer over CMU backup, or of brick veneer over stud backup. Masonry is erected by a relatively slow field assembly—the exact opposite of a concept behind metal building systems that stresses fast construction. Prefabricated masonry panels are available but not yet widely used.

7.3.1   Concrete masonry units (CMU)

Walls made of single-leaf CMU, 8 to 12 in thick, are gaining popularity in pre-engineered buildings because of their superior fire resistance and other masonry benefits mentioned above. When CMU walls act as shear walls, they eliminate the need for wall cross bracing. When, in addition, the walls are load-bearing, exterior building columns may be eliminated as well, and roof purlins can be supported on small bearing plates set inside wall pockets.

Exposed CMU walls need not be gray and nondescript; a wide variety of split-face, ground-face, and scored-block units is available in attractive colors.

Load-bearing concrete masonry units, or those in exterior applications where a possibility of freezing is present, should conform to ASTM C 90.[28] In the 1995 edition of this specification, the block is no longer differentiated by grade (N or S), but is still divided into type I, Moisture-Controlled Units, and type II, Nonmoisture-Controlled Units. Units of both types have the same maximum linear shrinkage, less than 0.065 percent. The units of type II, which are slightly less expensive, are suitable for most applications, except where an even smaller limit on residual drying shrinkage is desired.

Since masonry is brittle and very weak in tension, CMU walls used in seismic regions are required by codes to be reinforced in two directions to provide a measure of strength and ductility. Vertical reinforcement is placed in grouted cores and spliced as needed to extend the full height of the wall. Horizontal reinforcement can consist of rebars placed in bond-beam units, or of fabricated wire joint reinforcement placed in horizontal joints. Steel reinforcement allows some cracking to take place without leading to a total wall collapse; the amount of reinforcement is determined by analysis based on a local building code. Most local codes recognize ACI 530/ASCE 5/TMS 402 Specification[4] as an authoritative source.

In conventional construction, CMU walls are usually designed to span from top to bottom. The bottom support is provided by the foundations, the top by the perimeter roof beams. This approach is not well suited for a typical pre-engineered building where in place of perimeter roof beams one finds light-gage eave girts. The eave girts function best in providing lateral support for flexible roof siding, not brittle masonry. Whenever masonry is supported by the bottom flange of an eave girt, horizontal wall reactions from wind or earthquake produce torsional forces that a typical light-gage-channel eave section cannot resist. If standard eave sections must be used, the eave girt's bottom flange can be braced back to properly braced purlins. An even better solution is to substitute a closed steel section for the channel-type eave girt (Fig. 7.10): A closed section, such as structural tube, is better able to resist torsion.

In a different approach, CMU can be designed to span horizontally between the building columns as shown in Fig. 7.11. Under this scenario, primary structural reinforcement is contained in horizontal bond beams spaced at close intervals; masonry anchors transfer lateral loads from walls to columns. While this arrangement preserves regular eave girts and seems to be preferred by many in the industry, it presents obvious problems at doors and at wall control joints, if the latter do not coincide with the column locations. Furthermore, load-bearing and shear-wall capabilities of horizontally spanning CMU walls are not well established. And finally, the system becomes uneconomical with wide column spacing.

When wall height exceeds 20 to 24 ft, the strength of a standard 8-in block may become insufficient. (Note that the design thickness of ribbed blocks does not include the depth of ribs.) In this case, it is better to choose thicker masonry units than to attempt bracing the CMU with horizontal girts or wind columns. Bracing rigid masonry by flexible steel members does not work because of stiffness incompatibility, a consideration frequently missing in some sophomoric calculations that treat masonry no differently than metal siding.

This point brings us to the main problem with masonry use in metal building systems: Being rigid and brittle, masonry is inherently incompatible with light and flexible pre-engineered framing. When metal framing deflects horizontally under wind or seismic loading, it carries masonry with it, but masonry, unlike steel, cannot tolerate large lateral movements and may fail if displaced too much.

The most common method of safeguarding against cracking and failure of full-height masonry is to make the steel structure stiff enough—by using more metal than necessary for strength alone—to keep the building drift (lateral movement) within tight limits. For the frames that support vertically spanning

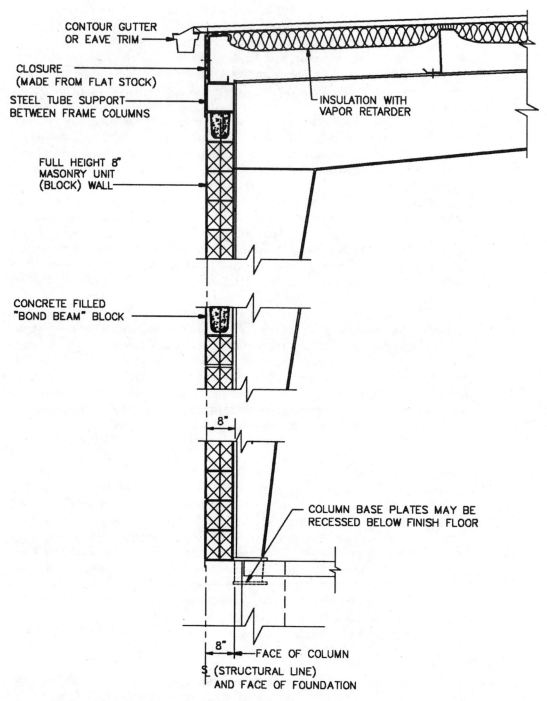

CONTOUR GUTTER
OR EAVE TRIM

CLOSURE
(MADE FROM FLAT STOCK)

STEEL TUBE SUPPORT
BETWEEN FRAME COLUMNS

FULL HEIGHT 8"
MASONRY UNIT
(BLOCK) WALL

CONCRETE FILLED
"BOND BEAM" BLOCK

INSULATION WITH
VAPOR RETARDER

8"

COLUMN BASE PLATES MAY BE
RECESSED BELOW FINISH FLOOR

8"   FACE OF COLUMN
(STRUCTURAL LINE)
AND FACE OF FOUNDATION

Figure 7.10 Vertically spanning full-height single-leaf exterior CMU wall. (*After a drawing by Butler Manufacturing Co.*)

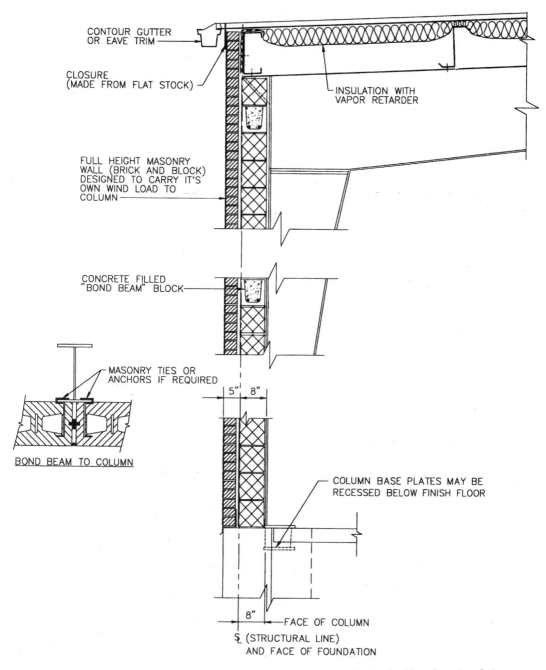

CONTOUR GUTTER OR EAVE TRIM

CLOSURE (MADE FROM FLAT STOCK)

INSULATION WITH VAPOR RETARDER

FULL HEIGHT MASONRY WALL (BRICK AND BLOCK) DESIGNED TO CARRY IT'S OWN WIND LOAD TO COLUMN

CONCRETE FILLED "BOND BEAM" BLOCK

MASONRY TIES OR ANCHORS IF REQUIRED

BOND BEAM TO COLUMN

5"   8"

COLUMN BASE PLATES MAY BE RECESSED BELOW FINISH FLOOR

8"

FACE OF COLUMN

℄ (STRUCTURAL LINE) AND FACE OF FOUNDATION

Figure 7.11   Horizontally spanning full-height brick and CMU cavity wall. (*Butler Manufacturing Co.*)

masonry, some engineers specify the allowable drift as the eave height divided by 600 (the $H/600$ criterion). This number has been criticized by some manufacturers, who decry it as wasteful and overly conservative. We revisit the issue of allowable frame drift values in a broader context in Chap. 11.

There are some other problems worth keeping in mind when specifying CMU walls in metal buildings. Not easily removed and reused, masonry is poorly suited for buildings where future expansion is likely. Also, masonry is heavy and requires continuous foundations for support. The insulation value of CMU walls is small; to increase it, furred interior walls containing insulation and finished with drywall or metal liner panels are needed. The blocks absorb moisture unless treated with a water-repellant admixture at the time of production or covered with a waterproof coating which needs to be periodically reapplied.

To reduce moisture penetration through mortar joints, the joints should be properly tooled, preferably to a concave shape. Like any exposed masonry, CMU requires control joints at close intervals such as 20 to 25 ft. (Luckily, this joint spacing coincides with popular building bay sizes.) With all the precautions taken, single-leaf CMU still does not enjoy the benefits of cavity construction, and any wall crack will allow moisture into the interior.

Single-leaf reinforced split-face CMU 12 in thick can cost $10 to $13 per square foot, not including any furring or drywall.

7.3.2 Brick veneer over CMU

This system combines the advantages of CMU walls such as durability and fire resistance with elements of a rain screen cavity wall. Figure 7.11 depicts a version of this wall as conceived by some major manufacturers, with CMU spanning horizontally between the columns. As was stated in the previous section, we would prefer a wall which spans vertically with a tubular member on top. Also, the cavity indicated should be at least 2 in wide and the base flashing should be firmly embedded into a block joint.

In this system, brick veneer and block are connected by horizontal joint reinforcement with integral or adjustable ties. The adjustable ties should be able to transfer the design lateral loads even at the outer limits of their movement. Brick's function is nonstructural, and all lateral loads are resisted by CMU; one can use 4-in split-face block or even stone instead of brick without changing the CMU design. All structural considerations discussed in Sec. 7.3.1 apply to the CMU of this assembly.

To improve insulating properties of masonry cavity walls, rigid insulation can be added into the cavity; it can also go on the inside surface of the CMU between furring channels and gypsum board.

Masonry cavity walls are rarely considered "systems" by either specifiers or code officials, in the sense that other curtain-wall systems such as EIFS (see Sec. 7.6.2) are conceived and tested. Instead, masonry is still thought of as an assembly of block and mortar; both components are specified and tested separately. As Kudder[5] points out, many masonry tests are available, but most predate cavity walls. For example, there is no standard watertightness test of the whole cavity-wall assembly with flashing and control joints. And yet, as specifiers well know, problems traced to improper flashing details and installation cause more harm than masonry materials which do not quite conform to the specifications. Other common occurrences that may compromise watertightness include mortar bridging the cavity and poorly filled mortar joints.

Brick and CMU are not identical in performance: After fabrication, clay- and concrete-based products tend to move differently with moisture changes. Oven-fired brick has practically no moisture at first and gains it, steadily expanding in the process. In contrast, CMU is born wet and is cured in a water-saturated condition, gradually losing moisture and shrinking.[6] It can only be hoped that, by the time the two materials are installed, most of the initial volume changes will have already taken place.

In a similar vein, brick and block react to temperature changes differently; normal-weight CMU may expand or contract up to 15 percent more than brick. Even more importantly, the exterior brick is directly affected by solar radiation and undergoes larger swings in temperature than the block separated by an air space.

Differential thermal movement can be mitigated by expansion and control joints spaced at close intervals, such as 18 to 25 ft. Expansion and control joints are often confused but are quite different in nature. Expansion joints in brick are filled with compressible materials, while control joints in CMU allow it to shrink and are normally filled with rigid inserts for stress transfer between the adjacent blocks.

A simple but often overlooked method of reducing brick thermal movement is to use light-colored brick.

The most common standard for face brick is ASTM C 216,[30] which covers about 93 percent of all the bricks sold in this country in 1994.[31] ASTM C 216 includes three types of brick: FBS for common use, FBX for applications requiring tight control of unit sizes, and FBA for deliberately nonuniform size and texture of units as is sometimes preferred by architects for a rustic look.

Mortar affects performance of brick and should be selected with care. The lower the mortar strength, the more displacement the wall can tolerate. A good overall choice for brick veneer, especially of that subjected to severe freeze-thaw cycles, is mortar conforming to ASTM C 270 type N. The mortar of type S has a higher flexural strength than that of type N; mortar of type M is reserved for load-bearing brick applications. Masonry cements are not recommended by the Brick Institute of America for use with brick veneer.

In contrast to brick, CMU backup walls usually contain steel reinforcing and require mortars of type S or M. With two different types of mortar often required for brick and CMU, proper supervision of the installation is essential.

There are many more fine points of masonry design that experienced architects and engineers learn throughout their careers. For good masonry wall performance, it is critical for design professionals to provide all pertinent details in the contract documents and to insist on good field supervision, even if the rest of the building system is "pre-engineered."

7.3.3 Brick veneer over steel studs

This wall type is commonplace in conventional construction, prized for its visual appeal, light weight, ease of insulation, and cost efficiency. The system, developed in the 1960s, consists of steel studs spaced 16 or 24 in on centers, interior gypsum wallboard, building paper on exterior-grade sheathing, and brick veneer separated from sheathing by an air space and attached to steel studs with adjustable metal ties (Fig. 7.12). The space between wall studs is often filled with fiberglass batt insulation.

Some argue that this complex wall system has become extremely popular too fast, before its weak points have been fully understood. Indeed, there are precious few sources of information on its long-term durability and on proper construction details. One of the best is *Technical Note* 28B by the Brick Institute of America.[8] The note contains detailed recommendations on structural design criteria, avoiding water penetration, minimum size of air space (2 in), maximum tie spacing, and other important aspects of the brick curtain-wall design.

It is not our intent to engage in a comprehensive discussion of this complicated system. Instead, we outline a few issues crucial to its performance in preengineered buildings.

A wall made of brick veneer over steel studs represents a simplified example of cavity-wall rain screen. It is assumed that some water will eventually get into the cavity. A proper functioning of waterproof building paper, flashing, and weep holes is therefore critical for a leak-free performance.

To be effective, flashing needs to extend through the wall and protrude at least ¼ in beyond its exterior face; the protruding point should form a drip by being bent at 45°.[7] Some architects loathe the look of the exposed flashing edge, yet flashing terminating within the brick invites water to travel under it back into the cavity, or worse—to freeze there. Flashing should be extended 6 to 9 in

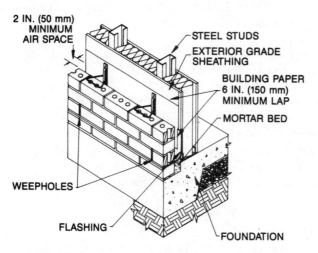

Figure 7.12 Brick veneer over steel-stud wall. (*Brick Institute of America.*)

above the penetration point and be embedded into the sheathing or a reglet. It is important to turn up and seal the edges of flashing at the sides to stop water from escaping sideways; prudent architects provide an isometric detail of this condition. The most durable flashing materials are stainless steel and lead-coated copper. Weep holes are located directly above the flashing at 16 to 24 in on centers.

The width of cavity needs to be at least 2 in, as mentioned above; anything less is difficult to keep free of mortar droppings which can carry water across the cavity. Why is this important? After all, is not water expected to get into the cavity anyway? The answer is yes, but the less water, the better, since water, or even moisture, lingering in the cavity attacks metal ties and their connections.

Adjustable ties anchor brick to steel studs and transfer lateral loads to them acting as minicolumns. The ties are best made of thick wire; those thin corrugated metal ties familiar to house builders are off limits in engineered construction. An adjustable tie is attached to the anchor connected to a steel stud; the anchor should permit a vertical adjustment of at least one-half height of a brick. Some common kinds of adjustable ties are shown in Fig. 7.13.

Most often, brick ties are attached to studs with self-drilling screws driven through sheathing. The screw-to-stud connection is a critical link holding up

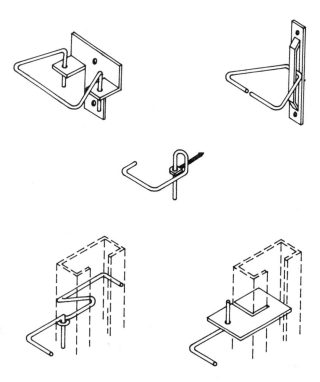

Figure 7.13 Brick tie assemblies. (*Brick Institute of America.*)

the veneer against the wind, which explains why moisture in the cavity is such a problem: Eventually, corrosion in this vital connection can lead to a brick-covered ground. The ties and studs might be made of galvanized steel, but the screws are normally protected only by a corrosion-resistant coating which can be easily damaged during installation. Stainless-steel fasteners offer a superior corrosion resistance of their own but can initiate corrosion when in contact with plain or galvanized steel.

At least one prominent engineer points out that all masonry contains some salt and laments that the brick "is literally hanging on the building by a thread, and the fine thin arris of the unprotected thread of a steel screw may be periodically bathed in a salt solution."[9] While others take a less apocalyptic view of the situation, the threat is clearly there. The author's practice is to specify galvanized studs of at least 18 gage, regardless of strength requirements, simply to provide a larger thickness of metal at the connection. This practice might be decried as a waste of metal by some, but the cost premium is minor in comparison with potential repair bills for a failed wall.

Many engineers, the author included, have sought ways to eliminate the screws from the anchors altogether; the products shown in Fig. 7.13d and e intend to do just that. While the objective is laudable in general, it is not clear how these particular designs affect long-term integrity of waterproofing, a critical component of cavity walls. Another possible approach is to use pop rivets, small bolts, or similar nonscrew fasteners, which completely eliminate the thread-corrosion problem. With any of those fasteners, one should think about special waterproofing requirements to prevent water penetration through the large resulting holes.

At present, the use of self-drilling screws and the anchors of the type shown in Fig. 7.13a and b is still prevalent. A special baked-on copolymer coating such as Stalgard* has been found more effective in corrosion protection and abrasion resistance than standard cadmium or zinc plating. To further restrict access of water to the joint, Gumpertz and Bell[10] suggest installing a piece of compressible gasket, made of EPDM or similar material, behind the base anchor. In addition, the screw should have either a built-in neoprene washer or a separate rubber grommet.

As if the issues involving flashing and brick ties are not overwhelming enough, a controversy involving lateral stiffness requirements is even more difficult to resolve. Recall that flexible steel studs are supposed to be a "structural" backup (lateral support) for "nonstructural" brittle brick veneer. The problem is, the brick will probably crack well before the flexible studs assume their deflected shape. An obvious solution is to stiffen the studs by using deeper sections made of thicker metal. But stiffen by how much?

BIA's Note 28B recommends that the maximum deflection of "steel stud backup, when considered alone at full lateral design load, [to] be $L/600$ to $L/720$."[8] There are those who consider this limit not stringent enough, pointing out that the brick will crack at deflections less than one-third of that—

---

*Stalgard is a registered trademark of Elco Industries Inc.

$L/2000$.[9] The steel-stud industry, on the other hand, insists that a limit of $L/360$ is adequate.[11]

To come to a solution, a rigorous finite-element analysis accounting for the stiffness of the veneer, steel studs, and brick ties needs to be made. Such analysis and investigation performed by Gumpertz and Bell[10] has found the BIA's limit of $L/600$ to be reasonable. The investigation has also discovered that the wind-force distribution among various brick ties along the height of the wall is not uniform: two or three ties located at the top and at the bottom of a stud carry a significant fraction of the total wind force. Thus the total wind load tributary to a stud should be divided into four or six, and the resultant should serve as the basis for design wind loading on all the ties.

The BIA[8] recommends that there be one tie for each 2 ft^2 (0.18 m^2) of wall area, and that the maximum tie spacing not exceed 24 in (600 mm). ACI 530/ASCE 5[4] is more stringent—it limits the area to 1.77 ft^2 per tie and tie spacing to 16 in on centers in every direction.

7.4 Combination Walls

Metal building systems allow plenty of opportunities to combine different materials in the same wall. For example, exterior masonry can be provided only at the bottom, where potential for physical abuse is the greatest. Masonry may also be selected as a wall-base material for aesthetic reasons, to add depth and interest to an otherwise flat or ribbed facade.

Partial-height masonry walls present some design challenges. As mentioned already, typical cold-formed wall girts are not appropriate for lateral support of masonry because of stiffness incompatibility. The best way to stabilize a partial-height ("wainscot") masonry wall is to cast vertical reinforcing bars into the foundation wall, so that a cantilever action is effected. Making such walls taller will result in a dramatic increase of the reinforcing bar size, so it is better to keep partial-height walls relatively short. The most expensive and difficult design is to run the CMU almost all the way up to the eave and terminate it in a ribbon window. If a situation like this arises, consider adding a few discrete windows at the top rather than a continuous one.

A common wall combination is that of a partial-height CMU with metal panels above. Panels can be attached to CMU with a sill channel or a sill angle (Fig. 7.14a and c), in which case the masonry should be structurally designed for an additional wind load imposed by the panel. In another possible solution, siding is laterally supported at the bottom by its own girt (Fig. 7.14b).

Metal wall panels of different profiles can be combined to create accent bands and to enliven the facade (Fig. 7.15). In this design, the bottom panel can be not only metal but also masonry (self-supporting, as noted above) or even precast concrete.

Occasionally, a combination of various wall materials can be avoided when metal panels are able to play a role of other building components, such as louvers. On a recent project in Puerto Rico, custom louvers were made of 18-gage metal liner panels bent to a louver-like configuration and spanning horizontal-

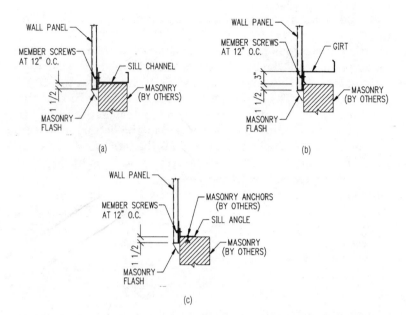

Figure 7.14 Details at top of partial-height masonry wall. (*a*) With sill channel; (*b*) with base girt; (*c*) with sill angle. (*Metallic Building Systems.*)

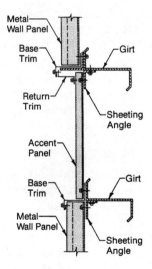

Figure 7.15 Detail of accent band. (*Star Building Systems.*)

ly. The resulting product not only blended in well with the surrounding wall panels but was also much less expensive than standard extruded metal louvers.[32]

7.5 Concrete Materials

7.5.1 Precast concrete panels

Precast concrete offers the same advantages as masonry, such as impact and sound and fire resistance, without the handicap of slow construction. Load-bearing and shear-wall applications of precast panels are especially popular in metal building systems. Nonstructural precast wall panels can add depth to the building (Fig. 7.16). To develop a distinctive wall texture, precast panels can be designed with deep horizontal or vertical grooves, perhaps with varying groove spacing; panels with projecting fins can also add visual interest. Even such unlikely elements as precast double-tee roof units have been successfully used as economical long-span curtain walls.

One of the main disadvantages of precast concrete is its low thermal insulation value, a problem that can be overcome with rigid insulation. Rigid insulation can be attached to the panel's interior face and covered with gypsum board, or can be enclosed as a sandwich between a thick structural part and a thin exterior course. For example, a 12-in insulated panel can consist of an 8-in solid or hollow-core structural part, 2.5 in of expanded polystyrene, and 1.5 in of exterior exposed-aggregate concrete. Such panels are commonly supplied in 8- to 12-ft widths; the cost might be $10 to $18 per square foot.

Figure 7.16 Precast wall panels add depth to the facade. (*Photo: Maguire Group Inc.*)

Many engineers neglect any structural contribution of the exterior layer and make it totally independent of the structural part. Interconnecting the two layers may ease concerns about panel delamination but may also lead to panel bowing and some loss of R value.

Insulated wall panels seem to be especially popular in food processing plants, where strict cleanliness requirements call for hard and smooth interior finishes without any crevices, pinholes, and horizontal ledges. For this application, panels are examined with binoculars during the erection to detect the most minute surface flaws which could become harbors for dust and bacteria.[29]

Designers increasingly opt to give the panels an exposed-aggregate finish. The recently released *Guide for Precast Concrete Wall Panels* by American Concrete Institute[12] differentiates among light, medium, and deep aggregate exposure for a variety of visual effects. The architect's arsenal of surface treatments for precast concrete also includes form liners, which can produce finishes ranging from wood board to split-face masonry, and the use of white and pigmented cements.

Structural design of precast concrete is governed by ACI 318[13] and is similar to cast-in-place concrete except as specified in ACI 533R.[12] Typically, concrete with a 28-day compressive strength of 5000 lb/in^2 or higher is specified for durability of precast panels. The reinforcement may consist of deformed bars, welded wire fabric, or prestressing tendons; epoxy coating or galvanizing is common to safeguard against corrosion.

A typical load-bearing panel spans the distance from the foundation to the roof. The bottom connection is made by welding embedded plates in the panel to those in the foundation, with any voids grouted and sealed. The top connection, which supports roof purlins, is made by field welding of clip angles to embedded plates (Fig. 7.17). Panel joints are handled in a similar manner.

The minimum thickness of conventionally reinforced precast panels ranges between one-twentieth and one-fortieth of the unsupported length.

Non-load-bearing precast panels can be laterally braced by heavy wall girts at their tops (Fig. 7.18). Although reinforced concrete exhibits a larger degree of ductility than masonry, careful consideration of the required girt and frame stiffnesses is required, especially if the bottom panel connection is close to fixity.

Design and detailing of precast concrete panels is a rather specialized field; it is best performed by the panel suppliers. Still, the information necessary to communicate the design intent should be specified in the contract documents. The extent of this information varies widely among design firms: at some, every panel is designed and detailed by the architect-engineer, at others, only the panel layout is shown on the building elevation drawings. We suggest that a typical panel be designed and detailed in-house, the design including the panel thickness, joint size, and a general method of attachment. The rest could be left to the precaster, who should be required to submit complete shop drawings and calculations, including structural analysis for handling, transportation, and thermal stresses.

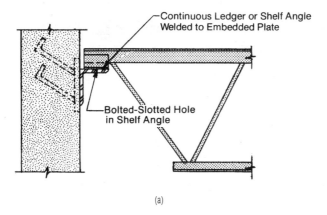

(a)

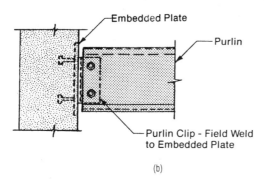

(b)

Figure 7.17   Joist (*a*) or purlin (*b*) connection to load-bearing pre-cast wall. (*Star Building Systems.*)

Architects are frequently concerned with acceptability of panel finishes and with casting or installation tolerances. The specifiers would be wise to review the relevant provisions of ACI 533R and *PCI Design Handbook*[14] and to be familiar with realistic visual and dimensional variations of plant-produced panels.

7.5.2 Tilt-up panels

While precast concrete is shop-formed and transported to the site, tilt-up panels are often cast right on top of the building's slab-on-grade. After a week's curing, panels are "tilted up"—lifted by a crane into proper locations—and braced. "Tilting up" normally occurs prior to erection of pre-engineered framing to avoid interference with the steel.

An obvious advantage of the tilt-up method is local fabrication, which avoids transportation costs and shipping limits on panel sizes; tilt-up panels can be as large as the available crane permits. A disadvantage is, of course, a loss of plant

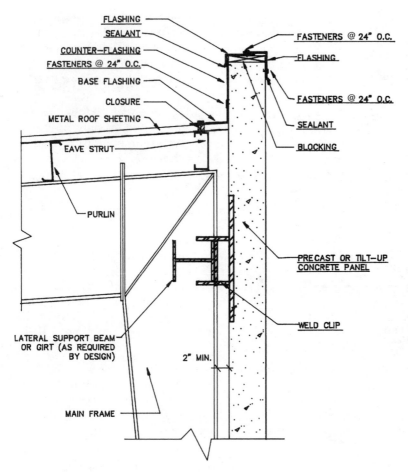

FLASHING
SEALANT
COUNTER—FLASHING
FASTENERS @ 24" O.C.
BASE FLASHING
CLOSURE
METAL ROOF SHEETING
EAVE STRUT
PURLIN
LATERAL SUPPORT BEAM
OR GIRT (AS REQUIRED
BY DESIGN)
MAIN FRAME

FASTENERS @ 24" O.C.
FLASHING
FASTENERS @ 24" O.C.
SEALANT
BLOCKING
PRECAST OR TILT—UP
CONCRETE PANEL
WELD CLIP
2" MIN.

Figure 7.18 Connection at non-load-bearing precast panel. (*Ceco Building Systems.*)

quality control and sophistication. While tilt-up panels are normally flat, a variety of reveals and surface treatments, such as exposed-aggregate and sand-blast finishes, expands the designer's choices. As in precast construction, wall liners can simulate the appearance of brick or stone. To facilitate panel removal, the casting slab is sprayed with a bond breaker; slab curing compound is usually adequate for this purpose. Most tilt-up panels emerge with an uneven or splotchy appearance that improves somewhat over time; but the panels are frequently painted nevertheless, often with acrylic-based coatings.[15] Some architects seize on this opportunity and create a trompe l'oeil effect, a painted pattern that looks from a distance like a three-dimensional structure.

The sides of tilt-up panels are usually formed with dimensional lumber, and panels $5\frac{1}{2}$ or $7\frac{1}{4}$ in thick are common. A rule of thumb limits panel thickness to one-fiftieth of the unsupported height. (This thickness is in addition to the depth of any reveals.)

In general, structural considerations for tilt-up panels parallel those for the precast; in both cases, stresses from lifting often govern the design. Calculations dealing with lifting loads and selection of embedded anchors are frequently performed by specialized engineers. A common panel design features a single reinforcement mat located at middepth with bars spaced 12 to 16 in on center.

Like precast panels, tilt-up concrete walls can be made with as much as 6 in of sandwiched-in insulation. A typical sandwich panel could be made of 2- to 3-in-thick exterior concrete layer, 2-in extruded polystyrene insulation, and an interior structural layer. Lately, fiber-composite connections between the panel layers, free from the disadvantages of thermal bridging endemic in the panels with traditional metal ties, have become popular.

Tilt-up construction imposes some specific requirements on the building slab-on-grade used for casting: The slab has to support the weight of the panels and the loaded crane (the erection is commonly done from inside the building). A minimum slab thickness of 5 to 6 in is recommended.[15] In addition, slab finishing requires special attention, as any surface irregularity will be reflected in the finished product. For best results, the slab joints are preplanned to coincide with the panel joints. The panel joints are usually ½ to ¾ in wide; they are sealed at both faces after erection.

For further engineering guidelines, readers are referred to ACI 551R, Tilt-Up Concrete Structures[16] and to Brooks.[17]

Tilt-up construction can be quite economical for medium-size buildings with high sidewalls (20 ft+) and a repetitive appearance. Most tilt-up panels are used as load-bearing elements and shear walls. The cost can range from $5 to $10 per square ft, depending on thickness and finish and excluding installation. Tilt-up is especially popular in areas with good climate such as Florida and southern California. According to the Tilt-Up Concrete Association, this wall system accounts for over 15 percent of new industrial building construction.[26]

7.5.3 Cast-in-place concrete

For some, concrete buildings with intricate plans or profiles panelization may not be an option. Cast-in-place concrete exterior may be needed to contain lateral pressures from the loose materials stored inside, as in most materials-recycling and resource-recovery facilities which require exterior concrete "pushwalls." Acting as cantilevered retaining walls, these pushwalls need to be rigidly connected to a horizontal base or to foundation walls; there are few alternatives to casting concrete in place.

Structural design and finishes of full-height cast-in-place concrete walls are similar to tilt-up construction, although cast-in-place walls are thicker to allow for concrete placement in a vertical position. For weather resistance of exterior walls, high-strength air-entrained concrete mixes are used. Control joints in cast-in-place walls are often made with rustication strips spaced 20 to 25 ft apart, perhaps coinciding with the bay spacing. To facilitate cracking at a joint, the amount of horizontal reinforcement passing through it should be halved.

Field-cast concrete might cost $12 to $16 per square foot for an uninsulated 8-in-thick wall.

7.6   Other Wall Materials

7.6.1   Glass fiber reinforced concrete (GFRC)

This relatively new material—it has been used in this country since the 1970s—is made by mixing glass fibers into a slurry of sand and cement and spraying the mixture onto molds. GFRC panels are thin (about $5/8$ in), lightweight, and durable, offering an appealing lightweight alternative to precast concrete and stone. Since a shape of the molds is limited only by the designer's imagination, a variety of striking forms can be achieved. GFRC affords an unmatched sharpness of detail due to smaller aggregate size. The available finishes include sandblasting and special decorative aggregate facing mixes ranging from $1/8$ to $1/2$ in in thickness.

Randomly dispersed glass fibers might account for only 5 percent of the panel weight, but the resulting increase in ductility is remarkable. GFRC skin is supported by lightweight steel frame with rigid ("gravity") and flexible ("wind") anchors embedded in thickened panel pads. The total panel weight is 8 to 20 lb/ft^2.

The installed cost is high—in the $20 to $50 per square foot range[18]—and restricts GFRC use to accent pieces, decorative parapets, column covers, elaborate cornices, and fascias.

While GFRC panels can add considerable interest to the project, they need to be specified with care, following the PCI's Recommended Practice.[19] The fabricator's and erector's qualifications should be checked carefully, since, owing to some loss of GFRC strength with age, it is possible that damage during fabrication or erection may not become noticeable for a long time. Nicastro[20] tells a story of some badly warped GFRC panels that were "straightened" in the field by a contractor and developed numerous cracks within a year after the erection. In the author's own experience, the panels are sometimes delivered cracked (mostly during transportation), discolored, and with delaminated anchor patches. It is wise, therefore, for architects to inspect the panels prior to erection and to insist that the manufacturer's operations comply with PCI's *Manual for Quality Control.*[21]

All these precautions do not guarantee success. GFRC simply has not been used long enough to reveal all its limitations. For example, GFRC skin of a 21-story office building in Texas mysteriously developed widespread cracking and eventually had to be replaced with another cladding system.[33] The problem was attributed to several factors. In some panels, "wind" anchors were made too rigid, restricting the panel's expansion and contraction. Other panels delaminated at the interface of the face mix and GFRC base; the delamination was blamed on differential thermal and moisture movement—and buildup of stress—between the two materials. Numerous workmanship deficiencies have also been observed.

We can only hope that, once the weak points of GFRC are clearly understood, this promising material will overcome its problems and become commonplace in metal building systems.

7.6.2 Exterior insulation and finish systems (EIFS)

Born in Europe, EIFS were introduced in this country in the late 1960s by Dryvit System, Inc., and for a while were called by that name. Today, this system is offered by many other companies represented by EIFS Industry Members Association (EIMA). The association publishes guideline specifications, technical notes, and other useful information about the product.[22]

The system most likely to be used in metal buildings includes steel studs spaced 12 to 32 in on center, exterior sheathing, rigid insulation attached to sheathing with adhesive, base coat, reinforcing mesh, and finish coat (Fig. 7.19).

There are two generic classes of EIFS: polymer-based (PB) and polymer-modified (PM). Class PB systems are made with thin flexible materials and are far more popular than the thicker class PM, or "hard-coat," cementitious products.

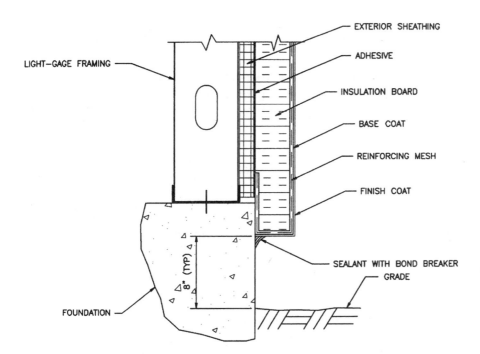

Figure 7.19 Detail at the base of EIFS panel.

The main advantages of EIFS are design flexibility, high insulation value, and low cost. EIFS make possible a variety of shapes and surface textures; the systems can be applied over existing surfaces in the field or manufactured in panels with light-gage steel framing.

EIFS have become incredibly popular, but the failure rate has been dramatic as well. The author was intimately involved with production of EIFS panels in the early 1980s and later witnessed failures of many of those applications. Recent reports indicate that the first EIFS introduced in the United States were not nearly as good as their European siblings.

Today's designs are much more advanced and will likely perform better. Still, the products that simply meet the minimum EIMA criteria might not be worry-free; some additional requirements will greatly improve the chances of obtaining a durable EIFS skin, although at an added cost. The following is a sampling of the experts' recommendations, after Piper and Kenney[23] and others.[24,25]

- The PB class coatings are flexible, but they cannot cover large gaps and defects in the substrate without cracking. A classic source of cracks is the sloppy joint between pieces of rigid insulation that, instead of being filled with slivers of insulation, is left unfilled or is filled with adhesive.

- Although exterior-grade gypsum sheathing is allowed by EIMA, it can absorb moisture and lead to failure. Instead, Piper and Kenney recommend using cement fiberboard or proprietary products conforming to ASTM C 1177 such as Georgia-Pacific's Dens Glass* or the newest Dens Glass Gold.

- Expanded polystyrene insulation must have a good bead fusion; its joints should not be aligned with edges of the openings, sheathing joints, or rustications.

- The base coat should not be more than 33 percent cement, even though many products on the market have a 50 percent cement content. The extra cement not only reduces the coating's flexibility but also increases its alkalinity, which can break down an alkali-resistant mesh coating and eventually corrode the mesh.

- A common base-coat thickness is $\frac{1}{8}$ in, which does not provide adequate moisture protection. For best results, the base coat should be at least $\frac{3}{32}$ in thick and be applied in two layers.

- Use low-modulus sealants applied to the primed base-coat areas, not to the finish coat as is commonly done. Silicone sealants will likely provide the best service.

- Avoid using EIFS on sloped surfaces with a pitch less than 1:1.

- To guard against delamination, attach EIFS to the substrate with both mechanical fasteners and adhesives.

---

*Dens Glass is a registered trademark of Georgia-Pacific.

While some of these recommendations might be argued with, they highlight the evolving state of the art in the EIFS industry as well as large differences between the available products. Do not treat this wall system as a commodity.

7.6.3 Thin-veneer panels on steel framing

Occasionally, architects wish to use a wall system that looks and feels like stone, brick, or precast concrete but weighs much less. Thin-veneer panels might provide an answer. A typical panel consists of thin sheets of veneer supported on light-gage steel studs or on structural-steel frames. Thin veneer, which can be stone, precast concrete, and thin-plate—or even full-size—brick, is attached to the framing by steel anchors and clips. The panels may be insulated.

Thin-veneer panels have been developed primarily for mid- and high-rise buildings and for building retrofit applications, where light weight and speedy erection of cladding are highly valued. The system is no less appropriate for metal building systems that value the same attributes.

The main disadvantage of thin-veneer construction—as well as of GFRC, EIFS, and any other single-leaf systems—is the absence of an air cavity. Not conforming to the rain-screen principle, all these systems depend on joint sealants for moisture protection and are vulnerable to fastener corrosion, thus requiring periodic inspections and maintenance.

7.7 Choosing a Wall System

Selection of exterior wall materials logically fits into an overall assessment of the available building systems outlined in Sec. 3.6. Throughout this chapter, advantages and disadvantages of various wall materials have been listed. Often, the surrounding environment will dictate the choice of finish, color, and texture; sometimes the client will have a major input by insisting on easy-to-maintain finishes, for example. The functional requirements will often dictate whether a "hard" wall is needed, at least at the base, and whether the wall should be insulated.

A building with a lot of forklift traffic may need "hard" walls unless some kind of wall protection is employed. Similarly, a warehouse storing valuable goods will not be well protected by metal siding. A food-processing plant, as already mentioned, requires a hard and smooth interior finish, best achieved with precast concrete. Fire-resistance criteria may also limit the choice to masonry or concrete. Beyond these considerations, the trade-off is between aesthetics and cost.

It is critically important to remember that exterior walls are a part of the metal building system; they must be compatible with it visually, structurally, and functionally. Nonferrous wall systems, and especially their connections, should be carefully reviewed for compatibility with metal framing. Unfortunately, low-rise curtain walls seem to have more problems than their high-rise counterparts, perhaps owing to the limited design budgets usually available. Yet the potential liability of the designer for a poor coordination or an unwise product selection knows no such limits.

A helpful reservoir of ideas on combining various materials to achieve the desired effect can be found in *Metal Architecture* and other publications of the metal building industry.

References

1. *Metal Building Systems,* 2d ed., Building Systems Institute, Inc., Cleveland, Ohio, 1990.
2. John A. Hartsock and W. J. Fleeman, III, "How Foam-Filled Sandwich Panels Work," *Metal Architecture,* May 1994.
3. Richard Keleher, "Rain Screen Principles in Practice," *Progressive Architecture,* August 1993.
4. Building Code Requirements for Masonry Structures (ACI 530/ASCE 5/TMS 402-95), American Society of Civil Engineers, New York, 1995.
5. Robert J. Kudder, "Deliver Masonry as a Cladding System," *Masonry Construction,* August 1994.
6. Stephen Szoke and Hugh C. MacDonald, "Combining Masonry and Brick," *Architecture,* January 1989.
7. Douglas E. Gordon, "Brick Basics," *Architectural Technology,* Fall 1986.
8. "Brick Veneer Steel Stud Panel Walls," *Technical Notes on Brick Construction* no. 28B, rev. II, Brick Institute of America, Reston, Va., 1987.
9. Clayford T. Grimm, "Brick Veneer: A Second Opinion," *The Construction Specifier,* April 1984.
10. Werner H. Gumpertz and Glenn R. Bell, "Engineering Evaluation of Brick Veneer/Steel Stud Walls," Part 1, "Flashing and Waterproofing," Part 2, "Structural Design, Structural Behavior and Durability," *Proceedings, Third North American Masonry Conference,* Arlington, Tex., June 1985.
11. *Here Are the Facts about Steel Framing—Brick Veneer Systems Design,* Metal Lath/Steel Framing Association, Chicago, Ill.
12. *Guide for Precast Concrete Wall Panels,* ACI 533R-93, American Concrete Institute, Detroit, Mich., 1993.
13. Building Code Requirements for Reinforced Concrete, ACI 318, American Concrete Institute, Detroit, Mich., 1995.
14. *PCI Design Handbook, Precast and Prestressed Concrete,* 3d ed., Precast Prestressed Concrete Institute, Chicago, Ill., 1985.
15. Ed Santer, "Tilt-up Basics," *Concrete Construction,* May 1993.
16. Tilt-Up Concrete Structures, ACI 551R-92, American Concrete Institute, Detroit, Mich., 1992.
17. H. E. Brooks, *The Tilt-up Design and Construction Manual,* 2d ed., H.B.A. Publication, Newport Beach, Calif., 1990.
18. Gordon Wright, "GFRC Provides Lightweight, Versatile Cladding," *Building Design and Construction,* December 1988.
19. Recommended Practice for Glass Fiber Reinforced Concrete Panels, MNL-128, Precast/Prestressed Concrete Institute, Chicago, Ill., 1994.
20. David Nicastro, "Uncovering the Reasons for Curtainwall Failure," *Exteriors,* Summer 1988.
21. *Manual for Quality Control for Plants and Production of Glass Fiber Reinforced Concrete Products,* MNL-130, Precast/Prestressed Concrete Institute, Chicago, Ill., 1991.
22. EIMA Guideline Specification for Exterior Insulation and Finish Systems, Class PB, and other publications, EIFS Industry Members Association, Clearwater, Fla., 1994.
23. Richard Piper and Russell Kenney, "EIFS Performance Review," *Journal of Light Construction,* Richmond, Vt., 1992.
24. Michael Bordenaro, "Avoiding EIFS Application Pitfalls," *Building Design and Construction,* April 1993.
25. Margaret Doyle, "Trends in Specifying EIFS," *Building Design and Construction,* August 1988.
26. Maureen Eaton, "Tilt-up Concrete Offers Economical Construction Option," *Building Design and Construction,* May 1995.
27. ASTM E 331, Test Method for Water Penetration of Exterior Windows, Curtain Walls, and Doors by Uniform Static Air Pressure Difference, ASTM, Philadelphia, Pa., 1993.

28. ASTM C 90-95, Standard Specification for Load Bearing Concrete Masonry Units, ASTM, Philadelphia, Pa., 1995.
29. Craig A. Shutt, "Food Processors Savoring Insulated Precast Wall Panels," *Ascent,* Summer 1995.
30. ASTM C 216, Standard Specification for Facing Brick/Solid Masonry Units Made from Clay or Shale, ASTM, Philadelphia, Pa., 1992.
31. Brian E. Trimble, "Building Better with Brick," *Building Design and Construction,* August 1995.
32. "Modified Panels Used for Manufacturing Facility's Wall Louvers," *Metal Architecture,* October 1995.
33. Gordon Wright, "Inappropriate Details Spawn Cladding Problems," *Building Design and Construction,* January 1996.

Chapter

8

Insulation

8.1 Introduction

Ever since the oil crisis of the 1970s, demand for energy-efficient buildings has been steadily rising. Model building codes have already adopted many energy conservation provisions; federal regulations such as the Energy Policy Act of 1992 also emphasize energy-efficient building construction. To set uniform design requirements for energy conservation, American Society of Heating, Refrigeration and Air-Conditioning Engineers (ASHRAE) has issued its Standard 90.1,[1] that is being rapidly adopted by building codes.

Historically, insulation issues have not been on the "front burner" of the metal building industry. Indeed, inadequate insulation is still among the most often heard complaints about pre-engineered buildings. This chapter reviews some available insulation products, systems, and details to help designers make educated choices about materials and installation methods. It will not delve too deeply into the domain of HVAC engineers, dealing with thermal loads, energy conservation, and equipment selection topics. It will also avoid the matters of mass consideration, annual heating loads, and life-cycle costing for metal building systems that are well addressed in *Metal Building Systems* by the Building Systems Institute.[2]

8.2 The Basics of Insulation Design

Heat loss or gain in a building can occur via three modes: radiation, conduction, and convection. Of these, conduction through the building envelope accounts for most heat transfer. Heat losses due to conduction can be reduced—but not

---

The author wishes to express his sincere thanks to the North America Insulation Manufacturers Association (NAIMA) Metal Building Committee for its contribution to this chapter and for permission to reproduce the illustrations from its publications.

eliminated—by additional insulation. Convection by air leaks can and should be prevented by "tightening" the building. Radiation affects mostly glass surfaces and can be minimized by reflective coatings. Since exterior walls contain openings and are otherwise thermally nonuniform, heat transfer via several *parallel heat flow paths* may be considered separately.

The building codes and ASHRAE 90.1 specify a certain required level of *thermal conductance (or transmittance)* U_0 for roof and wall assemblies in various locales, expressed in Btu/(h)(ft²)(°F). Thermal transmittance is a reciprocal function of the *thermal resistance R* of an assembly, which is a sum of R values of the components including those of the inside and outside air films and any air cavities.

Heat transfer is often accompanied by moisture movement, since warmer air contains more water vapor than cold air. Movement of vapor does not have to coincide with the actual movement of the air containing it. When warm air cools down or meets a cool surface, it loses some of its moisture, producing condensation. The air temperature at which condensation starts to occur is the *dew point*.

Condensation may lead to metal corrosion, growth of mold and mildew, loss of insulating properties, and ruined finishes. It can be minimized by a vapor retarder installed on the warm side of the wall (or, more precisely, on the side with the higher vapor pressure). Vapor retarder slows down moisture transfer toward a cooler surface. In winter months the warm side is of course the inner surface of the wall; during hot summer months it is the outside surface. Most roofing and siding is virtually vapor-impermeable and is quite able to act as a vapor retarder for hot-weather conditions; it's the cold-weather condensation protection that is normally needed for interior surfaces.

The term *vapor retarder* is more accurate than a frequently used *vapor barrier*, because building materials do not totally stop moisture movement and can only slow it down. Retarders may not be needed at all in moderate dry climates but are important in humid locales and in buildings where moisture is released.

8.3   Types of Insulation

Type and thickness of insulation have the largest influence on thermal efficiency of a building. Indeed, spending money on insulation could be among the best investments ever made by a building owner. Properly selected roof insulation does even more for thermal performance of a single-story metal building than for a multistory structure.

All insulation functions by entrapping still air, which slows down conductive heat transfer through the insulating medium. The various types differ mainly in *how* this is accomplished. Four basic types of insulation are available for metal building systems: fiberglass, rigid, spray-on, and foam core.

8.3.1   Fiberglass blanket insulation

Fiberglass functions similarly to a fur coat: Both trap air on the surface of numerous individual fibers. Fiberglass blankets are the most common kind of insulation used in roofs and walls of pre-engineered buildings because of their low cost, fire and sound resistance, and ease of installation. The *R* value of fiberglass insulation ranges between 3 and 3.33 per inch of thickness. The blankets are normally provided with a vapor retarder that is "laminated" on the fiberglass (Fig. 8.1*a*). In addition to serving its main purpose, vapor retarder often doubles as the only ceiling finish found in metal buildings; accordingly, the facing is usually white for better light reflectivity.

Fiberglass insulation for metal buildings is not quite the kind used by homeowners to insulate their attics. First, it is wider, corresponding to typical metal girt and purlin spacing (5 and 8 ft), rather than to that of wood studs and rafters (16 or 24 in). Second, it uses a different type of vapor retarder, as will be explained in Sec. 8.4.

The "official" insulation conforming to the North American Insulation Manufacturers Association (NAIMA) standard 202 must meet a set of stringent criteria, such as maintaining dimensional stability after application of a vapor

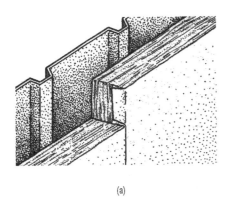

(a)

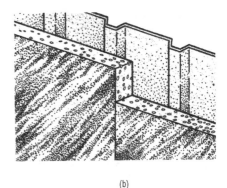

(b)

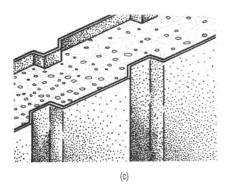

(c)

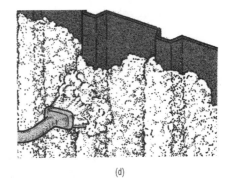

(d)

Figure 8.1 Four basic types of insulation used in metal buildings. (*a*) Fiberglass with laminated facing; (*b*) foam boards; (*c*) preinsulated panels; (*d*) spray-on cellulose. (*Courtesy of NAIMA Metal Building Committee.*)

retarder and during construction. It is tested for fire safety, as well as for its resistance to corrosion, mold, odor, and moisture.[3] (See Chap. 2 for information on NAIMA.) The NAIMA 202 certification is imprinted on the unfaced side of the fiberglass rolls. According to the association, such insulation has been tested with all UL-approved facings. In addition to NAIMA certification, some specifiers require compliance with ASTM E 553[4] as well.

Another type of fiberglass insulation for pre-engineered buildings is Metal Building Insulation (MBI) that does not have a vapor retarder and is intended to serve as a second layer over the faced insulation. MBI comes in both batts and rolls. Owens-Corning's Metal Building Insulation—Plus is one example of this kind.

Fiberglass insulation is produced by a few major companies (mostly NAIMA members) and is normally purchased by the builder from the laminators who apply vapor retarders. Some major metal building manufacturers such as Varco-Pruden and Butler offer a complete range of insulation products for one-stop shopping.

8.3.2   Rigid insulation

Rigid insulation, also known as foam board, functions by entrapping air in a multitude of individual foam cells. Most often, the foam is made with polystyrene, polyisocyanurate, and phenolic materials (Fig. 8.1*b*).

Rigid insulation offers excellent thermal efficiency (*R* value), vapor retardance, and dimension stability but lacks in acoustical performance. Rigid insulation is more expensive to make and install than fiberglass. The insulation can be applied both inside and outside the framing and can be faced with various surface treatments; it may require protective boards during construction.

Polystyrene insulation used to be an undisputed king of the foam boards. Recent CFC regulations have severely limited its production and, unless another equally good blowing agent is found, now threaten its existence. This insulation is available in two incarnations: expanded polystyrene ("bead boards"), an open-cell material sometimes specified for perimeter foundation protection, and extruded polystyrene used as wall and roof insulation. Extruded polystyrene, found among other things in disposable coffee cups, has a closed-cell composition with better vapor-retarding properties than an open-cell variety.

Major advantages of extruded polystyrene include an *R* value of 5.0 or slightly higher and a compressive strength of 30 to 40 lb/in². Being combustible counts as its chief disadvantage beyond the CFC problem.

Polyisocyanurate insulation, essentially a modified urethane foam, has a major advantage of being fire-resistant. It is more thermally efficient than polystyrene, with *R* values ranging from 5.8 to 7.2. However, it is also more compressible and friable than polystyrene and less water-resistant. Therefore, polyisocyanurates are normally covered with facing materials and are heavily dependent upon them for stability and moisture resistance. If the facers delaminate, the insulation is endangered. Unfortunately, this problem is far too common, especially with thicker boards.[5]

Polyiso insulation, as it is known, is less expensive to install than either extruded or expanded polystyrene and, also being more efficient, produces a much larger return on investment than polystyrene. This fact undoubtedly helps explain the fact that polyisocyanurate insulation is used in one-half of all new commercial construction.[12]

One of the most popular polyisocyanurate products is Thermax* by the Celotex Corporation. Thermax consists of a glass fiber–reinforced polyisocyanurate foam plastic core with aluminum foil facing on each side. The facings come in a variety of finishes: exposed facings often receive embossed white finish, while the concealed side normally has a reflective foil finish. The available insulation thicknesses range from ½ to 3 in; the standard panel width is 4 ft. Thermax was specifically designed to dampen the noise due to movement of standing-seam roofs.[6]

A new product called Nailboard, produced by NRG Barriers of Portland, Maine, uses polyiso insulation with bonded sheets of oriented-strand board $\frac{7}{16}$ or $\frac{5}{8}$ in thick. The product provides a nailable surface for roofing and eliminates damage to insulation during handling and erection.

Phenolic insulation boasts the highest R value of all foam boards, about 8.3 per inch,[5] as well as excellent fire resistance. It is more expensive than the other products and is reserved for the most demanding applications. Phenolic insulation may require an overlayment board for protection. Recently, phenolic foam insulation has been linked with accelerated corrosion of steel roof deck and should be specified with caution.

8.3.3   Foam-core sandwich panels

Preinsulated panels discussed in Chap. 7 can incorporate either rigid or fiberglass insulation. Rigid-insulation foam is usually sprayed between the metal faces and allowed to "bubble" and expand, filling all the corrugations (Fig. 8.1c). The fiberglass is simply inserted between the sheets. Either way, the panels offer excellent R values but obviously lack the acoustical performance of exposed fiberglass. A typical panel with 2 in of urethane or isocyanurate foam core may have a U factor of 0.06 and weigh only 2½ lb/ft^2; a 6-in-thick panel may have a U factor of about 0.02.[16]

8.3.4   Spray-on cellulose

Cellulose, a paper product often treated with fire retardants, is brought to a sprayable form by addition of liquid binders (Fig. 8.1d). Low cost and good noise absorption are about the only advantages of this material. Its disadvantages are many and include limited R value, lack of vapor retardance, and rough appearance; it can collect dust, absorb moisture and oily residues, and precipitate corrosion of metal surfaces.

---

*Thermax is a registered trademark of the Celotex Corporation.

Spray-on cellulose should conform to ASTM D 1042, type II, class (a)[7] and should not contain asbestos, crack, or lose bond with the substrate. The applicator's skill is critical for successful installation.

Recently, spray-in-place polyicynene foam insulation was introduced. This soft-foam insulation shows great promise, clearly outperforming cellulose. Spray-in-place foam can be trimmed after curing to provide a relatively flat interior surface.[16]

8.3.5 Choosing type and thickness

The process of insulation selection starts with determination of code-mandated U values and moves in concert with the wall and roof design. (A list of U values for several most popular wall assemblies is included in Sec. 8.6.) Apart from satisfying the minimum requirements, the insulation thickness should be substantial enough to prevent condensation on the facing by keeping its temperature above the dew point. The acoustical performance, appearance, and cost should also be considered. Fire-hazard classification rating determined by a flame-spread test could eliminate some insulation choices from the beginning.

Fiberglass blanket insulation with an appropriate facing often provides the best overall performance and is almost exclusively specified for metal building roofs. Selection of wall insulation is closely tied to wall system design and fire-rating requirements.

8.4 Vapor Retarders

As was mentioned above, the main function of a vapor retarder is to slow down the flow of moisture through a roof or wall assembly. (Another function is to reduce the flow of *air*, as explained below.) No known material is a true vapor barrier that completely stops passage of moisture, but some vapor retarders can slow the process better than others because of lower *permeability*, or permeance.

Permeability is measured in *perms* in the British system (a perm is 1 grain of water transmitted per hour per square foot for 1 inch of mercury vapor pressure differential) or in nanograms per pascal per second per square meter in SI. (One grain is $\frac{1}{7000}$ pound.) One perm equals 57.2 nanograms per pascal per second per square meter.

The rate of water vapor flow is equal to the vapor pressure difference on two sides of the material times its vapor permeance. The lower the permeance, the more effective the vapor retarder is. According to NAIMA, the perm ratings of commonly used vapor retarders range from 1.0 for a basic vinyl to 0.02 for such top-of-the-line composite materials as foil/scrim/kraft, metallized polypropylene/scrim/kraft, vinyl/scrim/foil, polypropylene/scrim/foil, and vinyl/scrim/metallized polyester.

In order to select an apppropriate facing material, one needs to assess the amount of moisture likely to be generated within the building and compare it with the annual temperature and humidity conditions in the area. Factors like appearance, resistance to abuse, and cost are then considered. Some occupancies may impose additional requirements on vapor retarders such as noise reduction or low emissivity. Wherever chemical fumes are released—chemical plants, laboratories, poultry farms, steel mills, and the like—a special chemical analysis may be needed. Consulting a manufacturer is recommended for critical applications. Among the industry leaders in vapor retarders are Lamtec Corporation of Flanders, N.J.; Vytech Industries, Inc., of Anderson, S.C.; Thermal Design, Inc., of Madison, Neb.; and Rexam Performance Products, Inc., of Stamford, Conn.

White-colored materials are popular because of their high light reflectance, a desirable quality for ceilings. Some abuse-resistant insulation facings such as those marketed by Lamtec Corp. tout their toughness; these products aim to compete with interior-wall panel liners and gypsum board.

Pure-vinyl vapor retarders were used almost exclusively earlier, but vinyl usage is declining. Despite its low cost, vinyl has very high permeance, low puncture resistance, and a tendency to yellow and crack with age. Any of the new materials can do the job much better. If the price is the only criterion, non-metallized polypropylene/scrim/kraft facing, often marketed as a replacement for vinyl, is among the cheapest white reinforced facings available. However, its high perm rating (0.09 or so) is 4.5 times larger than that of the metallized version, which is still moderately priced. It is worth noting that the difference between the cheapest and the most expensive product available could be as little as two-tenths of 1 percent of the total building cost.[15]

8.5   How to Maximize Thermal Performance

8.5.1   Avoiding condensation

Condensation and rusting problems in metal building systems are quite common. Says William A. Lotz, a noted consultant on building moisture and insulation: "In my experience, 'conventional' metal building insulation and facers do not protect the metal building from condensation and rust for very long when there is humidity inside a building located in a cold climate."[17] Yet even the best insulation and vapor retarder facing can be severely compromised if installed incorrectly. Specifying a vapor retarder with low permeability is only the beginning. As Ref. 13 and others point out, moisture diffusion through vapor retarders is a slow process which rarely causes actual problems. A much more practical danger is leakage of moisture-laden air into the insulated space caused by poorly sealed seams of vapor retarder and by unprotected penetrations. Unless all the facing joints and penetrations are properly sealed, moisture will seep into the insulation and eventually condense on the underside of metal surfaces, saturating the insulation and ruining its thermal performance.

Some of the latest insulation facing products offer adhesive or release-paper seams which could provide a better seal than the traditional fold-and-staple method. Thus a mediocre vapor retarder with good seams may perform better than a top-of-the-line product with leaky edges.

Permeability of interior vapor retarders is usually larger than that of metal roofing and siding. When humidity inside the building is high, moisture will eventually seep through the vapor retarder and condense on the metal. (For this reason, items like insulation retainers should be made of nonferrous materials or have a plastic coating to prevent rust stains.) At this stage, ventilation of insulated space between the metal and the vapor retarder becomes the only solution that can prevent moisture accumulation and hidden corrosion. Therefore, providing the cold side of the assembly—the exterior in this case—with some mechanism for moisture release, such as soffit and ridge vents, is beneficial. Unfortunately, the ventilated metal-building roof assembly is still a rare one.

Ventilation does not guarantee absence of condensation and the resulting damage to finishes, only a reduction in its severity. In cases where indoor humidity is expected to be extremely high, especially in cold climates, it may be wise to use a nonmetal roof system better able to dissipate moisture. In the words of Tobiasson and Buska:[14] "As an example, a vinyl vapor retarder/fibrous glass batt insulation/standing seam metal roofing system, ventilated or not, is probably not appropriate for a building housing an indoor pool in Minnesota."

Moreover, ventilation is sometimes a double-edged sword, inviting moisture into the roof during summer months. (Remember that to keep a house basement dry it is best *not* to ventilate it in the summer and to use a dehumidifier. Regrettably, dehumidification of an insulated roof cavity does not yet seem practical.)

A special problem occurs with composite metal panels that are factory sealed for the stated purpose of preventing water intrusion. Alas, no seal is ever perfect. It is much easier for the moisture to get inside this vapor trap than to evaporate, as the owners of many permanently clouded insulated-glass doors and windows can testify. The trouble is, while the glass does not corrode and the seal failure is clearly visible in a window, it is exactly the opposite for metal panels. The difficult task of simultaneously providing ventilation openings in the panel and preventing rain intrusion should go beyond simply making weep holes to drain the condensate.

8.5.2 Minimizing heat loss through fiberglass insulation

A well-known phenomenon of thermal bridging occurs when a piece of highly conductive material such as metal connects exterior and interior spaces, "short circuiting" the insulation. In cool weather it leads to cold spots, energy losses, and condensation problems. A great example of thermal bridging occurs when metal roofing or siding is fastened through fiberglass insulation into secondary framing. The insulation is compressed to less than $3/16$ in at supports and gains its full thickness near midspan, its shape resembling an hourglass (Fig. 8.2).

This "hourglass" method of installing insulation is the easiest, the cheapest—and the least energy-efficient. Chances are, this is what you will get unless you specify another design in the contract documents. Were you to take a bird's-eye view of such a roof covered with snow, you would immediately see where the purlins were: where the snow has melted.

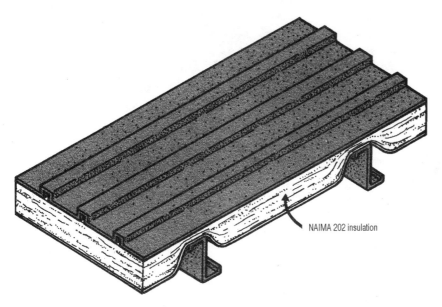

NAIMA 202 insulation

Figure 8.2 NAIMA system 1—insulation installed over purlins. (*Courtesy of NAIMA Metal Building Committee.*)

How to estimate the amount of heat lost through the framing? Some simply overlook it and pretend there is none—hardly an option for an enlightened designer. Others conduct a sophisticated parallel-flow analysis. Still others rely on the results of actual "hot box" testing. A simplified but practical way is to increase the supplied R values by 25 to 40 percent over those required by analysis.

The thicker the insulation, the more efficiency it loses in an hourglass installation, since the heat loss through a purlin or girt stays almost constant.[8] So, while an insulation with an R value of 10 would lose 25 percent of its efficiency in this kind of installation, the one with an R value of 19 would lose 42 percent.[16]

In addition to heat losses, the hourglass design encourages condensation on the roof purlins and wall girts during cold weather, since vapor retarder is located on the outside of the framing. Ironically, when humidity inside a building is high, the more effective the vapor retarder is, the more serious condensation may become—in some cases serious enough to be mistaken for roof leaks.

In an attempt to fit more insulation between the purlins, another system has been developed specifically for roofs. Here, the insulation fills almost all the space between the purlins, resulting in a slightly better installed R value (Fig. 8.3). The facing tabs on the sides of the vapor retarder are overlapped over the purlins for continuity. Still, the problems with thermal bridging and condensation are not resolved. This system requires insulation support bands running between the purlins, an extra-cost item.

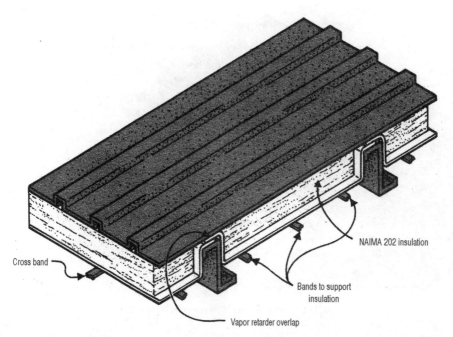

Cross band

NAIMA 202 insulation

Bands to support
insulation

Vapor retarder overlap

Figure 8.3 NAIMA system 2—insulation installed between purlins. (*Courtesy of NAIMA Metal Building Committee.*)

The third insulation-support system attempts to solve the problem of thermal bridging by relying on thermal blocks made of polystyrene-foam insulation and running on top of the purlins. The blocks have to stop at the standing-seam roof clips, but this design is clearly a major improvement over the previous ones. To further increase the system R value, a second unfaced filler layer of insulation may be added between the thermal blocks (Fig. 8.4). This design seems to have been developed with standing-seam roofs in mind, because through-fastened roofs would need longer fasteners penetrating the blocks and potentially shattering them. Also, some feel uneasy about using strips of foam for roof panel support. This system still does not address the purlin condensation problem.

The fourth available installation system combines the advantages of both the uncompressed filler insulation and the thermal blocks (Fig. 8.5). Its installed R value approaches that of the insulation alone, a major accomplishment. On the down side, the installation is rather laborious, still requires insulation support bands, and still does not solve the condensation problem.

The last system incorporates a new element: insulated ceiling board with an integral vapor retarder facing. The insulation board not only provides support for fiberglass blankets but also solves—at last—the problem of condensation by placing purlins within the insulated area. This is a premium system, in terms of both performance and price. For a still better performance, thermal blocks

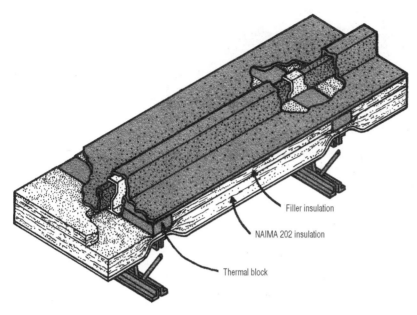

Figure 8.4 NAIMA system 3—use of thermal blocks. (*Courtesy of NAIMA Metal Building Committee.*)

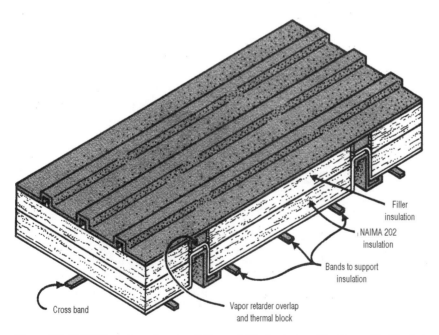

Figure 8.5 NAIMA system 4—use of thermal blocks in combination with blankets between purlins. (*Courtesy of NAIMA Metal Building Committee.*)

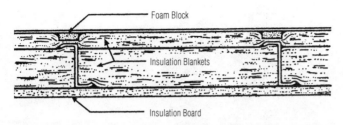

Figure 8.6   A premium system with insulated ceiling board. (*Courtesy of NAIMA Metal Building Committee.*)

can be used on top, resulting in installed R values of over 34 when 1.5-in-thick foam boards are used (Fig. 8.6).

Which system to select? The choice depends on the building use, the climate, and the budget. Figure 8.7, adapted from Ref. 3, may be used to find the R value of insulation for the first four systems in order to meet the U values prescribed by codes. (The fifth system is superior to the first four and need not be directly compared to them.)

Architects should become familiar with the new products and proprietary technologies entering the market. For example, the proprietary "Simple Saver System" by Thermal Design, Inc., is claimed to have improved on the traditional technology by using a continuous heavy-duty vapor retarder underneath the purlins and thus eliminating the purlin condensation problem. The liner is supported by a series of straps attached to the underside of the framing. Above the liner, insulation is installed similar to the fourth system.

Another proprietary product is the "Sky-Web System" by Butler Manufacturing Co., an open-web polyester scrim mesh with openings of about ½ in designed to visually blend with white-colored vapor retarders. In addition to insulation support, the mesh is intended to protect workers from falls and to catch falling objects.

Any insulation system that relies on materials being attached to the underside of purlins should be reviewed in light of potential interference with purlin bracing described in Chap. 5.

8.6 U_o Values of Various Wall Systems

Determination of the overall conductance values U_o for roof assemblies is relatively straightforward, given the fact that metal roofing and fiberglass blanket insulation are used almost exclusively. Our previous discussions have dealt with various fiberglass installation methods that will, more than anything else, make a difference in the thermal performance.

The walls, however, are another matter. A variety of both metal and "hard" (masonry or concrete) materials, coupled with either fiberglass or rigid insula-

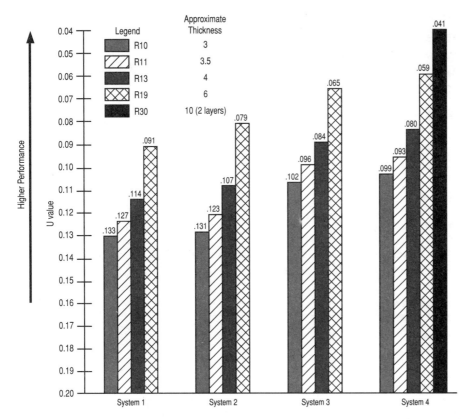

Figure 8.7 Determining fiberglass insulation thickness for various systems. (*Courtesy of NAIMA Metal Building Committee.*) Enter the required roof U_o value from the local code and proceed until intersection with various insulation thicknesses (only system 1 can be used for both walls and roofs.) Disclaimer: The overall thermal transmittance coefficients, U values, for the four systems were calculated using the zone method as described in ASHRAE's handbook *Fundamentals*. The use of this information does not ensure or guarantee code compliance. Consult local code authorities for approval before finalizing design.

tion, are available. Tables 8.1 and 8.2 are included to facilitate comparison between overall conductance values of the most common wall systems.

The R values for the tables were taken from ASHRAE *Handbook of Fundamentals*[9] and Ref. 10. To simplify comparison, the following R values have been assumed for all systems: outside air film in a 15-mi/h wind = 0.17; inside air film = 0.68; an air space $\frac{3}{4}$ to 4 in deep = 0.97. To compute the overall conductance value U_o one needs to add up all the R values of the components and take a reciprocal function of the sum as illustrated in the following example. R values for selected materials are indicated in Table 8.1.

U_o is the number of Btu that will flow through 1 ft^2 of the wall in 1 h when the temperature differential between the two sides is 1°F.

Example   Compute U_0 value for brick veneer over CMU insulated assembly:

Component	R value
Outside air film	0.17
4-in brick veneer	0.44
2-in air space	0.97
2-in polystyrene insulation	10.8
8-in CMU	1.51
$3/4$-in furring air space	0.97
$1/2$-in drywall	0.45
Inside air film	0.68
Total $R =$	15.99

Thus $U_0 = 1/R = 0.0625$.

Table 8.2 includes the results of similar computations for common wall systems, listed roughly in the order established in Chap. 7. For some assemblies,

TABLE 8.1   *R* Values for Selected Materials

	$1/2$	$3/4$	1	$1\frac{1}{2}$	2	3
4-in Clay brick			0.44			
4-in block (72% solid)			1.19			
6-in block (59% solid)			1.34			
8-in block (54% solid)			1.51			
10-in block (52% solid)			1.61			
12-in block (48% solid)			1.72			
Concrete, normal weight, in						
5.5			1.30			
6			1.33			
8			1.49			
10			1.64			
12			1.82			
$1/2$-in drywall			0.45			
Exterior air film (winter)			0.17			
Interior air film			0.68			
Dead air space ($3/4$ to 4-in) (winter)			0.97			
Air space at foil face			2.8			
Insulation-type thickness, in	$1/2$	$3/4$	1	$1\frac{1}{2}$	2	3
Polyisocyanurate (foil face)	4.0	5.8	7.7	11.5	15.4	23.1
Extruded polystyrene	—	4.05	5.4	8.1	10.8	16.2
Fiberglass blanket insulation (approx.), in						
3			10			
3.5			11			
4			13			
6			19			
10			30			

SOURCE: Compiled from Refs. 3, 8, and 9.

TABLE 8.2 U_o Values for Selected Wall Assemblies

Wall assembly	Illustration	U_o
Steel siding with 3-in fiberglass insulation*	Figs. 7.1 and 7.5	0.13
Steel siding with 4-in fiberglass insulation*	Figs. 7.1 and 7.5	0.12
Concealed-fastener panel, 3-in fiberglass blanket, metal furring, and $\frac{1}{2}$-in gypsum board†	Fig. 7.2	0.112
Concealed-fastener panel, 2-in Thermax board, wood furring, and $\frac{1}{2}$-in gypsum board†	Fig. 7.2 (similar)	0.047
Factory-insulated panel 2-in thick‡	Fig. 7.4	0.069
8-in uninsulated CMU	Fig. 7.10	0.424
8-in CMU + $\frac{3}{4}$-in furring + 1-in polystyrene + $\frac{1}{2}$-in glued-on gypsum board		0.109
4-in brick + 2-in air space + 8-in CMU + $\frac{1}{2}$-in drywall on furring	Fig. 7.11	0.193
4-in brick + 2-in air space + 2-in polystyrene insulation + 8-in CMU + $\frac{1}{2}$-in drywall on furring (see example)		0.0625
4-in brick + 2-in air space + $\frac{1}{2}$-in sheathing + $3\frac{1}{2}$-in fiberglass batt§ in $3\frac{5}{8}$-in steel studs + $\frac{1}{2}$-in drywall	Fig. 7.12	0.10
5.5-in concrete		0.465
5.5-in concrete + glued-on 1-in polystyrene insulation + $\frac{1}{2}$-in glued-on gypsum board		0.125
8-in concrete + $\frac{3}{4}$-in furring + 1-in polystyrene + $\frac{1}{2}$-in glued-on gypsum board		0.069
EIFS (2-in polystyrene + $\frac{1}{2}$-in sheathing, neglecting other materials)	Fig. 7.19	0.082
Insulated glass with $\frac{1}{4}$- to $\frac{1}{2}$-in air space		0.57

*As stated by Star Manufacturing Co. for Durarib or Starmark Wall, Panel to Girt Fasteners at 12 in o.c., girt spacing 5 ft 0 in o.c.

†As stated by Star Manufacturing Co. for Star CFW panel, 24-in furring spacing.

‡As stated by Star Manufacturing Co. for STARTHERM II panels.

§Insulation value decreased by one-third to account for parallel flows.

the manufacturers' hot-box test data are used. Thermal efficiency losses through metal studs are accounted for by decreasing the R values of fiberglass insulation by one-third.

For those seeking an analytic method of parallel flow analysis through metal girts or studs and through fully insulated areas, the following procedure is offered:

$$U_{average} = \frac{S}{100} U_{steel} + (1 - \frac{S}{100}) U_{insul}$$

where S = percent area taken by steel framing
U_{steel} = U value for the area taken by steel framing
U_{insul} = U value for insulated area between framing

For example, for steel studs spaced 16 in on centers, the S factor is about 20 percent; for studs 24 in on centers, it is around 15 percent.[11]

References

1. ASHRAE Standard 90.1, Energy-Efficient Design of New Buildings Except Low-Rise Residential Buildings, ASHRAE, Atlanta, Ga., 1989.
2. *Metal Building Systems,* 2d ed., Building Systems Institute, Inc., 1230 Keith Building, Cleveland, Ohio, 1990.
3. *Understanding Insulation for Metal Buildings,* NAIMA, Alexandria, Va., 1994.
4. ASTM E 553-92, Mineral Fiber Blanket Thermal Insulation for Commercial and Industrial Applications, Philadelphia, Pa., 1992.
5. Thomas Scharfe, "Specifying the Right Roofing Insulation," *Building Design and Construction,* September 1988.
6. "Board Insulation Designed for Metal Buildings," *Metal Architecture,* April 1994.
7. ASTM E 1042-92, *Acoustically Absorptive Materials Applied by Trowel or Spray,* Philadelphia, Pa., 1992.
8. *Facts about Metal Building Insulation and System Performance,* Booklet Information from NAIMA, Alexandria, Va., 1994.
9. *ASHRAE Handbook of Fundamentals,* American Society of Heating, Refrigerating and Air Conditioning Engineers, Atlanta, Ga., 1989.
10. *Cavity Walls,* Masonry Advisory Council, Park Ridge, Ill., 1992.
11. John H. Callender (ed.), *Time-Saver Standards for Architectural Design Data,* 6th ed., p. 4-49, McGraw-Hill, New York, 1982.
12. "Polyiso Insulation Outperforms Polystyrene in Tests," *Metal Architecture,* April 1995.
13. Wayne Tobiasson, "General Considerations for Roofs," ASTM Manual 18, *Moisture Control in Buildings.* Also available as *CRREL Misc. Paper* 3443.
14. Wayne Tobiasson and James Buska, "Standing Seam Metal Roofing Systems in Cold Regions," *Proceedings of 10th Conference on Roofing Technology,* 1993 (from CRREL Reprint).
15. Bob Fittro, "Manufacturers Stress Importance of Proper Vapor Barrier Selection," *Metal Architecture,* September 1995.
16. Bob Fittro, "More Ways Than One to Insulate a Metal Building," *Metal Architecture,* December 1995.
17. William A. Lotz, "Simple Rules Sidestep Insulation Failures," *Building Design and Construction,* April 1996.

9

The Process of Buying
a Metal Building

9.1 The Start

9.1.1 Before it all begins: a note to owners

In this chapter we talk about the many things that have to happen between your decision to build and getting the keys for a newly constructed facility.

Once the decision to expand is made, you need to establish building dimensions, shape, and clear height. No matter how well everything else goes, if these fundamentals are not properly thought through, the project will not be successful. If a critical piece of machinery does not fit by a few inches, what was the point of building?

We strongly recommend that you let an experienced architect perform programming and preliminary design of your building: Architects are trained to analyze owner's needs and to offer solutions. Many years spent with architects under one roof have convinced the writer of tremendous improvements these design professionals can make to the original plans conceived by their clients. An architect often comes up with a completely different—and better—building layout. Unless you need a small basic rectangle of a building or must suit a preestablished equipment layout, let designers, not contractors, help you make design decisions. (Of course, the architect you select should be experienced in specifying metal building systems or at least should have read this book....)

The architect will help you identify your immediate and future space needs, prepare a preliminary cost estimate, and propose a timetable for construction. On your part, you have to determine whether adequate financing is available, a budget appropriated or planned, and the members of your in-house planning team are in agreement on what needs to be done.

9.1.2 Selecting the site

After the programming phase, the project moves into schematic design. By this time a prospective location might have already been selected. If several sites are still being considered, it is best to focus on one or two choices before proceeding further, since many building parameters such as height, size, and type of construction may be affected by surrounding buildings and by local zoning codes. As usual in such transactions, prior to purchase a prospective buyer performs title and easement search, zoning check, site survey, and environmental investigation. The site should be large enough to allow for all required property setbacks, parking, access roads, and future expansion needs. If time is of the essence, it is best to stay away from protected areas such as wetlands.

Ideally, the site already has or can be economically served with all the necessary utility hookups. Sewer requirements might be tightly controlled by the community and need to be specifically investigated, and any site drainage problems addressed.

Now is the time to hire a reputable local engineer and a soil boring contractor to perform a soils investigation. Is the soil good enough to allow for economical shallow foundations? The question is not idle, because in many industrial areas the best land has long been developed, and only the less desirable parcels are left—often those deemed uneconomical to build upon by others. Poor soils may require expensive deep foundations such as piles and caissons; the added cost could push the budget beyond the acceptable limits. Some other site preparation costs include demolition of any present structures, lot clearing, excavation, fill, and paving.

9.2 The Architect's Role

9.2.1 The basic responsibilities

After schematic design is completed, either by in-house personnel or by an outside architect, and the site is selected, the owner can start thinking about methods of construction delivery. With a set of schematic plans and specifications in hand, an owner can pursue any one of the three basic construction methods discussed in Chap. 3: conventional (design and bid or negotiate), design-build, or contracting directly with a preselected manufacturer of metal building systems.

In both conventional and pre-engineered construction, the owner can be represented by an independent architect; in a design-build mode, the architect is a part of the builder's team. Since we are specifically interested in metal building systems, and since the manufacturer's staff rarely includes architects, the best course for the owner to follow is to hire an independent design team.

One of the first priorities of the design team is to develop a site-plan package for review and approval by a local planning and zoning board. The package will demonstrate how the owner intends to comply with federal, state, and local regulations. It may address such issues as wetland protection, increased traffic, pollution, sewage flow, parking, and appearance. While some localities are development-friendly, others might not be; occasionally, obtaining all the permits may take longer than the design and construction time combined.

In order to prepare the site package, the design team undertakes a comprehensive code review. (It goes without saying that the intimate knowledge of complex code provisions is a good enough reason to retain an architect in the first place!) By submitting a set of documents in compliance with the local code and all the local regulations, the owner can save a lot of valuable time and lower construction loan interest charges.

Design development and final design can proceed while the site package is being reviewed. The goal is to produce a set of contract documents that adequately communicate design intent without being overly specific and prescriptive.

In broad terms, the design professional is responsible for selecting the design criteria, for any items not normally carried by the metal building manufacturer, and for overall coordination. The items not commonly available from the manufacturers are listed in Sec. 2 of MBMA's *Common Industry Practices,* and include foundations, insulation, fireproofing, finishes, cranes, electrical and mechanical equipment, overhead doors, and miscellaneous iron.

The *Practices* specifically state that ventilation, condensation, and energy conservation issues are beyond the manufacturer's responsibility and therefore are to be included in the design professional's scope of services.

The design team should examine the effects of the proposed building on adjacent structures, such as a possibility of snow drifting onto a lower existing roof. The manufacturer should not be expected to perform this purely engineering task, because some smaller manufacturers might not even have an engineer on staff—only a technician who punches the numbers into a computer program. (Most owners are not aware of this fact, because the term "pre-engineered building" implies the presence of an engineer.)

9.2.2 What to specify

Some people still think that the architect's role in specifying metal building systems is selecting siding colors. It isn't quite so. While construction documents prepared for a pre-engineered building might not be as extensive or detailed as for conventional construction, they still need to communicate a great deal of information. Some of the items sought by manufacturers for proposal preparation are:[1]

- Information on the governing building code including, significantly, the edition. Avoid listing too many codes that may contain conflicting design criteria. While reputable manufacturers will use the most conservative criteria in cases of such conflict, some hungry upstarts might choose to do otherwise.

- Design loads to be used, such as collateral, snow, live, wind, and seismic. Some recurring problems with specifying snow vs. roof live loads are addressed in Chap. 10. Collateral (superimposed) dead load allowance should be carefully considered and its nature preferably identified. Rooftop

HVAC equipment needs to be located on the roof plan, its weight and required roof openings specified. Any other concentrated loads, such as from a suspended walkway, warrant a separate mention. It is important to research the *local* code, which might contain higher design load requirements than model codes. For example, the design wind speed might be specified by a local code as 110 mi/h, while a national code calls for only 70 mi/h. (Of course, the opposite could be true, too: The local code could be based on an obsolete edition of a model code.)

- Load combinations. In addition to the combinations listed in MBMA *Low-Rise Building System Manual,* the designers may wish to include some others, as discussed in Chap. 3.

- The structural scheme assumed in the design (e.g., multispan rigid frame with pinned supports).

- Building dimensions, including length, width (do not forget that, to a manufacturer, building width is the distance between outside flanges of wall girts, not between column centerlines), eave height, and clear height.

- Exterior wall materials, finishes, and insulation. Some specifiers choose to leave doors and windows out of a metal building package: By purchasing these items locally, it is often possible to buy sturdier products with better hardware, and to avoid transit damage. However, in this case the design wind pressures to be used for these important components of the building envelope must be conveyed to their suppliers.

- Locations where wall bracing is to be avoided, for aesthetic or functional reasons and, perhaps, where bracing is desired. Also, any open-wall locations.

- Corrosion protection requirements. The specifiers are well advised to mention a presence of any existing facilities within ½-mile radius which emit corrosive chemicals, a proximity to saltwater areas, and any other possible sources of corrosion. They should also evaluate a corrosive or moisture-producing potential of the operations within the building itself. In metal buildings, corrosion from the inside is difficult to protect against. While the exterior panel finishes might be quite good in fighting corrosion, interior steel framing is often protected only by a primer coat. Many manufacturers lack the facilities for high-quality surface preparation and for application of premium coatings; they send the steel to specialty shops if those coatings are specified, driving up the cost. For main framing, it is preferable to use a high-quality field-applied paint than to specify a galvanized finish: Hot-dip galvanizing tends to promote warpage and distortion of framing members made of thin built-up plates. A few manufacturers offer mill-galvanized C or Z girts and purlins.

- Any restrictions on framing sizes. The drawings should indicate the largest column depth that the foundations can accept. The rate of column taper should also be controlled if any equipment or interior walls are to be located near the columns.

- Lateral drift and vertical deflection criteria for both the main and the secondary framing. (This issue is critical enough to deserve its own chapter; see Chap. 11.)

- Crane requirements, if any are needed, including service levels, as is further explained in Chap. 15.

9.3   The Manufacturer's Responsibilities

The manufacturer is responsible for design and fabrication of the metal building, exclusive of the items mentioned above, down to the bottom of the column base plates. Using the owner-supplied design criteria, the manufacturer either selects a "pre-engineered" frame from a catalog of standard products or custom designs it. While the industry started out with the former approach (hence the buildings were called "pre-engineered"), presently the latter is the norm. Computers have revolutionized design of metal building systems by erasing the line between "standard" and "custom" choices; today, a vast majority of metal building systems are custom designed for a specific project.

Many larger manufacturers have developed extensive CAD libraries of details and connections which can help quickly assemble a computerized framing design. While each company tends to develop its own software with slightly different features, all the programs perform similar functions. Typically, computers help generate anchor bolt plans and details, frame elevations, structural computations for each member, and cost data. The newest graphic software can generate impressive-looking documents useful for presentations to owners and permitting agencies. Large metal building manufacturers view their advanced software capabilities as both technical and marketing tools differentiating them from the less-equipped competitors whose staff might not even include registered engineers.

9.4   The Builder's Role

Usually, the manufacturer does not contract with the owner directly. The entity that does is a local franchised builder, or dealer in MBMA terminology. Builders can act either as general contractors for the project with complete responsibility for it, or only as suppliers of the pre-engineered building. In either case, they may subcontract building erection to another firm.

Major manufacturers are quite selective in the kind of people they allow to become their dealers, seeking contractors who are financially stable, experienced, and dedicated to quality workmanship. A prospective dealer is often required to complete a course sponsored by the manufacturer and receive a renewable certificate.

The dealer does not simply take a set of the owner's contract documents and send it to the manufacturer; many manufacturers would not want to struggle through a thick set of plans and specifications to make a proposal. Instead, the dealer interprets the documents and distills them into the so-called order documents, a standard proposal form accompanied by sketches on graph paper, and other supporting data (Fig. 9.1).

The architect's contract documents are referenced only in an agreement between the dealer and the owner; a contract between the dealer and the manufacturer is based solely on the order documents. Given the fact that builders generally do not have any in-house engineering expertise, a potential for misinterpretation of some complex design requirements is quite real. Indeed, the task of condensing hundreds of pages of information into a simple form could stymie even the specifiers themselves. It is easy to see how some fine points of the design might be lost in translation and never make it to the manufacturer. A close examination of the manufacturer's design certification letters (Sec. 9.5) and shop drawings becomes extremely important in assuring the owner that the building will be constructed as conceived.

Some major manufacturers have created departments of "national accounts," reflecting a desire to allow for personalized service and a better communication with large interstate repeat customers who would otherwise have to deal with a variety of local builders.

9.5   Bidding and Selection

Excepting the cases of "captive" relationships, the project will in all probability be competitively bid or negotiated. Several manufacturers will be submitting their proposals via their dealers. Which one to select?

The lowest-priced proposal may of course deserve the best chance of being accepted, *if* it is in line with the others in all respects. It is not easy to make this determination. Sometimes a builder will not even mention which manufacturer will supply the building or whether roof and wall metal panels will be shop- or field-formed.

One way to ensure a level playing field is to require all the manufacturers to submit with their bids letters of design certification. A letter of this kind should clearly state the dealer's name, building configuration, governing codes and standards to be followed, and every load and load combination the building will be designed for. The letter should bear a seal and signature of a registered engineer employed by the manufacturer.

Another critical item to be checked in the letter is the design roof snow or live load, a common target of manipulation by some manufacturers seeking an advantage over the competition. (More on this subject in the next chapter.) Insist on seeing the actual design roof load, not the "ground snow load," which is only a starting point for further calculations. If roof live load, not snow, controls the design, the letter should indicate the actual loading used to design various members and any live load reductions taken by the manufacturer.

Also, design wind and seismic loads and, especially, collateral loads should be clearly stated in the letter. Verify that proper lateral drift and vertical deflection criteria will be used: A manufacturer who "overlooks" these will have a major cost advantage over the others. If too little—or too vague—information

PROPOSAL — CONTRACT

Submitted To: _____ Project: _____
[Owner(s) Name(s)]

Billing Address: _____ Project Location: _____
                                           [Street Address]
City & State _____ Telephone No.: _____

This Contract made this _____ day of _____, 19___ by _____
_____ (Contractor) _____
to _____ [Owner(s)]. Upon
acceptance, the Contractor agrees to furnish all necesssary labor and materials to make improvements at the named project and
Owner(s) agree(s) to pay in accordance with all terms and conditions set forth.

DESCRIPTION OF WORK TO BE PERFORMED:

_____
_____
_____
_____
_____
_____
_____
_____
_____
_____
_____
_____.

1. The parties agree that the Contract price is (Words and Figures) _____
_____ ($_____); the
terms of payment are as follows: _____
_____
_____.

2. Contractor does not include in this proposal any Work or items not specifically mentioned above.

3. Estimated completion time is _____ days from date Contractor is authorized and able
to proceed with the Work to be performed.

4. All terms and conditions printed on the back hereof are, by agreement of the parties, incorporated within this Contract as
fully as if printed and written on the front, and the parties acknowledge that they have read the foregoing provisions of this
Contract, both front and back.

Owner(s) hereby accept this proposal.

_____   _____
        [Owner(s) Name(s)]   (Contractor Name)
By: _____   By: _____

Title: _____   Title: _____

By: _____   IF OWNER(S) IS/ARE A CORPORATION:
        (Spouse if applicable)
Title: _____   _____
                                           PRESIDENT
Date: _____
                                           _____
                                           SECRETARY

Issued by the Systems Builders Association [logo: SYSTEMS BUILDERS ASSOCIATION]

©1988, Systems Builders Association, 28 Lowry Drive, P.O. Box 117, West Milton, OH 45383-0117; (513) 698-4127

Figure 9.1 Sample proposal-contract. (*System Builders Association.*)

TERMS AND CONDITIONS

1. The Owner hereby represents that he is the Owner in fee of the premises whereon the improvements herein specified are to be made, having a deed for the same. Owner, on request, shall furnish all surveys describing the physical characteristics, legal limitations, and utility locations for the site of the Project.

2. The Contractor agrees to use his best efforts to complete the Work promptly. He cannot be held responsible for delays due to strikes, fires, acts of god, unworkable weather conditions, or otherwise inability to meet the anticipated schedule. Should the Contractor not be so excused, then the Owner hereby agrees to receive as liquidated damages, $_____ per calendar day, as full compensation.

3. The Owner, without invalidating the Contract, may order extra Work (within the general scope of the Contract) or make changes by altering, adding to or deducting from the work. All such changes shall be in writing and executed by both parties. The Contract sum shall be altered accordingly. All such Work shall be executed under the conditions of the original Contract The value of any such changes shall be determined in one or more of the following ways: A.) By estimate and acceptance of a specified; amount. B.) Unit prices named in the Contract or subsequently agreed upon. C.) By actual cost plus _____ percent. If none of these methods are specified by the Change Order, the third method shall apply.

4. The Owner shall notify the Contractor of any complaints he has about the completion of this Contract within ten days after conclusion of Work by the Contractor and submittal by the Contractor of a notice of conclusion of the Work and absent such notification, the Work shall be deemed to have been completed in a satisfactory manner. Furthermore, the Owner must notify the Contractor within two (2) days of learning of any problems before any progress or final payment is withheld or reduced in value.

5. If tests (other than those that may have been made in initial design) shall be deemed necessary (by the Engineer or Architect or Owner) to establish bearing capacities, such tests and reports as required shall be paid for by the Owner.

6. It is expressly understood and agreed that the Contractor shall be entitled to liquidated damages in the amount of 25% of the Contract price in the event this Contract is terminated before the Work begins and without the consent of the Contractor. Should the Work have already begun, then the Contractor shall be due all costs incurred plus 25% of the Contract price.

7. Progress payments or final payment due and unpaid under this Contract shall bear interest from the date payment is due at the maximum rate allowed. Such rate to be that rate used for revolving credit at the locale of the Work.

8. Should either party employ an attorney to institute suit or demand arbitration to enforce any of the provisions hereof, to protect its interest in any matter arising under this Contract, or to collect damages for the breach of the Contract or to recover on a surety bond given by a party under this Contract, the prevailing party shall be entitled to recover reasonable attorney's fees, costs, charges, and expenses expended or incurred therein.

9. It is expressly agreed between the parties that in the unlikely event of a dispute of any nature relating to this Contract arising between them, that it will be submitted to the American Arbitration Association for binding arbitration, under the Construction Industry Rules.

10. WARRANTIES: The Contractor shall issue limited warranties for materials, products, and other related Work for differing periods of time depending upon the specific material, product, or Work, which shall be issued upon completion and payment for the Work specified herein upon request of the Owner and in conformance with the terms, conditions, and guarantee policies of the Contractor. THE FOREGOING IS IN LIEU OF ALL OTHER WARRANTIES EXPRESS, IMPLIED, OR STATUTORY AND CONTRACTOR NEITHER ASSUMES, NOR AUTHORIZES, ANY PERSON TO ASSUME FOR IT ANY OTHER OBLIGATION OR LIABILITY IN CONNECTION WITH THE WORK TO BE PERFORMED UNDER THE TERMS OF THIS CONTRACT.

11. Contractor's scope of work shall not include the identification, detection, abatement, encapsulation or removal of asbestos or any other hazardous substances.

12. Permits, licenses, fees, tests, inspections, or surveillances of any type which may be imposed under the building or zoning ordinances or by Cities, Counties, States, or other regulatory authorities are to be obtained by and paid by Owner and are not included in the Contract unless so stated.

13. All promises, understandings, or Contracts of any kind, to this Contract, not mentioned herein, are hereby expressly waived; and it is agreed that this instrument shall constitute the entire Contract between the parties, and shall not be modified in any manner, except in writing signed by both parties.

Figure 9.1 *(Continued)* Sample proposal-contract. *(System Builders Association.)*

is provided in the letter, do not hesitate to ask for a written clarification of any murky issues.

A sample letter of design certification is reproduced in Fig. 9.2 from Ref. 2. Note that the letter identifies the specific edition dates for the governing standards and codes, a fact especially important for the AISI Specification, as explained in Chap. 5. This and other items most commonly lacking in clarity in design certification letters are highlighted in boldface.

To help assure quality of both design and fabrication for a critical project, the owner may elect to deal only with the manufacturers whose facilities are awarded AISC Quality Certification program, Category MB. While it is true that there are plenty of capable manufacturers that are not so certified, limiting the pool of bidders to those with the designation greatly simplifies the comparison.

Carefully compare the warranties. Will the warranty called for by the contract documents be provided? A standard warranty for framing components is 1 year from the shipment date. Warranties for metal roofing are available up to 20 years and depend on the material.

Check the builder's qualifications, too. Since it is the builder who signs the contract with the owner, this check is at least as important as comparing the manufacturers. Have the builders worked on similar projects? Do their references confirm their ability to deliver on time and within budget? Are they satisfied with the quality of their work? Are they financially stable and bonded? (Who hasn't heard about a contractor going belly-up in the middle of a job?) Are they members of System Builders Association? Are they certified by the manufacturer? Investigate how the builder tends to approach any out-of-scope items. Are the "extras" priced reasonably or become a source of enrichment?

Find out if the builder (or the president in a large construction company) is personally involved with the projects: The best builders are. For example, the president of Span Construction and Engineering, the *Metal Construction News'* "1994 Top Metal Builder," personally inspects every major metal roof completed by the company, investing up to $\frac{1}{2}$ day in each such "walk." In his words, such inspection lends credibility to the 20-year weathertight warranty.[3]

The selection process ends with signing of the contract documents by the owner and the builder. The contract documents may include the agreement, general and supplementary conditions, drawings, and specifications. The contract should assign a clear responsibility for various facets of design, fabrication, erection, code compliance, and permitting. Except for very small and simple buildings, we recommend that the AIA contract forms be used, rather than a one-sheet proposal and contract form similar to that of Fig. 9.1. (Incidentally, the SBA contract requires owners to pay a penalty equal to 25 percent of the contract price if they fail to proceed with the work after signing the contract.) The contract may reference MBMA *Common Industry Practices* to establish a scope of work.

Some contract provisions may lead to protracted negotiations. If discussions over a truly important provision are deadlocked, dealing with the lowest bidder might be abandoned, if permitted by law, and the no. 2 bidder invited to

March 30, 1992

Mr. John Doe
XYZ Corporation
Grand Rapids, MI 49508

To Whom It May Concern:

12 x 21 x 5.5 m (40' x 70' x 18')
LMDS .25:12 slope
Warehouse expansion
Grand Rapids, Michigan
BMC Ord.Nos.052789,052790
Builder Order Ref: CB 8673

Please accept this letter as our certification that the Butler components of the subject building are designed in accordance with the **1989 Edition of the AISC Specification for the Design, Fabrication and Erection of Structural Steel and the 1986 Edition of the AISI Specification** for the Design of Cold-Formed Steel Structural Members. The basic loads of the subject building meet or exceed the County Climatic Data as published in the **1986 Edition of the MBMA Low Rise Building System Manual.**

The governing design code is the 1987 Edition of the BOCA National Building Code. The following loads are applied in accordance with the governing code:

Roof Snow Load	1436 Pa (30 psf)
Wind Speed	129 km/h (80 mph)
Wind Exposure	B
Seismic Zone	1

The building system is designed for a drift snow load applied in accordance with the Low Rise Building Systems Manual which meets or exceeds the governing code. Load combinations are in accordance with the governing code.

This building has been reviewed for an 363 kg (800 lb) concentrated load at the midspan of the 6 m (20 ft) long beam trusses in addition to the full 1436 Pa (30 psf) snow load.

These Butler components, when properly erected on an adequate foundation in accordance with the erection drawings as supplied and using the components as furnished, will meet the above loading requirements. The design of this building for wind load assumes that doors not supplied by Butler are designed to sustain the same wind pressures and suctions as the walls in which they are installed. This certification does not cover field modifications or design of material not furnished by Butler Manufacturing Company.

This building is produced in a manufacturing facility that is certified by the American Institute of Steel Construction—Category MB.

Cordially Yours,

Duane Miller, P.E.
Division Engineer

Figure 9.2 Sample letter of certification. (*Butler Manufacturing Co.*)

negotiate. In this tense situation, some owners in their zeal to build may rely on verbal promises instead of ironed-out written agreements, forgetting the Samuel Goldwyn quip about an oral contract that isn't worth the paper it is written on.

The manufacturers might be more open to negotiations if contacted during their slowest months of the year—November and December.

9.6   Shop Drawings and Construction

One of the first pieces of information the manufacturer should submit to the owner's designers is the value of column reactions. As mentioned above and discussed further in Chap. 12, metal-building foundations are often designed prior to receiving these data from the manufacturer and are rechecked later against the actual reactions. This process is on the critical path, and such submittal is required as early as possible, preferably at the bidding stage. In any case, it should not take the manufacturer longer than 2 weeks to process the order plus a week to generate the reactions. Thus the reaction report, such as that shown on Fig. 9.3, and perhaps even a complete approval set, might be ready in 3 weeks.

The approval set may include an erection plan, frame elevations, anchor bolt plan, wall elevations, and some details. (What the set includes should have been specified in the contract documents and, hopefully, in the order documents prepared by the dealer.) An example of frame elevation is shown in Fig. 9.4; an anchor bolt plan may be found in Fig. 12.20. The approval set may not be drawn to scale but is still a source of valuable information; it should be closely scrutinized for any hints of misunderstanding the design intent.

The submittal should also include detailed structural calculations sealed by the manufacturer's engineer. Some owners insist that the engineer be registered in the state where the building is located and include this requirement in the contract documents. MBMA's *Common Industry Practices* requires only that the engineer be registered in the manufacturer's home state.

While reputable manufacturers tend to submit calculations with clearly identified assumptions and input data, some others might try to overwhelm the reviewers with mounds of incomprehensible computer data. Those submittals may look as if they are, in Tom Clancy's words, written by computers to be read by calculators. If anything looks suspect, asking questions in writing and insisting on strict adherence to the design requirements is the only defense. (On a recent project, a manufacturer stated that the building complied with the project's strict lateral drift criteria. The calculations indicated otherwise.)

Do not be surprised to see that any marked-up comments on the approval set are construed as changes and greeted with a change order by the manufacturer. Some three-way complex negotiations might ensue. The reader is invited to review Secs. 2.2 and 3.3.3 of MBMA's *Common Industry Practices* which deal with such changes. In some unfortunate circumstances, the project might stop in its tracks right there and end in dispute.

Varco-Pruden Buildings
A United Dominion Company

Built On Superior Service

REACTIONS REPORT
(Project Name: Phoenix Associates Test)
(VP Quote No.: TQ940323085603)
(VP Order No.:)

DATE: 04/06/94
TIME: 09:43
PAGE: 03 of 05

FRAME PLANE   Version 2.1

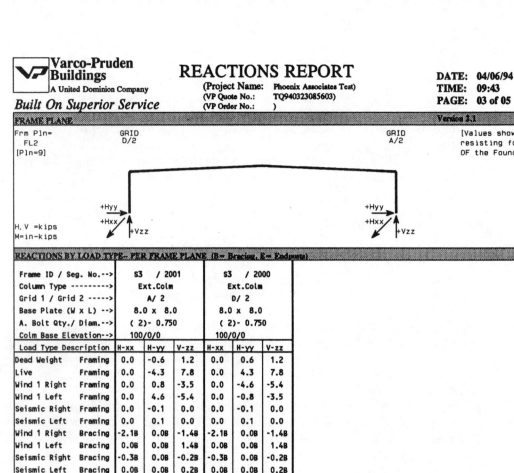

Frm Pln=
  FL2
 [Pln=9]

GRID
D/2

GRID
A/2

[Values shown are resisting forces OF the Foundation]

H, V =kips
M=in-kips

+Hyy +Hxx +Vzz   +Hyy +Hxx +Vzz

REACTIONS BY LOAD TYPE... PER FRAME PLANE (B= Bracing, E= Endpost)

Frame ID / Seg. No.-->	S3 / 2001			S3 / 2000		
Column Type -------->	Ext.Colm			Ext.Colm		
Grid 1 / Grid 2 ---->	A/ 2			D/ 2		
Base Plate (W x L) -->	8.0 x 8.0			8.0 x 8.0		
A. Bolt Qty./ Diam.-->	(2)- 0.750			(2)- 0.750		
Colm Base Elevation-->	100/0/0			100/0/0		
Load Type Description	H-xx	H-yy	V-zz	H-xx	H-yy	V-zz
Dead Weight Framing	0.0	-0.6	1.2	0.0	0.6	1.2
Live Framing	0.0	-4.3	7.8	0.0	4.3	7.8
Wind 1 Right Framing	0.0	0.8	-3.5	0.0	-4.6	-5.4
Wind 1 Left Framing	0.0	4.6	-5.4	0.0	-0.8	-3.5
Seismic Right Framing	0.0	-0.1	0.0	0.0	-0.1	0.0
Seismic Left Framing	0.0	0.1	0.0	0.0	0.1	0.0
Wind 1 Right Bracing	-2.1B	0.0B	-1.4B	-2.1B	0.0B	-1.4B
Wind 1 Left Bracing	0.0B	0.0B	1.4B	0.0B	0.0B	1.4B
Seismic Right Bracing	-0.3B	0.0B	-0.2B	-0.3B	0.0B	-0.2B
Seismic Left Bracing	0.0B	0.0B	0.2B	0.0B	0.0B	0.2B
Wind 1 Right Endpost	- -	- -	- -	- -	- -	- -
Wind 1 Left Endpost	- -	- -	- -	- -	- -	- -

REACTIONS BY LOAD COMBINATION... PER FRAME PLANE (B= Bracing, E= Endpost)

Grid 1 / Grid 2 ---->	A/ 2			D/ 2		
Ldg Comb (Strs Incrs)	H-xx	H-yy	V-zz	H-xx	H-yy	V-zz
Frmg Cmb No. 1 (1.00)	0.0	-4.9	9.0	0.0	4.9	9.0
Frmg Cmb No. 2 (1.33)	0.0	0.2	-2.3	0.0	-4.0	-4.2
Frmg Cmb No. 3 (1.33)	0.0	4.0	-4.2	0.0	-0.2	-2.3
Frmg Cmb No. 4 (1.33)	0.0	-0.7	1.2	0.0	0.5	1.2
Frmg Cmb No. 5 (1.33)	0.0	-0.5	1.2	0.0	0.7	1.2
Post Cmb No. 1 (1.33)	- -	- -	- -	- -	- -	- -
Post Cmb No. 2 (1.33)	- -	- -	- -	- -	- -	- -

Figure 9.3 Reaction report. (*Varco-Pruden Buildings.*)

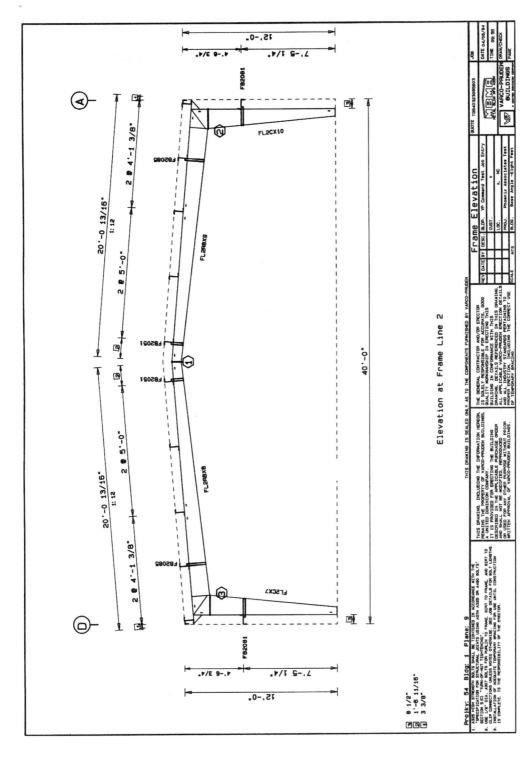

Figure 9.4 A manufacturer's shop drawing. (*Varco-Pruden Buildings.*)

The approval set, however schematic, is the first and usually the last occasion to review the manufacturer's shop drawings. The detailed work prepared afterward constitutes fabrication drawings which are not generally furnished by the manufacturer unless specifically required by contract. With all the shop drawing issues resolved, construction can, at last, start. Some "red flags" to be watched out for during construction are described in Chap. 16.

References

1. Alexander Newman, "Engineering Pre-engineered Buildings," *Civil Engineering,* September 1992.
2. Duane Miller and David Evers, "Loads and Codes," *The Construction Specifier,* November 1992.
3. Shawn Zuver, "Span Construction & Engineering Wins Fifth Top Builder Honor," *Metal Construction News,* May 1995.

Some Common Problems and Failures

Time after time, those who specify metal building systems complain about the same troublesome issues that cause more than their share of problems. These issues deserve close attention of the specifiers. Most troubles are rooted in misunderstanding or miscommunication between the owner and the owner's design team on one side and the manufacturer and the dealer on the other. Every one of this chapter's vignettes has been inspired by an actual not-so-pleasant occurrence. A brief review of some building failures—the ultimate problems—completes our discourse.

10.1 Specifying Buildings with Complex Shapes and Wall Materials

10.1.1 Building too small

Sometimes, metal building systems are specified for inappropriate applications, where their advantages cannot be fully utilized. The systems are best suited for large rectangular low-rise buildings, especially those that can benefit from metal panel walls and roofs. Still, time and again, pre-engineered structures are provided for small buildings with irregular layouts, complex roof shapes, and varied wall materials—with the results mixed at best. Whenever such conditions apply, a rule of thumb puts a lower floor-area limit of the buildings suitable for pre-engineered construction at about 3000 ft^2. While some smaller buildings have been successfully constructed, they might have been built even more economically with some other framing systems identified in Chap. 3.

To be sure, there are manufacturers that specialize in production of simple rectangular stand-alone metal buildings at very competitive prices. Our focus, however, is on custom-designed structures, perhaps with some special architectural features.

Manufacturers often find that small buildings with complex layouts require careful engineering and no less effort than the large simple boxes. This engineering and detailing time, as well as mobilization and transportation costs, are difficult to amortize on small structures. For all these reasons, major manufacturers rarely pursue minor construction in distant areas, leaving such buildings for their smaller competitors who, unfortunately for the owners, might not be experienced with custom applications of metal building systems. A common result is a host of engineering and coordination issues which confront the owner and builder, the issues that a more experienced manufacturer could have easily resolved.

10.1.2   Complex configuration

Large buildings of complex shapes can present a problem, too. Most design programs used by manufacturers are geared toward rectangular structures. A C-shaped building, for example, can be broken up into three rectangular "units" by the manufacturer, who could assign the framing design of each to a different engineer. Not surprisingly, each unit might have its own set of columns, with a double line of columns at the unit interface.[1] Unless the owner's design team has anticipated this turn of events by providing expansion joints at the interface—with a double set of columns and foundations—it is in for a shock when the shop drawings come in.

The unexpected double set of columns could wreak havoc with the assumed column sizes and could require new expansion joints in the exterior walls and roof, complicating the appearance of the building. It will also require much larger foundations than those shown in the contract drawings. At this stage of construction, the foundation contract might already have been awarded, or, in a most nightmarish scenario, the concrete has already been placed. In either case, the change won't be easy—or cheap.

We recommend that the owner's designers either divide L-, C-, and Z-shaped buildings into rectangles from the beginning or carry a warning on the contract drawings against an introduction of any columns not shown there.

10.1.3   Nonmetal wall materials

As Chap. 7 has demonstrated, metal building systems are increasingly designed with masonry, concrete, and other nonmetal materials. These applications expand the limits of the systems' acceptance and will undoubtedly increase in the future. Still, to paraphrase the famous Calvin Coolidge's saying, the business of metal building manufacturers is metal. The industry has accumulated substantial know-how in the design and construction of metal-clad structures, but the same level of expertise might not be necessarily found in its dealing with masonry or concrete, let alone with GFRC or EIFS.

To ensure a successful project, the owner's design team has to provide as much detailing of the nonmetal exterior as it would for a conventionally framed structure. An interaction between exterior walls and pre-engineered frames is a common source of problems. Chap. 7 has already mentioned and Chap. 11 will address further an inherent conflict between the "hard" wall materials and flexible metal framing.

For example, a quite substantial eave member is needed for lateral support of vertically spanning masonry and concrete walls or for gravity support of heavy parapet and fascia panels. And yet pre-engineered buildings do not normally have a large perimeter roof beam found in conventionally framed steel structures. Instead, the usual feature is a light-gage eave strut that works well with metal roofing and siding but is nearly useless for "hard" walls. Most major manufacturers are quite capable of modifying their standard framing for "hard-wall" conditions, but some others might not be so knowledgeable. If the wall design requires a hot-rolled steel section at the eaves or elsewhere, it is better to design it in-house and show it on the drawings. To delineate responsibility, such members designed by the owner's team could be excluded from the design scope of the manufacturer.

Similarly, the architect-engineer team has to decide how to support any partial-height "hard" walls. The preferred method recommended in Chap. 7 is to "fix" any such wall at the bottom and forgo any lateral support from flexible metal structure. If not shown this solution, manufacturers would probably think in terms of Z girts as they would for a metal siding.

CMU and concrete walls are often expected to act as load-bearing or shear-resisting elements. While this use is certainly rational and cost-effective, it introduces another set of problems. Similarly to the building foundations, the walls have to be designed by the architect-engineer team, but for which loads? In theory, the manufacturer could supply the horizontal and vertical loads on the wall upon completion of the framing design. In practice, wall design needs to be completed in advance, based on the specifier's own analysis and later compared with the manufacturer's numbers. The manufacturer might even expect the architect-engineer team to design connections between the wall and metal framing.

A point often forgotten: The manufacturer's standard wall accessories are designed for metal siding and are not necessarily adaptable to nonmetal walls. Items like doors, windows, and louvers, which could otherwise be available from the manufacturer, might need to be purchased outside. As we have already suggested in the previous chapter, it could be a good idea in any case.

10.2 Fixed-Base versus Pinned-Base Columns

Most experienced structural engineers will agree that it is rather difficult to achieve a full column-base fixity in metal building systems; some of the reasons were mentioned in Sec. 4.8.4. Accordingly, the "outside" designers routinely assume the column bases to be pinned, that is, not transmitting any bending moments to the foundations. Unfortunately, not everybody remembers to put this assumption in writing—in the contract documents.

Without being "nailed down," the pin-base assumption exists only in the minds of the specifying engineers. Manufacturers may feel free to propose a flagpole-type system with fixed-base design, which is a much cheaper solution—for them. Since manufacturers do not see the added foundation costs, they may sincerely believe that the fixed-base solution is truly economical.

By this time, however, the foundations may have already been designed with the pin-base assumption in mind and could even be partially constructed by, or at least awarded to, a concrete contractor. Since the fixed-base column design results in an unanticipated bending moment applied to the foundations, a redesign is almost always required.

At this point, the owner's choice is between accepting a large change order from the foundation contractor or a protracted battle with the metal building manufacturer. Both could have been easily avoided with one sentence inserted in the contract drawings and specifications: "The pre-engineered building columns shall have pinned bases and shall transfer no moments to the foundations."

10.3 Anchor Bolts

Anchor bolts (Fig. 10.1) intersect a demarcation line between the design responsibilities of the metal building manufacturer and the structural engineer of record, who is responsible for the whole project. An inexperienced specifier often assumes that anchor bolts are provided by the manufacturer, who, after all, submits the shop drawings that include an anchor bolt plan, with the bolt sizes and locations clearly indicated. Moreover, shop drawings of column base plates also include the bolt sizes; the submitted calculations spell out how many and what kind of bolts are needed.

In reality, the manufacturer does not normally supply the bolts. This fact is clearly stated in MBMA *Common Industry Practices,* Sec. 2.1.3. The manufacturer does not even determine the bolt length, which depends more on foundation construction than on parameters of the metal building.

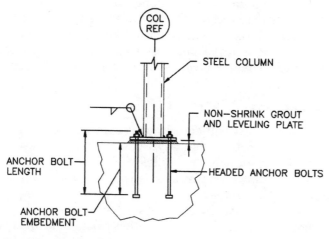

Figure 10.1 Anchor bolts.

As explained in Chap. 12, holding strength of anchor bolts is controlled by one of the two factors: tensile capacity of the steel section and strength of concrete. The manufacturer determines the former, the design professional the latter. A case in point: anchor bolts embedded in an isolated pier are likely to be longer than those in a large thick footing. The reason: tension capacity of the pier concrete may be less than that of the bolts, a fact that requires a certain bolt length to engage the pier's reinforcing bars.

We suggest that the structural engineer of record review the anchor bolt sizes and layout indicated in the shop drawings for consistency with foundation design. If the submitted information is acceptable, he or she should then determine the required length of the anchor bolts and advise the builder who actually provides them. If, however, the proposed size and layout cause a problem, the engineer could try to resolve the issue with the manufacturer's engineer, either directly by a phone call, or by going through the proper channel—the builder.

10.4 Framing around Overhead Doors

Pre-engineered building manufacturers share common methods of making openings in metal-clad exterior walls. Normally, a window or a door is framed by extending the jamb channels to the nearest horizontal girt; header girts are used if needed (Fig. 10.2). This detail works fine for the doors and windows of small to moderate sizes but not, in the writer's opinion, for large industrial overhead doors.

Overhead doors are generally analyzed as horizontally spanning members carrying wind loads to the jambs. The jambs, in turn, span from the floor to the roof eave strut. Recall an earlier discussion about a standard eave-strut section that works reasonably well with metal siding but is not appropriate for lateral support of masonry. The eave strut of Fig. 10.2 suffers from the same problem of a potential torsional overstress caused by horizontal wind reactions from the jambs.

In addition to wind loading, the overhead door's weight needs to be supported when the door is raised. For coiling doors, it means a substantial weight hung eccentrically from the jambs or—worse—from the eave strut. Light-gage sections are ill equipped to resist such loads and often need to be reinforced or superseded by hot-rolled steel members. Still, some manufacturers continue to use cold-formed channels that are less than satisfactory around large doors.

The framing around overhead doors has a very important role to play: its failure during a hurricane lets the wind into the building, leading to a loss of its contents. Unfortunately, this is exactly what often happens. For example, in surveying the damage from two 1970 hurricanes, Ref. 2 observes that the wind has caused light framing around large overhead doors to deflect excessively, which led to derailment of roller supports and a subsequent door failure.

We recommend that the framing around large overhead doors (wider, say, than 16 ft) be designed in-house and indicated in the contract drawings. An example of such framing is shown in Fig. 10.3. Here, the heavy hot-rolled chan-

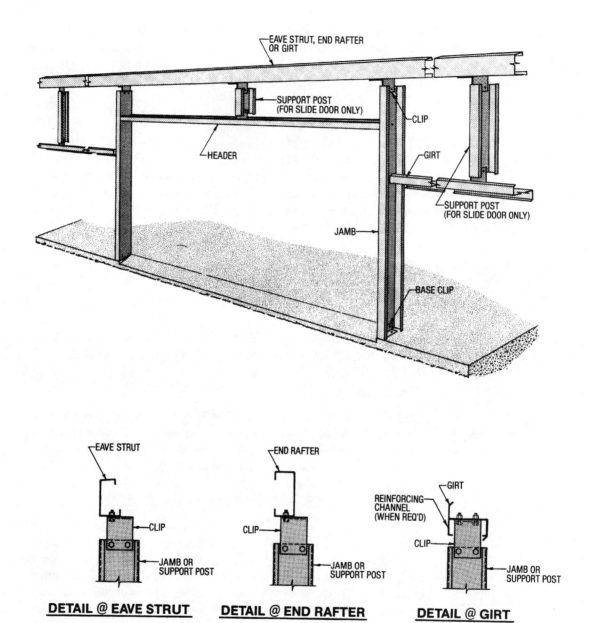

Figure 10.2 Typical framed opening for a sliding door. (*Star Building Systems.*)

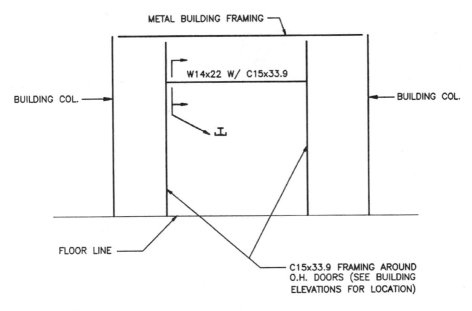

METAL BUILDING FRAMING

W14x22 W/ C15x33.9

BUILDING COL.

BUILDING COL.

FLOOR LINE

C15x33.9 FRAMING AROUND
O.H. DOORS (SEE BUILDING
ELEVATIONS FOR LOCATION)

Figure 10.3 Typical framing at large overhead doors.

nels are designed to resist a wind loading from the door and from the adjacent tributary areas; the wide-flange beam is provided to support the weight of the door and the siding above.

Incidentally, regardless of the detail used, it is important to require the builder to provide *some* framing around doors. The MBMA *Common Industry Practices* considers such framing an accessory and states that it will be supplied only if expressly required. The door itself is not included in the system either and needs to be specified separately; it should be designed for at least the same wind loading as the building walls.

10.5 Supports for Rooftop HVAC Equipment

Rooftop-mounted or suspended HVAC equipment may include anything from small fans and unit heaters to large air-handling units. While mechanical equipment is not a part of metal building systems and is not provided by building manufacturers, equipment supports need to be integrated with roof structure. Unless specifically addressed in the contract documents, responsibility for these supports tends to fall through the cracks.

To be sure, the design profession has well realized that equipment does not belong on roofs at all: Equipment restricts movement of roofing panels and invites leaks through the penetrations. Many roofs have been ruined by careless equipment installation and maintenance. Unfortunately, there are few economic alternatives to the rooftop location. Mechanical penthouses, common in high-rise construction, are a rarity among pre-engineered buildings. It seems that roof-mounted HVAC equipment will continue to detract from appearance and function of metal roofs for years to come.

There are two basic methods of supporting rooftop equipment: a continuous curb and an elevated steel frame on legs. A properly designed and installed curb includes sheet flashing and is less prone to leakage than discrete penetrations at frame legs. The best curbs are custom manufactured to the specific roofing profiles and fit nicely in the panel corrugations (Fig. 10.4). Curbs for standing-seam roofs may consist of two pieces: a light flashing curb premounted on a base panel of the same thickness and configuration as the roofing, and a heavier (such as 10-gage) unit support frame connected directly to roof purlins. The two parts move independently; any difference in their movement is absorbed by the counterflashing.

Butler Manufacturing Co.[3] recommends that the two-part design be used for the equipment weight of 2000 to 4000 lb. For lighter equipment, a single-piece curb typically made of 14-gage steel may be adequate. The manufacturer suggests that the curb assembly be made by a specialized curb supplier. Still, many specifiers include the curbs in a scope of work for the metal building manufacturer to increase the chances for better fit and coordination.

Despite its sophistication, even this curb system may be vulnerable to leaks if it is simply laid on top of the roofing and depends solely on the sealant in between. For best results the curb and the roofing should be installed at the

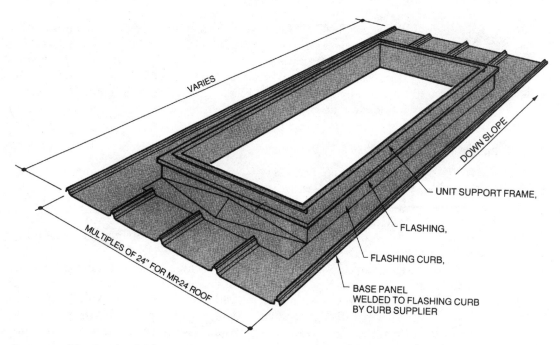

Figure 10.4 Metal roof curb for standing-seam roof. (*Butler Manufacturing Co.*)

same time, allowing the curb sheet to be placed under the upslope roof panel and over the downslope panel, so that water does not run into the joint. In addition, Buchinger[4] recommends that the curb's base sheet be large enough to provide at least 1 ft of space between the end of an upslope panel and the closest edge of the curb or its cricket to avoid water backup at that critical spot.

An elevated steel frame concentrates the equipment weight on four legs (Fig. 10.5). These point loads often exceed a load-carrying capacity of the purlin and require an additional purlin—or even a wide-flange steel beam—for support. For this reason, it is better to bear the legs on the main building frames wherever possible.

As with metal curbs, the issue of weathertightness arises any time a frame support leg penetrates the roof. A common solution is to use an elastomeric boot pipe flashing (also known as roof jack) covering the penetration. Available in several sizes for various pipe or column diameters, boot flashing must be able to accommodate the differential movement between metal roofing and structural supports below. Otherwise it is certain to invite leaks. The boots are commonly but incorrectly installed through the roofing corrugations, as shown in Fig. 10.6. This detail invites water leakage through difficult-to-seal panel seams made unprotected by the penetration. A better detail is to locate roof penetrations in the flat part of the panel where more effective waterproofing can be made.[4]

Who designs the rooftop structural frames? Some architect-engineer firms prefer to design the frames in-house and have them provided separately from the pre-engineered framing. Some others prefer to ease the coordination by

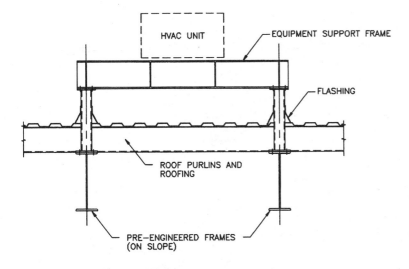

Figure 10.5   Elevated frame on legs.

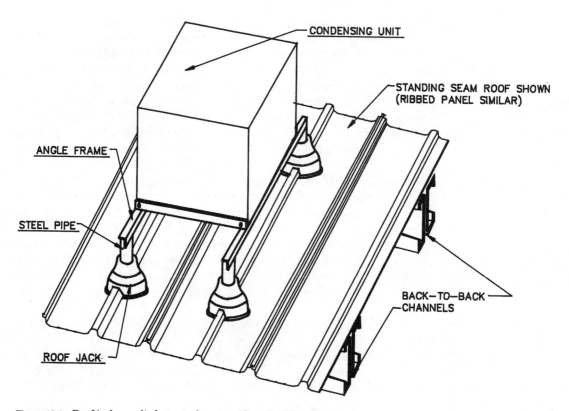

CONDENSING UNIT

STANDING SEAM ROOF SHOWN
(RIBBED PANEL SIMILAR)

ANGLE FRAME

STEEL PIPE

BACK—TO—BACK
CHANNELS

ROOF JACK

Figure 10.6 Roof jacks applied at panel seams. (*Ceco Building Systems.*)

requiring the manufacturer to design and provide the frames. That way, the roof openings are likely to be incorporated into the metal building design, any interferences between the purlins, purlin bracing, and support beams noticed and addressed, the weights of the units included in structural loading on the roof, and flashing provided for all the penetrations.

In any event, the location and weight of HVAC equipment needs to be closely coordinated with the building manufacturer to ensure proper structural support. It is easy to see how a lack of such coordination can easily result in "extras" if any additional framing has to be provided after the roof is in place.

10.6 Confusing Roof Loads

As readers might remember from Chap. 3, there are two externally applied uniformly distributed roof loads: snow load and roof live load. The nature of these loads and the differences between them have been explained already; our present discussion deals with problems of specifying them.

During the last two decades, application criteria for these loads have changed dramatically, confusing the manufacturer and the specifier alike; even the building officials are sometimes unclear on this subject. The owner has to be assured that all the manufacturers interested in the project apply the same design loads and would supply buildings of similar strength. Otherwise, a party claiming to conform to the project design criteria but employing legerdemain to arrive at the design loads which are lower than those used by others will gain a major competitive advantage. As Miller and Evers[6] put it:

> Competition may prompt some companies to look for any way to gain a pricing advantage. For the hard-pressed competitor, this may mean using questionable methods of interpreting code and design loads to achieve the lowest price. It is normally less a question of outright cheating than of clever interpretation of a sometimes allowable reduction, or failure to consider a recent code update.

Experienced structural engineers know that the magnitude of design loads is among the most important factors affecting construction cost. As Ruddy[5] has found for single-story steel-framed buildings, the cost of structure increases at a rate of 2 cents per square foot for each additional pound per square foot of superimposed load. For example, if the roof framing, columns, and foundations cost \$5 per square foot when the design load is 20 lb/ft^2 the same building may cost \$5.20 per square foot when designed for a 30 lb/ft^2 roof load.

A common problem is the designer's failure to differentiate between roof live and snow loads. The specifiers should compare the design values for both, as listed in the local building code, and clearly understand which one controls the design for the project's location. For northern regions, it is usually snow, for the south—roof live load.

The code could specify the snow-load value as either "ground snow" or "roof snow." Often, the ground or "basic" snow load can be converted into the roof snow load by a multiplication coefficient of 0.7. To eliminate a potential for confusion, the contract documents should list a value of the design roof snow load (if snow controls the design) and clearly call it snow. A careful check of the design certification letters (see Sec. 9.5) should be made to verify that no bidder has mistakenly assumed the design load to be a ground snow load that could be further reduced by another 30 percent. Such a reduction alone could cut the amount of steel by as much as 5 percent.[6]

Roof live loads present another complication: live load reduction allowed by the codes for large areas—usually in excess of 200 ft^2—supported by a roof structure member. In many codes, including MBMA's, low-slope roof live load is taken as 20 lb/ft^2 for tributary areas of up to 200 ft^2, 16 lb/ft^2 for tributary areas between 201 and 600 ft^2, and 12 lb/ft^2 for tributary areas over 600 ft^2. Therefore, if the contract documents refer to "roof live load" of 20 lb/ft^2, this load will be reduced by all the bidders in the same fashion. However, if the documents specify a "roof live load" of 30 lb/ft^2, although what was really meant is a *snow* load of 30 lb/ft^2, an expensive problem is invited. Some manufacturers will have understood that snow load was meant and will design all their roof members for 30 lb/ft^2, while others might reduce it proportionally to the above-listed numbers for roof live load. A load reduction from 30 to 18 lb/ft^2 may save as much as 8 percent from the cost of primary frames.[6] The manufacturer taking this reduction will clearly be positioned to win—unfairly to the competition—because of ambiguous contract documents.

10.7   Specifying Exposed Framing

Some architects of a "high-tech romantic" streak like to incorporate exposed steel framing into building exteriors. Exposed framing can look dramatic on sketch paper, but the real-life structure might not turn out so good, for several reasons.

First, as was mentioned in the preceding chapter, there could be a problem of obtaining a high-quality shop-applied paint or color-galvanized finish for primary framing. Exposed steel with a mediocre coating may not survive for long in a corrosive atmosphere.

Second, steel framing is used most efficiently when column and rafter flange bracing is available; such bracing, as well as bolted member splices, may not look particularly attractive when exposed to view.

Third, some architects forget that pre-engineered framing is built up from relatively thin plates. The web-to-flange welding usually occurs on one side of the web only. Apart from raising conceptual concerns in some structural engineer's minds, this kind of welding does not approximate a familiar smooth fillet line offered by hot-rolled steel and does not provide a good weather barrier.

Fourth, the framing-to-wall connections are difficult to weatherproof; the wall integrity depends solely on sealants. Owing to fabrication and erection tolerances, the gaps between framing and siding are rarely uniform in width and may require massive amounts of sealants. The results are rarely pretty. On a recent project where exposed steel was used, the only party totally satisfied with the building's appearance was the caulking salesman.

10.8   Failures of Pre-engineered Buildings

Like any other construction system, pre-engineered buildings can, and do, fail. The majority of known failures is probably due to a simple overload, but a few dramatic failures might have been avoided if only the designers had known what to safeguard against.

Some metal building failures can arguably be traced to improper purlin and girt bracing, as was discussed in Chap. 5. The unbraced purlins and girts are likely to fail in a real hurricane or during a major snow accumulation because of buckling of their laterally unsupported flanges. Zamecnik,[7] for example, has investigated several pre-engineered buildings with evident failures of roof purlins suffered under the snow loads well *below* the design values. Some of those roofs have partially collapsed. He places much of the blame on inadequate purlin bracing. (His conclusions are disputed by others who insist that the failed buildings were of older vintage, perhaps improperly engineered, and therefore not representative of the modern practice.)

Another suspected mode of failure in cold regions involves water freezing in the purlin-supported gutters. The resulting overload of the first one or two purlins causes their rotation into a near-horizontal position. The roofing then pulls on the rest of the roof purlins, which also "lay over" and fail. A proper eave and ridge bracing assembly illustrated in Chap. 5 can help prevent this type of progressive collapse.

The insurance industry is well aware of the fact that some metal building roofs can fail very quickly under a major snow overload. Such failures occurred in March 1993 in northwest Georgia and in 1994 in eastern Pennsylvania. A magazine published by Factory Mutual notes that "metal building systems fared the worst in the Storm of '93, perhaps because of their design....When a collapse occurs, it usually is total."[8] One building in Pennsylvania reportedly collapsed in 10 seconds, start to finish. Again, the apparent cause of that failure was a sheer amount of snowfall.

After Hurricane Iniki had hit the Hawaiian Islands in September 1992, the Structural Engineers Association of Hawaii prepared a damage survey. The survey found that, along with residential structures, some pre-engineered buildings had been affected. Among the report's conclusions: "Pre-engineered metal buildings appeared to have proportionally more damage than other types of engineered structures."[9] Again, collapse of some buildings was total (Fig. 10.7).

The troubling aspect of such metal building failures is not that some of the buildings were overloaded—this can happen to any structure—but that there was so little ductility and reserve strength when the overload came. The reports about building failures under the loads which were less than the design

Figure 10.7 Total collapse of a metal building. (*Photo: Structural Engineers Association of Hawaii.*)

values are especially disconcerting. In general, however, properly designed metal building systems should provide safe and sound shelter that can withstand all code-mandated structural loads without undue deformations. It should also be noted that metal building systems have a good record of resisting earthquakes. The troubles are likely to occur not in well-engineered metal building systems but in structures put together from metal building components without proper engineering.

References

1. Alexander Newman, "Engineering Pre-engineered Buildings," *Civil Engineering,* September 1992.
2. Joseph E. Minor, et al., "Failures of Structures Due to Extreme Winds," *ASCE Journal of the Structural Division,* vol. 98, ST11, November 1972.
3. *Butler Roof Systems Design / Specifiers Manual,* Butler Manufacturing Co., Kansas City, Mo., 1993.
4. Ken Buchinger, "Proper Design of Roof Penetrations in Trapezoidal Standing Seam Roofs," *Metal Architecture,* June 1994.
5. John L. Ruddy, "Evaluation of Structural Concepts for Buildings: Low-Rise Buildings," BSCE/ASCE Structural Group Lecture Series at MIT, 1985.
6. Duane Miller and David Evers, "Loads and Codes," *The Construction Specifier,* November 1992.
7. Frank Zamecnik, "Errors in Utilization of Cold-Formed Steel Members," *ASCE Journal of the Structural Division,* vol. 106, no. ST12, December 1980.
8. "Preventing Roof Collapse from Snow Loading," *Record,* P.O. Box 9102, Norwood, Mass., Fall 1994.
9. *A Survey of Structural Damage Caused by Hurricane Iniki, September 11, 1992,* Structural Engineers Association of Hawaii, Honolulu, Hawaii.

Lateral Drift and Vertical Deflections

11.1 The Main Issues

The discussion in Chaps. 7 and 9 has already highlighted an importance of specifying correct design criteria for lateral drift and vertical deflections. This chapter is specifically devoted to this critical and controversial topic that occupies the minds of many structural engineers.

First, the definitions: *Lateral (story) drift* is the amount of sidesway between two adjacent stories of a building caused by lateral (wind and seismic) loads (Fig. 11.1). For a single-story building, lateral drift equals the amount of horizontal roof displacement. *Horizontal deflection of a wall* refers to its horizontal movement between supports under wind or earthquake loading. *Vertical deflection* of a floor or roof structural member is the amount of sag under gravity or other vertical loading.

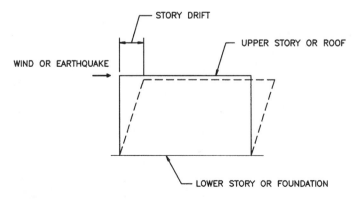

Figure 11.1 Story drift caused by lateral loads.

Why is any of this controversial? Drift and deflection criteria, along with some other issues such as vibrations, deal with *serviceability,* or functional performance, of buildings under load. Model building codes traditionally prescribe the desired levels of strength and safety, leaving the more nebulous topics of satisfying the occupants' perceptions of comfort and solidity up to the designers. The designers' criteria for achieving these goals are necessarily subjective, as the building which seems "flimsy" to one person may feel fine to another.

Various design firms tend to specify similar, although not identical, limits on horizontal and vertical building displacements for medium and high-rise structures. On the other hand, stiffness requirements for low-rise pre-engineered buildings remain a mystery to many engineers. These squat structures have been traditionally clad in flexible metal siding and roofing which could tolerate large amounts of framing movement, obviating any concerns about serviceability. The specifiers became concerned only when brittle wall materials such as masonry and concrete began to find their way into metal building systems. For those cases, some engineers continued to specify the same strict drift and deflection criteria used in conventional construction—only to be rebuffed by many manufacturers denouncing such rigidity requirements for metal buildings as unnecessarily expensive and impractical. Should metal buildings with hard walls be granted special privileges?

Our journey through these emotionally charged waters will begin with topics of lateral story drift and horizontal wall displacement and will then continue to the subject of vertical deflection criteria.

11.2   Lateral Drift and Horizontal Wall Displacement

11.2.1   Should it be mandated by codes?

Lateral drift and deflection limits are commonly expressed in terms relative to the story, or wall, height (H), such as a story drift of $H/400$. Another frequently used term is *deflection index,* an inversion of the drift limit; e.g., for a drift limit of $H/400$ the deflection index is 0.0025.

With many building codes being silent about specific horizontal deflection and drift criteria, building designers are left to decide for themselves. A "sturdy" building costs more to build than a "flimsy" one and may bring a higher resale value. Two buildings designed for the same structural loads but used for different occupancies could have different stiffness requirements. For example, a high-tech research lab or a hospital would probably impose much stricter limits on allowable building movement than a typical office would. Many designers feel that horizontal deflection and drift criteria relate to quality of building construction and should not be code-mandated.

Still, the codes may specify a maximum drift threshold for the sake of preserving structural integrity of buildings and their brittle components. Indeed, excessive lateral displacements may lead to unanticipated overload of building columns due to P-delta effect; large movements could also damage exterior cladding and interior finishes.

The codes seem to be more concerned with controlling building sway caused by violent earthquakes than by strong hurricanes, and justifiably so. A case in point: After the 1995 Kobe earthquake, some buildings seemed relatively undamaged from the outside but were still unusable because of buckled partitions and jammed doors caused by violent movements.

11.2.2 Provisions of model codes for drift limits from seismic loads

The drift criteria listed below are found in the code sections dealing with seismic loads. The Uniform Building Code[1] limits calculated story drifts to $0.04/R_w$, or 0.005 times the story height for structures with fundamental periods of under 0.7 s. For all other structures, it limits calculated story drifts to $0.03/R_w$, or 0.004 times the story height. The values of the coefficient R_w depend on the type of the main structural system. For metal building systems with rigid frames and cross bracing, the drift index of 0.005 (or $H/200$) usually controls.

The code makes two important exceptions to this rule. The first one states that the limit can be exceeded if it can be shown "that greater drift can be tolerated by both structural...and nonstructural elements that could affect life safety."

The second exception specifically exempts from any drift limitations single-story steel-framed buildings used for factory, manufacturing, storage, business workshop, and some other occupancies. To qualify for this exemption, a building cannot have any frame-attached equipment, unless it is detailed to accommodate the drift. To avoid damage from large frame movements, the code further requires that the walls laterally braced by the steel frame be designed to accommodate the drift. This goal is to be achieved by a deformation compatibility analysis and by meeting certain prescribed requirements for wall anchorage and connections.

In the 1990 BOCA Code[2] and 1988 Standard Building Code,[3] the lateral-drift limit is also specified as 0.005 times the story height (or $H/200$), "unless it can be demonstrated that greater deformation can be tolerated." The 1993 BOCA Code[11] takes a different tack. It describes the maximum allowable story drift as a function of several variables such as the building's seismic hazard exposure group, number of stories, and presence of interior finishes. The code significantly relaxes the allowable drift values: its most restrictive drift limit is $H/100$; the least restrictive is "none." The latter applies to those single-story buildings where all interior components are designed to accommodate the story drift and where no equipment is attached to the frame. Similar requirements are contained in ASCE 7-95.[12]

It is clear that, under seismic loading, none of the model codes seeks to impose any flexibility limitations on metal building systems with metal-only cladding. Such conditions occur mostly in industrial and warehouse buildings. For other structures, the drift limit under seismic loading may be spelled out in the governing building code. In any case, seismic loading and its drift limits rarely govern the design of pre-engineered buildings.

11.2.3   Drift limits for wind loading

For lightweight metal building systems, seismic loading rarely controls the design of lateral-load-resisting framing—wind loading usually does. The model building codes are silent about lateral drift limits due to wind, a fact that may reflect a lack of consensus on the matter and an understanding that such limits relate to building quality and should not be code-mandated. The guidelines are available elsewhere, however.

Since as early as 1940, a lateral deflection limit of $H/500$ has been recommended for tall buildings.[4] The authoritative *Structural Engineering Handbook* suggests that limits on "lateral deflection of 1/300 to 1/600 of the building height are reasonable and acceptable."[5]

The *Building Structural Design Handbook*[6] reflects that a 0.0025 drift index ($H/400$), even from a 25-year storm, "may be appropriate for a speculative office building. On the other hand, it may be completely inappropriate for a hospital, library, or any other type of high-quality building project." It goes on to suggest that the issue may be addressed by specifying a strict limit on the drift, say $H/500$, but for the drift to be computed using a smaller design wind loading than that imposed by a 50-year storm. For example, the loading from a 10- or 25-year windstorm might be used.

A survey of structural engineers around the country by ASCE Task Commitee on Drift Control of Steel Building Structures of the ASCE Committee on Design of Steel Building Structures[7] has found that the design practices with respect to wind drift vary considerably. Most designers, however, specify drift indexes of 0.0015 to 0.003 (corresponding to the limits of $H/666$ to $H/333$) caused by a 50-year mean wind recurrence interval for all types of structures. The most commonly used wind-drift limit for low-rise structures is, again, 0.0025 ($H/400$) caused by a 50-year wind. The task committee felt that wind-induced drift limits should not be codified.

11.2.4   Drift limits in AISC
Design Guide No. 3

Recognizing a dearth of serviceability criteria for metal building systems under wind loading, MBMA and AISC have published a design guide entitled, *Serviceability Design Considerations for Low-Rise Buildings.*[8] The guide's eminent authors, James M. Fisher and Michael A. West, have undertaken a major effort to stimulate discussion on various serviceability topics, including drift and deflections. The guide should be read by everybody involved in structural design of low-rise buildings.

Reflecting a subjective nature of serviceability criteria, the guide's authors base many of its recommendations on their own judgment and experience. They admit that the criteria are controversial and envision the guide as a catalyst for the debate rather than a final word in the discussion. (Some metal building manufacturers, however, seem to think exactly the opposite—that no further questions remain.)

The guide uses a 10-year mean recurrence interval wind speed loading for its drift-limit criteria, rather than a 50-year loading used for strength calculations. The rationale is that 50-year storms are rare events that have little in common with day-to-day experience of buildings. Furthermore, the consequences of serviceability failures are "noncatastrophic" and should be weighted against high up-front costs required to prevent the failures. The guide states that 10-year wind pressures can be reasonably approximated by using 75 percent of the 50-year wind pressure values.

(Some other sources have also questioned the common practice of basing wind-drift calculations on the wind loads likely to return only once in 50 years. Galambos and Ellingwood,[14] for example, advocate using a reference period of 8 years, which represents the average period of one tenancy in an office building.)

For several types of walls, the guide proposes certain maximum limits on the magnitude of bare-frame lateral drift, horizontal deflection, and racking (lateral movement parallel to the wall). Reproduced below are the criteria for foundation-supported cladding; the guide also considers criteria for column- and spandrel-supported panels. In the following expressions, H stands for the wall height and L for the length of a supporting steel member.

The maximum recommended story drift for various materials is

$H/60$ to $H/100$ for metal panels

$H/100$ for precast concrete

$H/200$ for reinforced masonry (can be reduced to $H/100$ with proper detailing)

Where interior partitions are used, bare-frame story drift is limited to $H/500$.

The maximum recommended horizontal deflections of girts or wind columns supporting metal or masonry walls are

$L/120$ for metal panels

$L/240$, but not over 1.5 in, for masonry walls

A limit on racking of $H/500$ is recommended for column- and spandrel-supported curtain walls. Again, all these criteria are for a 10-year wind loading.

The limitations on lateral drift and horizontal deflections proposed by the guide are more liberal than those of other sources. Some engineers find it counterintuitive that the guide seems to offer a larger degree of protection to interior drywall partitions than to brittle exterior walls.

11.2.5 The lateral drift provisions of MBMA Manual

The *MBMA Manual*[9] in the Appendix Section A6 briefly addresses the issue of lateral deflections due to wind. (Lateral drift due to seismic loading is not discussed.) The *Manual* stays away from recommending any specific limits for lateral drift and horizontal deflections, pointing out that such limits are highly dependent on the building occupancy, type of wall materials, presence of cranes, and other factors. "There is simply no way to set a limit that would be appropriate for all conditions," the *Manual* concludes.

The *Manual* suggests that lateral drift need not be computed using the full wind loading and points instead to the already-mentioned 10-year wind load, adding that this practice is recommended by the National Building Code of Canada. A further reduction in the frame stiffness for drift calculations, the *Manual* continues, could be achieved by using a design wind load that is less than the full tributary load on a frame.

11.2.6   How lateral drift is computed

Prior to a discussion of the various criteria listed above, it is necessary to briefly examine how drift and horizontal deflections are calculated and what the numbers actually mean.

The total story drift is a sum of two components—the frame drift and the diaphragm displacement between the frames (Fig. 11.2). For a typical pre-engineered building with rigid frames spaced 20 to 30 ft apart and a horizontal-rod roof bracing, the diaphragm deflection component might be insignificant. At another extreme, in buildings where no roof bracing is present at all, and wind loading is distributed to frames by eave struts, the diaphragm deflections could be larger than the frames' drift. Unfortunately, the diaphragm deflection computations are occasionally neglected by some inexperienced metal building designers.

Both BOCA[2] and Standard[3] Building Codes require the horizontal displacement computed for a frame under seismic loading to be increased by a factor of $1/K$ to obtain the design drift value. Metal buildings with ordinary rigid frames have a K value of 1.0.

The actual frame drift can be readily determined by most pre-engineered building software. For preliminary calculations, any general structural analysis computer program can be used. The approximate formula of Fig. 11.3 could

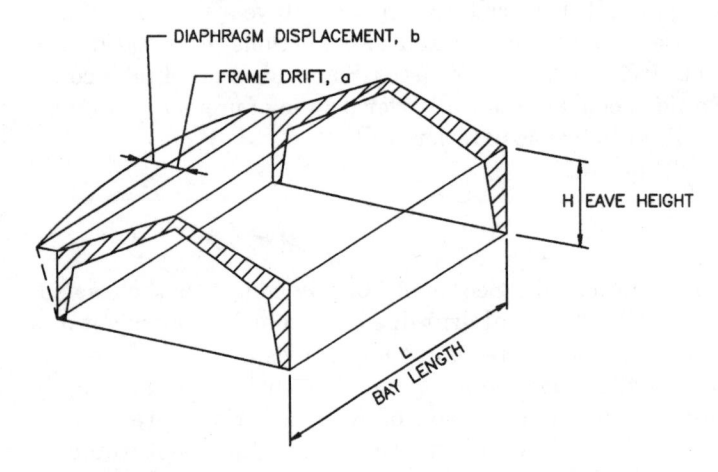

TOTAL STORY DRIFT = a+b

DIAPHRAGM DISPLACEMENT, b

FRAME DRIFT, a

H EAVE HEIGHT

L BAY LENGTH

Figure 11.2   Components of story drift.

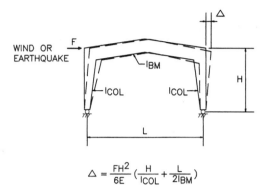

$$\Delta = \frac{FH^2}{6E} \left(\frac{H}{I_{COL}} + \frac{L}{2I_{BM}} \right)$$

Figure 11.3 An approximate formula for computing frame drift in a two-hinged frame with constant member sections.

be handy for rough checks of two-hinge frames with constant member sections. Naturally, the process is much more complex for rigid frames with tapered columns and beams, in which case computers are a must.

11.2.7 Lateral drift from gravity loads

A discussion focused solely on the lateral drift resulting from wind or seismic loading misses one important point: frame sidesway is caused not only by lateral loads but by gravity loads also. Many structural engineers used to design of conventional buildings do not realize that a gable frame can have a substantial amount of "kicking out" at the roof level when loaded with snow or roof live load (Fig. 11.4). Lateral displacements from large snow loads could exceed story drifts caused by winds. The codes do not address the issue, probably because gable frames are largely endemic to metal building systems.

We recommend that a drift limit established for the building apply to all sources of displacement, be it wind, snow, or earthquake. The computed values of lateral drift from some fraction of snow or roof live load should be combined with those caused by wind. In fact, it is wise to compute a total story drift from

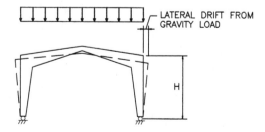

Figure 11.4 Lateral displacement of rigid frame under gravity load.

all the design load combinations, probably excluding only the dead and collateral loads applied prior to the installation of the finishes.

Excessive structural distortion due to snow is real and should not be ignored. Ruddy[13] tells of two pre-engineered buildings, used as a school and an office, that had suspended ceilings. After some snow had accumulated on the roofs, the ceilings in both buildings became noticeably displaced. The local building officials called on the scene insisted that both buildings be vacated, snow removed, and structural capacity of the roofs rechecked. After an occurrence like this, building designers tend to use rather strict displacement limits for future projects.

Lateral movement caused by gravity loads is less pronounced in multiple-span rigid frames than in their single-span siblings and is virtually absent in tapered-beam and truss systems.

11.2.8 Lateral deflections: The discussion

Are large story drifts always harmful? What concerns the designers is not only an absolute value of the story drift but also an angle of curvature assumed by the wall.

In elastic theory, the larger the slope of a deflected shape of a flexural member, the larger the stresses are. Brittle materials, such as unreinforced masonry and glass, can be simply broken up by large stresses; ductile materials, such as reinforced masonry and concrete, can tolerate some cracking without failure. Large cracks, however, especially in single-wythe masonry and concrete walls, are likely to become gateways for water intrusion, which can damage interior finishes and hasten wall deterioration caused by freeze-thaw cycles. The overall degree of wall curvature depends on a magnitude of three deflection components (Fig. 11.5):

1. Story drift, a sum of the frame drift (D_f) and diaphragm deflection D_{diaph}
2. Horizontal deflection of supporting girts and wind columns, if any, D_{girt}
3. Horizontal deflection of the wall itself D_w

Of these, the first two depend on a stiffness of the metal building and the third one is a function of the wall's stiffness. The second component, D_{girt}, occurs only when intermediate girts provide lateral support for the wall, as is often needed for tall walls made of steel studs and brick veneer. In this case, the wall curvature changes—points of inflection—occur near the girt locations (Fig. 11.5a). Obviously, horizontal deflections of the walls spanning from foundation to roof without any intermediate girts, such as full-height CMU or precast, do not include D_{girt} (Fig. 11.5b).

The critical issue in this discussion is whether the wall functions as a simply supported or continuous member. Stresses and deflections in simply supported beams are not affected by movement of supports. In contrast, continuous members are statically indeterminate and are influenced structurally by yielding supports.

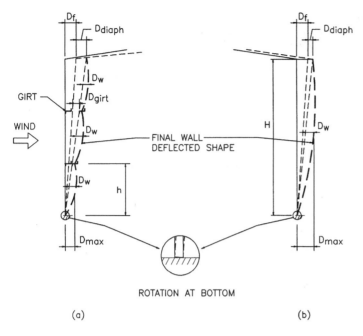

ROTATION AT BOTTOM

(a)   (b)

Figure 11.5 Components of total horizontal wall displacement. (a) Walls with girts; (b) walls without girts.

The ends of any simply supported member must be able to rotate freely. Therefore, a wall may be considered simply supported only if its ends at the base and at the roof are free to rotate. The maximum horizontal deflection of a simply supported wall may be taken as D_{wall}, not D_{max}, since the movement of supports is irrelevant. If, however, the wall is fixed at the bottom—a CMU wall doweled into the foundation is one example—the end rotation at the base is prevented, the simple-span model does not apply, and D_{max} must be taken as the maximum displacement. (The walls are rarely fixed at their tops, of course.) It is evident that the wall's vulnerability to cracking is influenced more by its own rigidity and by details at the base than by the magnitude of a story drift.

But is it possible to make doweled CMU walls behave as simple beams? For reinforced CMU, AISC Guide no. 3[8] suggests that, instead of a common CMU base detail of Fig. 11.6a, a detail similar to that of Fig. 11.6b be used to facilitate end rotation. In Fig. 11.6b, a continuous through-the-wall sheet flashing installed at the base, with mastic around the vertical bars, is provided to introduce a plane of weakness in the wall and make end rotation possible. It seems, however, that CMU with vertical bars and dowels will still develop a substantial fixity moment at the base (Fig. 11.6c), which puts the whole theory of pinned-base CMU into question.

The base rotation may become possible if the dowels are omitted but the flashing kept. Unfortunately, such a detail is likely to produce a wall without enough "grip" on the foundation, a wall that could shift under lateral loading. In a bet-

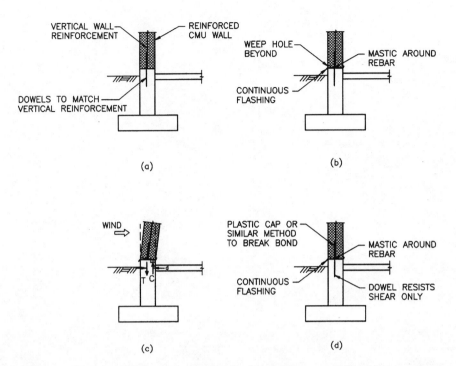

Figure 11.6 Overcoming fixity at the bottom of a doweled masonry wall. (*a*) A common construction detail; (*b*) introducing a plane of weakness to facilitate rotation; (*c*) forces resisting rotation and providing fixity; (*d*) detail of a possible true "pin" connection.

ter detail, the dowels are kept but the dowel length above the flashing is encased in a bond-breaking sleeve which allows the dowel to slide inside the wall (Fig. 11.6*d*). In this case, the dowels resist no tension but are able to transfer shear. To the best of our knowledge, this detail has not yet been tried in the field.

Another important issue in this discussion concerns building corners. A wall exposed to the wind, its top attached at the eave, will move with the frame, while the perpendicular wall will not (Fig. 11.7). By introducing a control joint

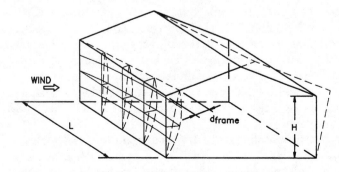

Figure 11.7 Deflected shape of the wall near a building corner.

in the wall near the corner, one hopes to avoid wall cracking. The joint, however, may not survive large wall rotations without failure—and leakage.

A problem with excessive deflections of the exterior walls normal to the direction of lateral loads is important, but so is *racking* of walls parallel to the load (Fig. 11.8). Racking affects, for example, interior drywall partitions attached at their tops to main building framing. A drywall partition can undergo significant deflections normal to its surface but is vulnerable to displacements along its plane. While it is possible to overcome this problem with special "sliding" connections at the top of the partition, such connections are best designed by structural engineers who, regrettably, are rarely involved with architectural details.

Exterior masonry and concrete walls in metal buildings also experience racking. These "hard" walls are much more rigid than any wall bracing that might be located along the same column line; they tend to act as shear walls, rendering the bracing ineffective. Unless completely separated from the frame movement—a rare scenario—these walls should be intentionally designed and reinforced as shear walls in lieu of wall bracing.

11.2.9 Choosing a lateral drift limit

It is clear that lateral drift and deflection criteria depend on many project-specific factors including the type of occupancy; presence of partitions, cranes, and structure-supported equipment; type of exterior wall materials; and expectations of the owner.

For "pure" metal buildings without any partitions, brittle exterior walls, or frame-mounted equipment, story drift limits in the range of $H/60$ to $H/120$ might be adequate.

For masonry or concrete exterior walls attached to metal frame, it is reasonable to use the story drift limits of conventional construction, such as $H/200$ for seismic loading and $H/400$ for wind. Somewhat more liberal limits for wind-induced drift ($H/200$ to $H/300$) may be justified if an excellent cooperation between the architects and engineers allows for a development of custom-engineered details for base, top, and corners of both exterior and interior walls. The prospective contractor's sophistication, experience, and quality of supervision count a lot, too.

A presence of drywall partitions without custom-designed connections—with any type of exterior walls—limits the design story drift to $H/500$.

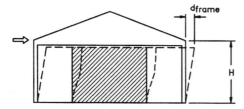

Figure 11.8 Wall racking.

Whether to base these drift limits on 50-year or 10-year wind loads is left to the specifying engineer's judgment. While there is merit in recommendations favoring a 10-year wind loading, we hesitate to embrace them for every condition. The use of a 10-year wind loading in tandem with relatively strict drift limits ($H/500$ or $H/400$) might be justified, but not if more liberal criteria are used.

11.2.10   Choosing a horizontal deflection limit

Horizontal deflection limits for some common exterior wall systems are discussed in Chap. 7. Building codes may or may not include requirements for maximum allowable horizontal deflections of walls. For example, both the 1990[2] and the 1993[11] edition of BOCA National Building Code impose an $H/180$ limitation on the horizontal deflections of *interior* walls and partitions, but only the 1993 edition specifically calls for a limit of $H/240$ for structural elements of *exterior* walls caused by the design wind load. (The 1990 BOCA Code required instead that deflections of "all other structural members" be confined to 1/240th of the span, which presumably applied to both horizontal and vertical deflections.) Based on the advance information, the 1996 edition of the BOCA code will include a limit of $H/120$ for exterior walls supporting metal siding and for those exterior walls designed to accommodate this wind-induced deflection.

The Standard Building Code[3] limits horizontal deflections to $H/240$ for exterior and interior walls and partitions with brittle finishes, and to $H/120$ for those with flexible finishes, except that $H/60$ may be used for metal siding.

Again, the issue of whether to base the horizontal deflection computations on 50-year or 10-year wind loading is left up to the specifier to decide, although the model codes cited above do not even mention 10-year wind loading.

Whenever exterior walls depend on the intermediate girts for lateral support, horizontal deflection criteria are influenced by the wall construction details. As was already demonstrated, a wall that can freely rotate at the base can tolerate larger girt deflections than a fixed-base wall. For the former, the provisions of AISC Guide no. 3, which limit girt deflections to $L/120$ for metal panels and to $L/240$ (but not over 1.5 in) for masonry, seem sensible. For the latter, a stricter limit is justified.

11.3   Vertical Deflections

11.3.1   Provisions of model codes

Vertical deflection criteria are less controversial than those for lateral drift: Everybody realizes that the sight of a sagging beam overhead does not make for a happy mood. Apart from such visual impact, uncontrolled vertical deflections can cause damage to interior partitions, windows, and plaster ceilings. Large deflections of low-slope roofs can lead to water ponding and, ultimately, to roof collapse.

Over the centuries, the design profession has concluded that a deflection limit of 1/360th of the member's length ($L/360$) is adequate to avoid cracking of plastered ceilings. The deflection limit is applicable to live load, snow, or any other superimposed load acting after the ceiling is constructed. This criterion has been widely adopted by the building codes; the specific model code provisions are listed below.

- Uniform Building Code.[1]

   Steel members supporting roofs with slope of less than ½:12: The deflections caused by snow load are limited to $L/180$.

   Steel roof members supporting plaster and all floor members: The deflections caused by live load are limited to $L/360$, the deflections from dead plus live load, to $L/240$.

- 1990[2] and 1993[11] BOCA National Building Code.

   Roof members supporting plastered ceilings: $L/360$

   Roofs with slope over 3:12 without ceilings: $L/180$

   All other members: $L/240$

   (Based on the advanced information, 1996 BOCA code will include a limit of $L/150$ for structural members supporting metal sheeting and without a ceiling load, provided that the minimum design slope is maintained.)

- Standard Building Code.[3]

   Floor and roof members supporting plaster: $L/360$ from live load, $L/240$ from dead + live.

   Roof members supporting ceilings without plaster: $L/240$ from live load, $L/180$ from dead + live.

   Roof members not supporting any ceilings: $L/180$ from live load, $L/120$ from dead + live.

   Exposed metal roofing: $L/60$ from dead + live.

11.3.2 Other recommended criteria

AISC specification[10] limits maximum live-load allowable deflection of roof and floor members supporting plaster to $L/360$. MBMA *Manual*[9] is silent on specific deflection criteria. AISC Design Guide no. 3[8] tabulates deflection limitations for various elements of roof construction including those required to satisfy ponding and drainage considerations. For example, it recommends the familiar deflection criteria of $L/360$ for roofs supporting plastered ceilings and $L/240$ for roofs supporting other ceilings. The Guide points out that some "maximum absolute value must also be employed which is consistent with the ceiling and partition details," and suggests a range of ⅜ in to 1 in. The Guide further recommends that the deflection of roof purlins be checked under a combination of dead and one-half design snow load (or a minimum of 5 lb/ft²) to verify that positive drainage still exists when the members are deflected under load. These Guide's criteria are based on the design live load or a 50-year snow.

The Guide states that the above-mentioned deflection criteria are truly important largely along the building perimeter, and that the maximum purlin deflection in the field of the roof from snow load could be limited to $L/150$. Presumably, the last number applies only where no ceilings or partitions are present. We feel that the $L/150$ deflection limit is too liberal for finished spaces. For a typical 25-ft span, it translates into a 2-in deflection under the design snow loading, a noticeable sag which may be large enough to make the occupants concerned.

The Guide makes an important point about localized deflections from concentrated loads being probably of larger importance than those from uniform loads. Indeed, a common complaint of pre-engineered building users is that a light fixture or a pipe suspended from a purlin deflects the purlin too much in relation to its neighbors. The only safeguard against such occurrence, short of designing each purlin for every minute load—an impractical task—is to use more rigid purlins, that is, to apply a rather strict deflection limit throughout the roof. The traditional $L/360$ and $L/240$ deflection criteria for plastered and nonplastered ceilings, respectively, should be adequate for most finished spaces.

References

1. Uniform Building Code, International Conference of Building Officials, Whittier, Calif., 1994.
2. The BOCA National Building Code, Building Officials & Code Administrators International, Inc., Country Club Hills, Ill., 1990.
3. Standard Building Code, Southern Building Code Congress International, Inc., Birmingham, Ala., 1988.
4. "Wind Bracing in Steel Buildings," Final Report, Subcommittee 31, Committee on Steel, Structural Division, *Transactions ASCE,* vol. 105, pp. 1713–1739, 1940.
5. Edwin H. Gaylord, Jr., and Charles N. Gaylord (eds.), *Structural Engineering Handbook,* 2d ed., p. 19-3, McGraw-Hill, New York, 1979.
6. Richard N. White and Charles G. Salmon, (eds.), *Building Structural Design Handbook,* p. 583, Wiley, New York, 1987.
7. "Wind Drift Design of Steel-Framed Buildings: State-of-the-Art Report," by ASCE Task Committee on Drift Control of Steel Building Structures of the Committee on Design of Steel Building Structures, *ASCE Journal of Structural Engineering,* vol. 114, no. 9, September 1988.
8. *Serviceability Design Considerations for Low-Rise Buildings,* Steel Design Guide Series no. 3, AISC, Chicago, Ill., 1990.
9. *Low-Rise Building Systems Manual,* MBMA, Cleveland, Ohio, 1986.
10. Specification for Structural Steel Buildings, Allowable Stress Design and Plastic Design, AISC, Chicago, Ill., 1989.
11. The BOCA National Building Code, Building Officials & Code Administrators International, Inc., Country Club Hills, Ill., 1993.
12. Minimum Design Loads for Buildings and Other Structures, ASCE 7-95, American Society of Civil Engineers, New York, 1995.
13. John L. Ruddy "Evaluation of Structural Concepts for Buildings: Low-Rise Buildings," BSCE/ASCE Structural Group Lecture Series at MIT, 1985.
14. Theodore V. Galambos and Bruce Ellingwood, "Serviceability Limit States: Deflection," *ASCE Journal of Structural Engineering,* vol. 112, no. 1, January 1986.

12

Foundation Design for Metal Building Systems

12.1 Introduction

Foundations, contrary to the dreams of building owners, do not come prepackaged with metal building systems. The concept of single-source responsibility for pre-engineered buildings is qualified by the fact that the foundations are usually designed by outside engineers.

In this chapter we look into the differences between foundations for metal building systems and those for conventional construction and examine some common solutions. We will not deal with the basics of foundation design, a subject that should be familiar to any practicing structural engineer and, hopefully, to most architects. Similarly, we will not delve into the complex topic of establishing allowable bearing pressures for various soils, a task best left to geotechnical engineers.

As discussed in Chap. 9, poor soils found at the site might call for deep foundations, an expensive item that could explode the project's budget and suddenly make a competing site much more appealing. On a more positive note, an experienced geotechnical engineer might be able to justify a much larger allowable soil-bearing value than could be learned from the necessarily conservative tables of presumptive bearing pressures contained in the building codes. This recommendation could lead to substantial cost savings. As Ruddy[1] has stated, "An increase in an allowable bearing pressure from 3 ksf to 6 ksf can result in a savings of $0.08/s.f. for a shallow spread footing foundation system in a one-story facility."

12.2 Soils Investigation Program

The results of a geotechnical exploration program are of interest to all parties of the construction project. The owner and the local building official need to be reasonably assured that the proposed design can be safely accomplished without endangering the building's occupants and adjoining properties. The engineer of record needs to know soil type, stratification, and water table location to determine the most appropriate foundation type and its bearing depth. The contractor seeks much of the same information to select an excavation support system and to determine dewatering needs. All of the participants are eager to know whether any unsuitable materials such as organic silt or peat are found.

A soils exploration program might uncover a presence of abandoned foundations, buried utilities, and occasionally, an archaeological site. Any one of those "finds" can adversely influence the project's cost and schedule.

Subsurface investigation usually includes several soil borings or test pits, the number, nature, and location of which are determined by the local codes and by experience. The BOCA National Building Code,[2] for example, requires at least one soil boring for every 2500 ft^2 of the building area for buildings over 40 ft, or more than three stories, in height bearing on mat or deep foundations. Normal practice calls for one boring at each building corner, one in the center, and the rest, if needed, near the critically loaded foundations. For large buildings founded on poor soils, soil borings should be spaced not over 50 ft apart and perhaps much closer.

How deep to carry the borings? The BOCA Code requires the borings to be carried to rock or to "an adequate depth below the load-bearing strata." For low-rise buildings, some engineers specify a depth of borings to be 20 ft below the anticipated foundation level, with at least one boring continuing deeper, perhaps to a lesser of 100 ft, the least building dimension, or refusal. If the longer boring encounters no unsuitable materials deep down, the rest of the borings could be stopped at the originally planned depth. A boring should never be terminated in an unsuitable material. Since the extent—and the cost—of the program can change so much during the field operations, it is advisable that a competent engineer be present at the site to observe the process and modify it if circumstances warrant.

Sometimes, instead of soil borings or in addition to them, test pits are excavated. Test pits are especially appropriate for lightly loaded foundations supported by good soil at some depth but by a questionable material near the surface. A test pit can provide a clear visual picture of the soil condition at a shallow depth, up to about 10 ft. Test pits are fairly inexpensive, since no specialized equipment is needed, and are often useful in supplementing the information provided by soil borings. For example, a refusal encountered by the boring rig could be a sign of a rock ledge or of a large boulder. A test pit can quickly provide an answer.

The end result of subsurface exploration is a soils investigation report prepared by the geotechnical engineer. The report describes soil conditions at the site and recommends the maximum allowable bearing pressure and other pertinent engineering characteristics of the soil.

12.3   What Makes These Foundations Different?

Three main factors distinguish pre-engineered building foundations from the rest: substantial horizontal column reactions, large column uplift, and a common need to design the foundations before column reactions are determined. Experienced structural engineers can easily spot an improper foundation design, because the first two issues are often overlooked by the uninitiated.

12.3.1   Horizontal column reactions

Lateral loads act on all buildings and naturally result in both vertical and horizontal column reactions. In "conventional" buildings with moderate footprints and relatively closely spaced columns, horizontal reactions are distributed by the first floor slab to a number of column and wall foundations. It's a rare case when the column foundation has to resist large horizontal loads. The situation is quite different in pre-engineered buildings.

Rigid frame, a staple of metal building systems, generates a large horizontal thrust from gravity loads (Fig. 12.1a), as well as horizontal reactions from lateral loads (Fig. 12.1b). Assuming that the frame columns are pin-connected, column reactions on a typical foundation are shown in Fig. 12.2a. For fixed-base columns, the fixity moment M (Fig. 12.2b) is added.

Horizontal column reactions tend to produce two modes of foundation failure—overturning and sliding—that will be elaborated on in Sec. 12.5.3.

12.3.2   Uplift

Uplift—an upward force—is a natural result of wind acting on gable-frame buildings (Fig. 12.3). In two- and multistory conventional buildings, wind uplift rarely exceeds the combined roof and floor dead loads and thus almost never governs the foundation design. Single-story metal building systems, on the other hand, have extremely lightweight roofs with a total weight of only 3 to 10 lb/ft^2, often not nearly heavy enough for uplift prevention.

Fortunately, column uplift can be resisted not only by the roof dead load but also by the weight of the foundation and the soil on top of it (Fig. 12.4). So, instead of making the roof heavier, it is better to increase the foundation weight, or better still, that of the overlying soil, the most economical way of doing which is to lower the footing. While this solution requires additional excavation and backfilling and thus is not quite "dirt cheap," it is often less costly than increasing the foundation footprint.

In Fig. 12.4, the uplift U is resisted by the soil weight W_1 and W_2 and the foundation weight W_3. To lift cohesive soil, such as clay, the uplift force must first overcome its shearing resistance; the plane of soil failure will generally be inclined from the vertical. In cohesionless soils such as sand the failure plane is close to the vertical line. Most engineers use a conservative approach and

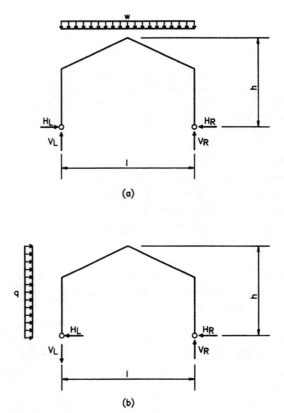

(a)

(b)

Figure 12.1 Column reactions of a rigid-frame structure. (a) From gravity loads; (b) from lateral loads.

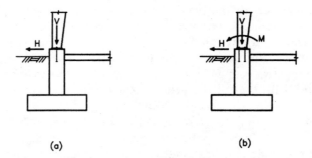

(a)   **(b)**

Figure 12.2 Forces acting on foundations supporting rigid-frame columns. (a) Pin-base columns; (b) fixed-base columns.

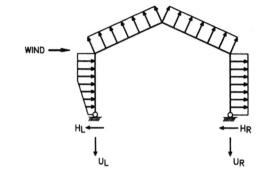

Figure 12.3 Uplift and horizontal column reactions caused by wind acting on gable-frame buildings.

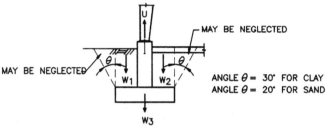

Figure 12.4 Development of uplift resistance.

neglect the inclined soil segment, as well as any shear resistance of the soil.[10] If included, the angle of incline may be taken as 30° for cohesive soils and 20° for cohesionless soils per Department of the Navy Criteria.[7] Building codes normally require a minimum factor of safety of 1.5 against wind uplift.

In the areas subjected to flooding, the "beneficial" dead load of the foundation is reduced by the water buoyancy pressure. It is not inconceivable that a flood and a hurricane will happen simultaneously, although the probability of such an occurrence should be carefully evaluated.

For deep foundations, additional uplift resistance can be mobilized by using friction piles; foundations on ledge can be anchored into the rock with drilled-in rods.

12.3.3 Foundations designed before the building

In conventional construction, including stick-built single-story buildings, the process of structural design normally follows a load path from the roof to the foundations. Load reactions determined for the structure at the top are applied to the lower members and, eventually, to the foundations, which are among the last items designed.

The situation is turned on its head in pre-engineered buildings. Excepting captive relationships, where the owner deals directly with a favorite manufacturer, a typical project requires preparation of complete contract documents—including the foundations—for competitive bidding or negotiations. The documents are prepared *before* the manufacturer runs a frame analysis and develops the reactions report described in Sec. 9.6. To add fuel to fire, some developers insist on an early "foundation contract" set of drawings to be released before the rest of the documents are ready.

One way or another, foundations may have to be designed before final building reactions are obtained from the selected manufacturer. This unfortunate situation is the bane of structural engineers specifying metal building systems, who often have to design foundations based on mere estimates of the column reactions.

12.4   How to Estimate the Magnitude of Column Reactions

12.4.1   Manufacturers' tables

For a single-span building with standard width, roof slope, and eave height, the values of column reactions may be obtained directly from the manufacturers' design manuals for a given set of building dimensions and loads. Typical tables of column reaction are provided here for common clear-span rigid frames (Fig. 12.5), multispan rigid frames (Fig. 12.6), and tapered beams (Fig. 12.7). For other types of framing, different sources must be tapped, some of which are described below in order of decreasing precision of the results they can provide.

Those who follow any approximate methods should be forewarned that final reactions supplied by the manufacturer are based on the actual sizes of tapered members and therefore will differ from the results of simplified analysis that assumes member cross sections to be constant.

12.4.2   Specialized software

A design firm frequently engaged in estimating reactions may consider investing in specialized software for design and analysis of metal buildings. This kind of software essentially duplicates the design process of manufacturers; it is especially appropriate if both the column reactions and the member sizes need to be known in advance. Any of several programs available on the market will be more than adequate for this purpose. There is a fair chance, however, that reactions supplied by the manufacturer will still be different from those determined by the software because of slight variations in member sizes and construction details.

12.4.3   General frame analysis software and frame formulas

Most frame analysis software programs are acceptable for determination of approximate column reaction values, especially when multiple-span rigid frames with unequal spans are involved.

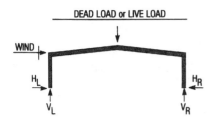

BAY SPACING — 25 ft.
ROOF SLOPE — 1:12
LIVE LOAD — 20 psf
WIND LOAD — 80 mph

REPRESENTATIVE FRAME REACTIONS 1:12 ROOF SLOPE

NOTES:
1. Dead load equals self weight of members.
2. Wind load is applied in accordance with MBMA (1986).
3. Negative value of reaction indicates direction opposite to that shown on sketch.
4. Reactions shown are approximate only and are not exact submittal values.
5. Reactions for various load combinations may be obtained by adding or subtracting the appropriate values.
6. Forces on the foundation will act in the opposite direction to the direction of the frame reactions.

FRAME REACTIONS (KIPS)										
SIZE		DEAD LOAD		LIVE LOAD		WIND LOAD				
SPAN	E.H.	V_L V_R	H_L H_R	V_L V_R	H_L H_R	V_L	H_L	V_R	H_R	
30	10	0.8	0.3	7.5	3.0	-4.9	-2.7	-3.2	-0.2	
	14	0.8	0.2	7.5	1.7	-5.3	-2.7	-2.8	1.1	
	16	0.8	0.2	7.5	1.4	-5.7	-3.0	-2.5	1.6	
	20	0.8	0.1	7.5	1.1	-6.8	-3.8	-2.0	2.6	
40	10	1.2	0.7	10.0	5.8	-6.3	-4.1	-4.5	-1.7	
	14	1.2	0.5	10.0	3.9	-6.6	-3.6	-4.2	0.1	
	16	1.2	0.4	10.0	3.2	-6.9	-3.5	-4.1	0.6	
	20	1.2	0.3	10.0	2.4	-7.9	-4.3	-3.8	1.8	
50	12	1.5	1.0	12.5	8.3	-7.8	-5.7	-5.7	-2.8	
	16	1.5	0.7	12.5	6.0	-8.2	-5.0	-5.5	-0.9	
	20	1.5	0.5	12.5	4.4	-9.2	-4.9	-5.5	0.6	
	24	1.5	0.4	12.5	3.4	-10.2	-5.6	-5.2	2.0	
60	12	1.8	1.6	15.0	13.2	-9.3	-8.3	-7.0	-5.4	
	16	1.8	1.2	15.0	9.9	-9.6	-7.0	-6.8	-3.1	
	20	1.8	0.9	15.0	7.8	-10.6	-6.7	-7.0	-1.4	
	24	1.8	0.8	15.0	6.3	-11.6	-6.7	-6.9	0.3	
80	12	2.6	3.2	20.0	24.4	-12.2	-14.3	-9.5	-11.5	
	16	2.6	2.5	20.0	18.8	-12.5	-11.8	-9.4	-8.0	
	20	2.6	1.9	20.0	14.7	-13.6	-10.7	-9.8	-5.5	
	24	2.7	1.6	20.0	12.1	-14.7	-10.2	-10.1	-3.4	
100	16	3.7	4.0	25.0	27.6	-15.5	-16.6	-12.0	-12.9	
	20	3.6	3.2	25.0	22.1	-16.7	-15.0	-12.6	-9.8	
	24	3.7	2.7	25.0	18.2	-17.9	-13.9	-13.1	-7.2	
	30	3.6	2.1	25.0	14.5	-19.4	-13.1	-13.4	-4.2	
120	16	4.8	6.2	30.0	38.8	-18.4	-22.7	-14.5	-19.0	
	20	4.8	4.9	30.0	31.0	-19.8	-20.1	-15.3	-15.1	
	24	4.8	4.1	30.0	25.7	-21.1	-18.5	-16.0	-12.0	
	30	4.7	3.3	30.0	20.8	-22.8	-17.1	-16.5	-8.3	

MODIFYING FACTORS:
To obtain approx. reactions for other bay sizes, live loads, and/or wind loads use the following rules:

BAY SIZE: (up to 30')
Divide all reactions shown by 25 then multiply by the bay length required.

LIVE LOAD: Divide live load reactions shown by 20 then multiply by the live load required.

WIND LOAD: Multiply the wind load reactions shown by the applicable factor:
70 mph use 0.8
90 mph use 1.3
100 mph use 1.6
110 mph use 1.9
120 mph use 2.3

Figure 12.5 Typical column reactions for clear-span rigid frames. (*Ceco Building Systems.*)

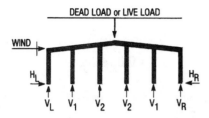

DEAD LOAD or LIVE LOAD

WIND

H_L H_R

V_L V_1 V_2 V_2 V_1 V_R

BAY SPACING — 25 ft.
ROOF SLOPE — 1:12
LIVE LOAD — 20 psf
WIND LOAD — 80 mph

REPRESENTATIVE FRAME REACTIONS 1:12 ROOF SLOPE

NOTES:

1. Dead load equals self weight of members.
2. Wind load is applied in accordance with MBMA (1986).
3. Negative value of reaction indicates direction opposite to that shown on sketch.
4. Reactions shown are approximate only and are not exact submittal values.
5. Reactions for various load combinations may be obtained by adding or subtracting the appropriate values.
6. Forces on the foundation will act in the opposite direction to the direction of the frame reactions.

FRAME REACTIONS (KIPS)

SIZE		DEAD LOAD				LIVE LOAD				WIND LOAD					
SPAN	E.H.	V_L V_R	H_L H_R	V_1	V_2	V_L V_R	H_L H_R	V_1	V_2	V_L	H_L	V_R	H_R	V_1	V_2
2 at 40	10	1.1	0.4	3.0	-	8.6	2.8	22.8	-	-6.0	-2.5	-3.4	-0.3	-12.3	-
	14	1.0	0.2	3.1	-	8.2	1.5	23.5	-	-5.9	-2.5	-3.1	1.0	-12.6	-
	16	1.0	0.1	3.2	-	8.0	1.1	24.0	-	-6.0	-2.7	-2.9	1.6	-13.0	-
	20	1.0	0.1	3.3	-	7.9	0.7	24.2	-	-6.6	-3.5	-2.9	2.6	-13.9	-
2 at 60	10	2.2	2.4	3.7	-	16.5	17.9	27.0	-	-10.8	-10.6	-7.2	-8.5	-14.5	-
	14	2.0	1.4	4.0	-	15.2	11.0	29.5	-	-10.2	-7.2	-6.4	-4.2	-15.8	-
	16	2.0	1.3	3.9	-	15.6	10.3	28.8	-	-10.6	-7.2	-6.6	-3.5	-15.7	-
	20	1.9	1.0	4.2	-	14.8	7.4	30.4	-	-11.0	-6.4	-6.5	-1.4	-17.6	-
3 at 40	12	1.1	0.4	2.6	2.6	9.4	3.0	20.6	20.6	-6.3	-2.7	-4.0	-0.1	-13.1	-9.0
	16	1.1	0.2	2.8	2.8	9.0	1.6	21.0	21.0	-6.4	-2.8	-3.6	1.2	-13.2	-9.7
	20	1.1	0.1	2.8	2.8	8.7	1.0	21.3	21.3	-7.1	-3.5	-3.5	2.3	-13.7	-10.9
	24	1.1	0.1	2.8	2.8	8.9	1.0	21.1	21.1	-8.0	-4.5	-3.4	3.1	-13.7	-12.1
3 at 60	12	2.2	2.0	3.9	3.9	16.7	15.4	28.3	28.3	-10.6	-9.4	-7.7	-7.1	-18.8	-11.6
	16	2.1	1.5	3.9	3.9	16.3	11.4	28.7	28.7	-10.6	-7.7	-7.5	-4.3	-19.1	-12.1
	20	2.1	1.1	4.1	4.1	15.8	8.4	29.2	29.2	-11.3	-6.9	-7.5	-2.2	-20.4	-13.6
	24	2.0	0.9	4.3	4.3	15.5	6.7	29.5	29.5	-12.0	-6.7	-7.5	-0.5	-21.2	-14.9
4 at 40	12	1.1	0.3	2.8	2.5	8.9	2.5	21.8	18.5	-5.9	-2.3	-3.9	0.1	-14.6	-9.8
	16	1.1	0.2	2.9	2.5	8.7	1.4	22.0	18.6	-6.0	-2.6	-3.6	1.3	-14.6	-10.0
	20	1.0	0.1	3.0	2.6	8.4	0.8	22.4	18.4	-6.6	-3.3	-3.5	2.3	-15.4	-10.6
	24	1.1	0.1	3.0	2.6	8.5	0.7	22.3	18.5	-7.5	-4.2	-3.4	3.2	-15.7	-11.2
4 at 60	16	2.1	1.4	4.2	4.0	15.8	10.6	30.1	28.1	-10.1	-7.1	-7.4	-4.0	-21.0	-15.1
	20	2.0	1.0	4.3	4.2	15.3	7.8	30.5	28.4	-10.7	-6.4	-7.4	-1.9	-22.4	-16.3
	24	2.0	0.8	4.5	4.3	15.1	6.4	30.7	28.4	-11.4	-6.4	-7.5	-0.5	-23.5	-17.1
	30	1.9	0.6	4.6	4.4	14.9	4.9	30.8	28.6	-12.4	-7.1	-7.4	1.6	-24.4	-18.3
5 at 60	16	2.1	1.4	4.2	4.0	15.6	10.3	30.6	28.8	-9.9	-6.7	-7.2	-3.9	-21.1	-18.5
	20	2.0	1.1	4.3	4.3	15.3	8.0	30.7	29.0	-10.6	-6.3	-7.4	-2.1	-22.3	-19.9
	24	2.0	0.9	4.5	4.3	15.1	6.5	30.7	29.1	-11.4	-6.2	-7.5	-0.6	-23.3	-21.2
	30	1.9	0.6	4.6	4.5	14.9	4.9	30.8	29.3	-12.4	-6.9	-7.3	1.5	-24.0	-23.0

MODIFYING FACTORS:

To obtain approx. reactions for other bay sizes, live loads, and/or wind loads use the following rules:

BAY SIZE: (up to 30')
Divide all reactions shown by 25 then multiply by the bay length required.

LIVE LOAD: Divide live load reactions shown by 20 then multiply by the live load required.

WIND LOAD: Multiply the wind load reactions shown by the applicable factor:
70 mph use 0.8
90 mph use 1.3
100 mph use 1.6
110 mph use 1.9
120 mph use 2.3

Figure 12.6 Typical column reactions for multispan rigid frames. (*Ceco Building Systems.*)

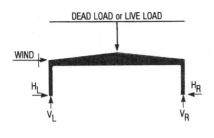

DEAD LOAD or LIVE LOAD

WIND

H_L H_R

V_L V_R

BAY SPACING —   25 ft.
ROOF SLOPE —   1:12
LIVE LOAD   —   20 psf
WIND LOAD   —   80 mph

**REPRESENTATIVE
FRAME REACTIONS
1:12 ROOF SLOPE**

NOTES:
1. Dead load equals self weight of members.
2. Wind load is applied in accordance with MBMA (1986).
3. Negative value of reaction indicates direction opposite to that shown on sketch.
4. Reactions shown are approximate only and are not exact submittal values.
5. Reactions for various load combinations may be obtained by adding or subtracting the appropriate values.
6. Forces on the foundation will act in the opposite direction to the direction of the frame reactions.

SIZE		DEAD LOAD		LIVE LOAD		WIND LOAD			
SPAN	E.H.	V_L V_R	H_L H_R	V_L V_R	H_L H_R	V_L	H_L	V_R	H_R
20	10	0.6 0.1	5.0 0.4	-3.7	-1.7	-1.7	1.2		
	12	0.6 0.0	5.0 0.3	-4.0	-2.0	-1.4	1.6		
	14	0.6 0.0	5.0 0.3	-4.3	-2.3	-1.1	1.9		
	16	0.6 0.0	5.0 0.3	-4.8	-2.7	0.7	2.2		
30	10	0.9 0.2	7.5 1.5	-4.9	-2.0	-3.2	0.6		
	12	0.9 0.1	7.5 1.1	-5.1	-2.2	-3.0	1.1		
	14	0.9 0.1	7.5 0.9	-5.3	-2.5	-2.8	1.6		
	16	0.9 0.1	7.5 0.7	-5.7	-2.8	-2.5	2.0		
40	10	1.2 0.4	10.0 3.4	-6.3	-2.9	-4.5	-0.5		
	12	1.2 0.3	10.0 2.7	-6.4	-2.7	-4.4	0.3		
	14	1.2 0.3	10.0 2.2	-6.6	-2.9	-4.2	0.8		
	16	1.2 0.2	10.0 1.8	-6.9	-3.1	-4.1	1.4		
50	12	1.6 0.6	12.5 4.9	-7.8	-3.8	-5.7	-0.9		
	14	1.6 0.5	12.5 3.8	-8.0	-3.5	-5.5	0.1		
	16	1.6 0.4	12.5 3.2	-8.2	-3.5	-5.5	0.6		
	18	1.6 0.3	12.5 2.6	-8.7	-3.8	-5.5	1.3		
60	12	2.1 1.0	15.0 7.0	-9.3	-5.0	-6.9	-2.1		
	14	2.1 0.8	15.0 5.6	-9.4	-4.4	-6.8	-1.1		
	16	2.1 0.7	15.0 4.6	-9.6	-4.2	-6.8	-0.2		
	18	2.1 0.6	15.0 3.9	-10.2	-4.2	-6.9	0.5		

FRAME REACTIONS (KIPS)

MODIFYING FACTORS:
To obtain approx. reactions for other bay sizes, live loads, and/or wind loads use the following rules:

BAY SIZE: (up to 30')
Divide all reactions shown by 25 then multiply by the bay length required.

LIVE LOAD: Divide live load reactions shown by 20 then multiply by the live load required.

WIND LOAD: Multiply the wind load reactions shown by the applicable factor:
70 mph use 0.8
90 mph use 1.3
100 mph use 1.6
110 mph use 1.9
120 mph use 2.3

Figure 12.7 Typical column reactions for tapered beam–straight column system. (*Ceco Building Systems.*)

Reactions of statically determinate, but relatively rare, three-hinge gable frames can be readily computed by statics equations. Two-hinge frames, which are statically indeterminate to one degree of freedom, are much more common. Vertical reactions of a two-hinge frame are the same as for a simply supported beam. Horizontal reactions of a single-span rigid frame with nontypical roof pitch, which is not covered by the manufacturers' tables, can be estimated by standard frame formulas found in Kleinlogel[10] and elsewhere.

12.4.4   Uplift check

Wind uplift, rather than downward loading, often controls footing sizes for metal building systems. A check for uplift involves taking the tributary area of a column, multiplying it by the vertical component of wind uplift force, and comparing the result with the counteracting weight of roof and foundations. For multispan rigid frames the computed uplift load may be increased by 10 to 20 percent to account for the effects of continuity. If the dead load does not provide a required factor of safety, the foundation size or depth is increased.

12.4.5   Other scenarios

As an alternative to the methods of estimating reactions described above, it might be wise to establish a good working relationship with a few pre-engineered building manufacturers. Many such companies would be glad to run a proposed framing scheme on their computers and print out the column reactions (and even indicate some preliminary member sizes). An additional benefit of this involvement might include good advice on constructibility of the project.

Occasionally, despite best efforts of the engineers in estimating column reactions, the final numbers provided by the manufacturer will differ substantially from the assumed values. Smaller numbers are obviously acceptable, but larger column reactions may lead to a foundation redesign. If a schedule-driven "early foundation package" has already been awarded—or worse, built—a change order from the foundation contractor is sure to follow. One such experience is usually enough to open the engineer's mind to the perils of such guesswork, however educated, and to the advantages of using large safety margins in such circumstances.

12.5   Methods of Resisting Lateral Reactions

After column reactions from various loads are determined, they must be combined into loading combinations to arrive at the most critical values for both inward and outward loads. At the very least, the three common loading combinations indicated in Fig. 12.8 should be checked (substitute roof live load for snow whenever applicable). Once the worst-case combination of reactions is known, a method of resisting the forces must be chosen. There are several foundation designs capable of resisting horizontal loads, some of which are discussed below.

12.5.1   Tie rods

A single-span rigid frame under uniform gravity loading will produce two equal horizontal reactions acting in the opposite directions (Fig. 12.1a). The most direct way of "extinguishing" both is to connect the opposing frame columns with a tie rod. Tie rods are suited best for large horizontal forces (40 to 50 kips,

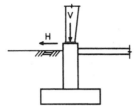

(a) CASE 1 DEAD+COLL+SNOW
(NORMALLY PRODUCES MAXIMUM OUTWARD LOAD)

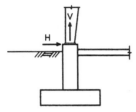

(b) CASE 2 WIND LEFT−DEAD
(NORMALLY PRODUCES MAXIMUM UPLIFT AND
INWARD LOADS ON LEFT FOUNDATION)

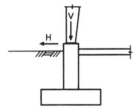

(c) CASE 3 DEAD+COLL+1/2 SNOW+WIND RIGHT
(MAY PRODUCE MAXIMUM OUTWARD LOAD)

Figure 12.8 Loading combinations that frequently control rigid-frame foundation design.

and more) and are not usually cost-effective for minor loads. The required cross-sectional area of a tie rod is determined by dividing the tension force by an allowable tensile stress in the rod, which may be taken as 60 percent of its yield strength.

Whenever the column base is recessed below the floor, a tie rod may be attached directly to the column by clevis and pin connected to the base plate (Fig. 12.9a). If the column base is at the floor level or higher, the clevis approach obviously does not work. Instead, the tie rod is embedded into the column pier (Fig. 12.9b), a solution best accomplished using common deformed reinforcing bars.

Naturally, steel bars hidden from view but exposed to moisture in the soil should be protected from corrosion. Tie rods should be galvanized or epoxy-coated and, as an extra measure of protection, encased in a plastic sheath filled

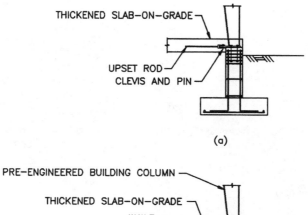

THICKENED SLAB-ON-GRADE

UPSET ROD
CLEVIS AND PIN

(a)

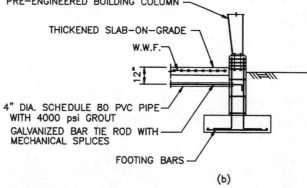

PRE-ENGINEERED BUILDING COLUMN

THICKENED SLAB-ON-GRADE

W.W.F.

12"

4" DIA. SCHEDULE 80 PVC PIPE
WITH 4000 psi GROUT
GALVANIZED BAR TIE ROD WITH
MECHANICAL SPLICES

FOOTING BARS

(b)

Figure 12.9 Foundations with tie rods. (*a*) Tie rod connected to base plate; (*b*) tie rod embedded in column pier.

with grout. Since building codes do not allow lap-spliced connections for tension-tie members, mechanical splices are required. Tie rods made of several reinforcing bars should have their splices staggered a minimum of 30 in.

Some engineers prefer to embed tie rods into thickened slabs instead of sheaths (Fig. 12.10). This solution is simpler but riskier, as the slab, thickened or not, could be removed during any future installation of underground piping and utilities. A sheath, on the other hand, has a better change of survival, since workers are trained not to cut underground cables.

While very effective in resisting gravity loads, tie rods are out of place whenever both column reactions act in the same direction, as from a wind load. Also,

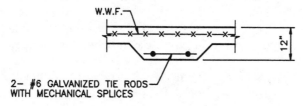

W.W.F.

12"

2- #6 GALVANIZED TIE RODS
WITH MECHANICAL SPLICES

Figure 12.10 Tie rods in thickened slab.

tie rods obviously cannot be used in buildings with deep trenches, large equipment pits, and similar discontinuities in the floor.

12.5.2 Hairpin rebars

Hairpin rebars utilize the same principle as tie rods, but instead of connecting two opposite columns by a steel rod, the hairpin system relies on the floor slab to function as the tie. Concrete itself cannot resist much tension, of course, but steel reinforcement within the slab can. The function of hairpin bars is to transfer horizontal column reactions into slab reinforcing bars or welded wire fabric, essentially by lap splicing. The required area of slab reinforcement—and that of hairpin bars—is determined by dividing the horizontal column reaction by an allowable tensile stress of the reinforcement—24 kips/in^2 for grade 60 reinforcing bars and 20 kips/in^2 for welded wire fabric. The length of hairpin bars depends on the amount of slab reinforcement to be engaged. Hairpin bars are commonly hooked around the outer anchor bolts and extended into the slab at 45° (Fig. 12.11). Hairpins should be long enough for the assumed failure plane to intersect the desired number of slab rebars or wires and to allow for their proper development.

Hairpin rebars function best when embedded in slabs containing properly spliced deformed bar reinforcement, which is common in structural slabs but not in slabs-on-grade. Use of hairpins in slabs-on-grade raises several troubling questions about the slab's ability to transfer tension.

First, slabs-on-grade are frequently unreinforced or contain only short fiberglass or steel fibers clearly unsuitable for tension transfer. Second, even if slab is reinforced, it is often with welded wire fabric, which tends to receive less attention in terms of placement and splicing practices than deformed bars. Third, slabs-on-grade usually have construction and control joints where all or

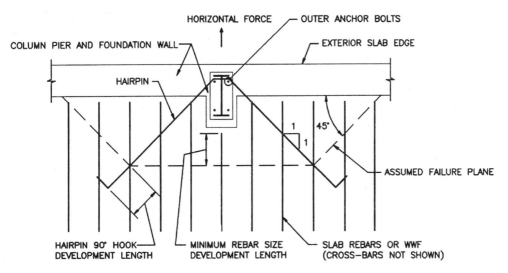

Figure 12.11 Hairpin rebars.

most of its reinforcement is interrupted—along with tension transfer. (A conscientious engineer who is determined to preserve the load path and chooses to continue the welded wire fabric through the joints may pay dearly by having a cracked floor—hardly a good alternative.) Even if the slab contains no joints at all, as is possible with shrinkage-compensating cement, there is a fourth, and biggest, problem—a possibility of the slab's being cut in the future to replace an underground cable or pipe. The building's lateral load-resisting system could be destroyed in one cut—without anybody realizing it! Such debacle is much less likely to happen to a structural slab.

Some engineers contend that a slab-on-grade need not be continuous between opposite ends of the building, because lateral loads can be transferred directly into soil by underslab friction; others respond that common use of polyethylene vapor barriers would severely reduce any friction potential. In any case, a sole reliance on friction to transfer lateral loads is unsettling. Still another unanswered question is: How does the slab, being weakened by joints, perform *in compression* against horizontal inward loads? Will the slab hold or will it buckle as a sheet of ice near a bridge pier?

As the author has recommended in the past,[4] it is better not to rely solely on hairpin rebars for tension transfer in slabs-on-grade. This suggestion goes against the persistent recommendations of many in the metal building industry, who see the inexpensive hairpins as an easy solution for a complex issue. Still, until some realistic tests of this system are made under various conditions, it is better to limit the use of hairpin bars to the slabs with continuously spliced deformed bar reinforcement.

A caveat about floor pits and trenches that was mentioned for tie rods applies equally well to hairpins.

12.5.3 Moment-resisting foundations

Column foundations can be designed to resist vertical and lateral loads in a cantilevered retaining-wall fashion, completely independent of any floor ties. The design methodology is well developed and widely available; one good source is *CRSI Design Handbook*.[5] Unlike retaining walls, however, moment-resisting foundations have sizable vertical loads applied to them, and it is often advantageous to proportion their footings with a longer toe than heel, instead of the other way around. In this configuration, the downward column reaction helps counteract the outward horizontal load (Fig. 12.12).

This solution offers many advantages: It allows for future cutting or even a total removal of the slab-on-grade without jeopardizing foundation integrity; it can accommodate any number of floor pits, trenches, and slab depressions; it can resist both inward and outward lateral loads. This method, however, normally results in foundations that are larger than those designed by any of the previous two methods, although some extra weight could be needed in any case for a wind uplift prevention. Design of moment-resisting foundations is rather time-consuming.

Horizontal column reactions can fail a moment-resisting foundation by overturning, sliding, or both. A minimum factor of safety against both overturning

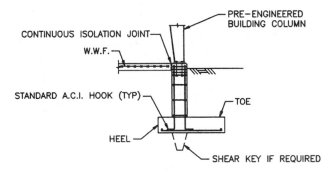

Figure 12.12 Moment-resisting foundation.

and sliding caused by transient loads should be at least 1.5; it should be increased to 2.0 if the loading is caused by gravity.

While column uplift is counteracted by the usual means such as adequate "ballast," resistance to lateral loads is achieved by a combination of soil friction and passive soil pressure. Sliding resistance can be developed by soil friction and, if needed, by a concrete shear key that protrudes below the bottom of the footing into undisturbed soil (Fig. 12.12).

Moment-resisting foundations depend on some degree of wall rotation under load to mobilize active and passive soil pressures, as most cantilevered retaining walls do. The tilt occurs because soil pressure under the footing is not uniform (Fig. 12.13). This rotational movement may endanger brittle exterior wall materials. Such rotation can be prevented by a floor slab, as discussed below, but in that case a much higher "at-rest" soil pressure coefficient must be used instead of active pressure. Basement walls, for example, are commonly designed for at-rest pressures.

How much rotation has to occur to allow the use of active and passive soil pressures instead of at-rest pressures? Common practice allows the use of active pressure coefficients for movements of walls or piers as small as one-tenth of 1 percent of their height.[6]

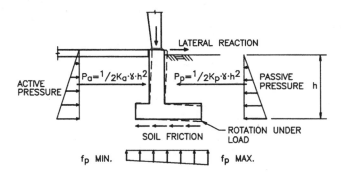

Figure 12.13 Load transfer in moment-resisting foundations.

12.5.4   A combination of methods

Some methods of resisting lateral loads can be combined to produce an economical and practical foundation design. The most common system involves a floor slab in combination with a moment-resisting foundation or with a tie rod. This system is often provided unintentionally by the engineers who routinely specify dowels extending from foundation walls into slabs (Fig. 12.14). The design function of the dowels, and of a foundation seat, is to support the slab, allowing it to span over poorly compacted areas near the walls; an unintended function is to restrain the movement of the wall and of the column pier under horizontal loads. In effect, the dowels act as well-distributed hairpins—at no additional cost.

The main advantage of this "belt-and-suspenders" system is its redundancy: While the moment-resisting foundations or the tie rods serve as primary means of resisting horizontal loads, the slab helps by acting as a limited horizontal diaphragm and by bracing the tops of the foundations. Were the slab continuity to be violated at some point, the foundation would not lose all of its lateral load-resisting capacity; at worst, the structure might suffer some minor serviceability problems, but not a complete breakdown in the load path.

Similarly, a tie-rod building foundation could rely on slab dowels to transfer into the slab the *inward* forces, against which tie rods are ineffective. In fact, design of tie-rod and moment-resisting foundations against horizontal forces acting inward and accompanied by uplift (Fig. 12.8b) would be quite difficult without any reliance on the slab. While the objections raised in Sec. 12.5.2 still stand, it seems easier to justify the slab contribution when the forces act inward rather than outward.

12.5.5   Mat foundations

Mat foundations may be specified for sites with uniform but poor soils extending to a considerable depth. A situation like this usually calls for deep foundations—piles, for example—unless the loads are so light that, when distributed over a large area, they result in acceptably small soil-bearing pressures. A light pre-engineered building with multispan rigid frames and flexible walls often qualifies. Mat foundations are considered cost-effective if the sum of the indi-

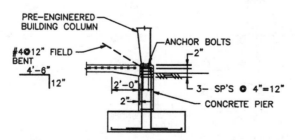

Figure 12.14 Moment-resisting foundation combined with slab dowels.

vidual column footing areas would cover more than one-half of the total building footprint. The biggest economy is achieved when the top of the mat is at the floor level, making a slab-on-grade unnecessary.

A special kind of mat foundation occasionally encountered in low-rise buildings with basements is a so-called floating mat. A floating mat is located well below the surface, usually at the basement level, and is designed not to exert any more pressure on soil than there had been prior to excavation. Even some very weak soils (excluding peat and organics, of course) might be able to support a floating mat.

A rigorous design of mat foundations considers mat stiffness, edge conditions, and variability of column loads. Most accepted methods follow the model of a slab on elastic foundation subjected to concentrated loads and bending moments. One source of information on mat design is the report of ACI Committee 436.[8] Another excellent method of mat analysis, complete with formulas and charts, can be found in Ref. 7. In addition, several computer programs dealing with mat design are available.

For buildings of modest size and orderly layout, it might be possible to simply locate the geometrical center of the mat at the centroid of all the column vertical loads and to assume soil pressure under the mat to be uniform.[2] This approach can run into problems if the soil properties are variable enough to cause significant local variations in pressure, the only safeguard against which is a generous and conservative factor of safety used in the design. In fact, good practice is to provide more reinforcement than required by analysis and to specify the same reinforcement at the top and bottom of the mat.[9] Local stiffening of the mat might be needed to span over any known isolated areas of unsuitable material.

In mat foundations, outward-acting horizontal column reactions are canceled out within the mat, while inward-acting reactions are resisted by mat-to-soil friction. Wind uplift is rarely a problem because of a large mass of concrete involved. With proper design, differential settlements between various columns are minimized. These advantages of mat foundations are counterbalanced by disadvantages which include design complexity, a need for heavy reinforcing, difficulties in accommodating deep pits and trenches, and, in northern climates, potential damage from frost heave for mats located at grade.

12.5.6 Pile foundations

Pile foundations are commonly utilized for weak and unsuitable soils underlain by good material. Piles made of wood, precast concrete, steel pipe, and steel H shapes are hammer-driven into the ground; concrete piles can also be cast in place.

Pile engineering is a complex subject well beyond the scope of this book; whenever piles are involved, geotechnical engineering assistance is a must. Our main interest is primarily in the way piles resist lateral loads and uplift.

Piles supporting building columns are normally installed in groups of three or more, regardless of the column loading. The "tripod" is stable even if the piles are not driven with perfect precision—common driving tolerances can easily result in a pile 3 in, or more away from the planned position—and the column ends up being eccentric to the centroid of the pile group. One- and two-pile groups require other methods of relieving unplanned eccentricities such as rigid pilecap bracing on each side.

A pile acts as a laterally braced column which extends through weak into suitable soil. Column loads are transferred into the soil by end bearing, skin friction, or both. Friction piles offer a superior uplift resistance as well as comparable compression and tension capacities, if tension transfer is provided for by proper splices and tension reinforcing. (Some engineers restrict the pile's tension capacity to two-thirds of that in compression.) End-bearing piles, on the other hand, can offer only their own weight against uplift forces. Uplift capacity of the piles resisting loads by a combination of skin friction and end bearing is computed as the pile's friction capacity plus the pile weight; it can also be determined by testing.

Piles resist lateral loads by bending. A simple common model assumes a cantilever-type behavior with a point of fixity some distance below the surface (Fig. 12.15a); the stiffer the pile and the soil, the smaller the depth to fixity. A more sophisticated model assumes piles to behave as beams on elastic foundations. However calculated, lateral resistance of vertical piles often ends up being 5 to 10 kips per pile. One problem with either approach is that the piles must undergo substantial displacements at their tops in order to engage the bending

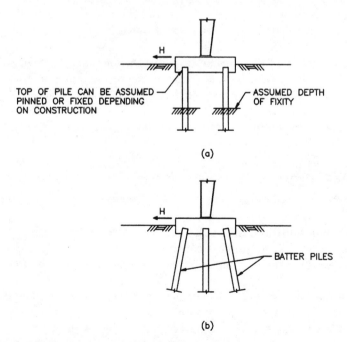

Figure 12.15 Lateral-load resistance of pile foundations. (*a*) Vertical piles for moderate lateral loads; (*b*) batter piles for major loads.

mechanism; such movements might prove damaging to buildings with brittle finishes. For major lateral loads, battered piles are appropriate (Fig. 12.15*b*).

Most earthquake codes require pilecaps to be interconnected by ties capable of transmitting tension or compression forces equal to 10 percent of the column loading. This bracing can be provided by reinforced slab-on-grade or by tie beams attached to foundation walls or grade beams, allowing for load transfer to the elements with substantial surface areas and passive-resistance capacities. Passive pressure on piles themselves is commonly neglected because of their small contact area and disturbance of soil during pile driving.

Pile installation is difficult to conduct next to existing buildings. Another limitation of pile foundations is cost: Building codes often require load tests for piles with capacities over 40 tons; each load test might cost $10,000 to $20,000, often adding a considerable enough expense to make other solutions worthwhile.

12.5.7 Drilled piers (caissons)

Drilled piers, or caissons as they are commonly called now, are somewhat of a cross between spread footings and piles. Some caissons are formed in an open excavation, but the majority are drilled by a special rig. Similar to piles, caissons are best suited for heavily loaded buildings bearing on poor soils. A typical pier has a vertical round shaft, with or without a flared bottom, and works by end bearing, like a massive plain concrete column with a deep footing. Caissons with flared bottoms bear on soil; caissons bearing on ledge are straight and are socketed into the rock.

Caissons can also rely on skin friction, as friction piles do. Such caissons are limited in their load-carrying capacities and are rarely used; a combination of end bearing and some skin friction is more common.

Drilled piers are often preferred to piles not only because of their low cost but also because they are free from such pile disadvantages as noise, vibration, and soil heave—important considerations for urban construction. Another advantage held by caissons over piles is the fact that the bearing stratum can be inspected prior to concrete placement. In contrast, pile driving is conducted blindly. In one humorous case, piles became so distorted during driving that they assumed a semicircular shape and penetrated the ground from below, damaging the cars parked nearby! All the while, the pile drivers were under an impression that everything was fine.

Caissons' substantial size—usually 2 to 6 ft in diameter plus the bell—and large weight makes them uplift-resistant. Some additional uplift resistance is provided by soil on top of the bell and by skin friction. When an uplift force exceeds the tension capacity of a plain-concrete shaft, a full-length reinforcement cage can help. Similarly, a partial-length reinforcement cage may be specified to improve the bending capacity of the shaft's upper part.

Caissons resist lateral loads as piles do—by shaft bending and by load transfer to grade beams which engage passive soil resistance (Fig. 12.16). The depth and thickness of the grade beams are determined by analysis.

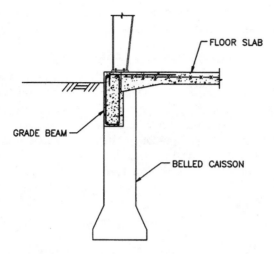

Figure 12.16 Belled caisson and grade beam foundation.

12.5.8   Downturned slab

The cheapest imaginable foundation—short of simply bearing columns on the ground—consists of slab-on-grade turned down at the perimeter (Fig. 12.17). This design dispenses with foundation walls and is commonly used by residential builders in warm frost-free climates.

This kind of foundation, whether in its basic or "heavy-duty" (slightly widened) design version, is inappropriate for anything larger than a house or a temporary structure. A common "footing," being only 12 to 18 in deep, provides little dead weight to counteract wind uplift and does not engage enough soil to take advantage of passive pressure. This design assumes some nebulous slab contribution to help resist uplift and lateral loads—with little justification. As a rigorous analysis would indicate, the thin unreinforced, or lightly reinforced with welded wire fabric, slab becomes overstressed at the point of thickness change. As Ref. 3 states, "A crack will almost surely occur in the floor slab at the point where the `grade beam' starts." Once the slab cracks, the thickened portion becomes a separate—and probably inadequate—foundation for the building column.

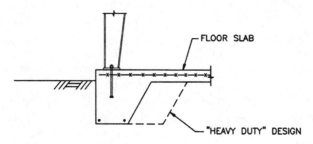

Figure 12.17 Downturned slab.

A better solution is to design the grade beam independently from the slab. If a combined foundation is desired, a mat may provide the answer.

12.6 Anchor Bolts and Base Plates

Anchor bolts transfer uplift and shear forces from building columns to foundations, acting as a crucial link between the domains of the metal building manufacturer and the engineer of record. Some of the pitfalls and misconceptions in specifying anchor bolts for metal building systems are addressed in Chap. 10; here, our focus is primarily on the technical aspects.

12.6.1 Anchor bolts: function and types

Anchor bolts serve two basic functions:

1. Positioning the column in the proper place and keeping it stable during erection. Thus the minimum number of anchor bolts is two.

2. Transferring lateral and uplift loads from the column to the foundation.

Most project specifications require that a template be used to set the bolts. Ordinarily, anchor bolt steel conforms to ASTM A 307, although sometimes high-strength bolts conforming to ASTM A 325 are specified. ASTM A 307 requires the bolt material to comply with the ASTM A 36 specification.

Anchor bolts derive their holding strength from bearing of their bent or enlarged ends against concrete. Excepting the cases of postinstalled adhesive and grouted-in anchors, any contribution of bond between the bolt shank and concrete is neglected. Depending on the configuration of their embedded ends, anchors are called L bolts, J bolts, headed bolts, and bolts with bearing plates. Of these, L and J bolts used to be most popular until it was demonstrated that L bolts were less effective in resisting slip than headed bolts.[12] Hooked anchors tend to fail by pulling out of concrete—an unsettling mode of failure. The bolts with a bearing plate at the end, once the darlings of structural engineers, also fell out of favor once it was recognized that the larger the plate, the bigger the plane of weakness it introduces into concrete.

Shear lugs—short flat bars welded to the underside of column base plates—are used to transfer shear forces to concrete without reliance on anchor bolts. To be effective, shear lugs need to be properly confined[13] and require special formwork inserts. Shear lugs can be used most effectively in resisting very large lateral loads and are not commonly encountered in pre-engineered buildings except on the West Coast.

As the evidence accumulated that properly embedded and confined headed anchor bolts were able to fully develop the tensile capacity of regular and high-strength bolts,[14] headed bolts gradually became the anchors of choice.

12.6.2 Headed anchors: design basics

A headed anchor located far away from an edge of concrete and subjected to pullout will develop a "concrete failure cone" (Fig. 12.18a). Whenever several

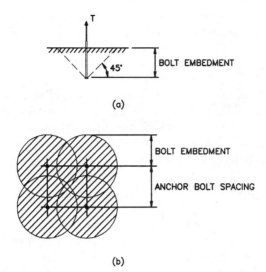

(a)

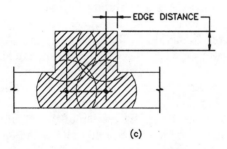

(b)

(c)

Figure 12.18 Concrete failure cones for headed anchors. (*a*) Single full concrete failure cone; (*b*) overlapping failure cones; (*c*) partial failure cones for bolts embedded in pier and walls.

closely spaced anchor bolts are used, as is the case with most practical designs, their failure cones partially overlap (Fig. 12.18*b*). The headed anchors located near an edge of concrete wall or pier will develop only a partial concrete failure cone (Fig. 12.18*c*). The theory and experimental data for all these models may be found in Refs. 13, 14, and 15.

Tension capacity of a headed anchor depends on two factors: the tensile strength of the steel shank and that of the concrete failure cone. It is easy to increase the former by simply using as many and as large bolts as required by design, but it is rather difficult to increase the latter. To enlarge the size of a concrete failure cone, longer anchor bolts are needed or the bolts must be spaced farther apart. To ensure some measure of ductility and prevent brittle failure of concrete, a standard practice is to make the concrete failure cone stronger in tension than the anchor bolt steel. With large loads, anchor bolt length can reach several feet.

A common complication involves anchors located too close to the edge of concrete. To avoid a splitting failure of concrete, adequate edge distance has to be provided. References 15 and 11 recommend a minimum edge distance equal to five bolt diameters or 4 in, whichever is larger.

As can be seen, it is much easier to determine the size, number, and spacing of anchor bolts—a task normally performed by a metal building manufacturer—than to develop those bolts in concrete, a task left to a foundation engineer.

12.6.3 Reliance on pier reinforcing

A practical approach to anchor bolt design relies on vertical pier reinforcement which of course should be adequate for the task—to transfer tensile loads into the foundation. In this model, closely spaced hoop ties transfer shear and safeguard against concrete splitting. The bolts must be reasonably close to the pier reinforcement and be long enough to allow for a proper development length of the pier rebars. The available development length is measured from an intersection of the rebars and the concrete failure cone (Fig. 12.19). The minimum required length of bolt embedment equals the bar development length plus horizontal distance between the bar and the bolt plus concrete cover above the end of the bar.

Design of anchor bolts for combined tension and shear loads is quite complex; simplified interaction curves have been developed for this purpose in Refs. 11 and 15.

12.6.4 Bolt tensioning

Should anchor bolts be tightened? Some say yes, stating that a clamping force resulting from bolt tightening helps prevent slippage at the column base. Most pre-engineered buildings, however, can tolerate some base slippage without ill effects; such slippage could be even beneficial, for the following reason.

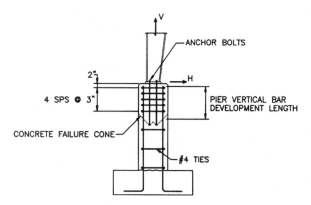

Figure 12.19 Transfer of tension and lateral loads into column pier by engaging vertical rebars.

Column base plates are normally provided with oversize holes to accommodate tolerances of anchor bolt placement, and chances are that only one bolt ends up actually bearing against an edge of the plate hole. As Fisher[16] points out, if the base is able to slip, perhaps another anchor bolt will come into bearing and thus help in load transfer; not more than two bolts in a cluster should be relied upon to transfer the base shear, however. As was already noted, pre-engineered building manufacturers and contractors commonly omit grouted leveling plates under columns. Regardless of what one thinks of this practice, it facilitates base plate slippage.

In pre-engineered buildings, anchor bolts are commonly pretensioned only to a "snug-tight" condition, which results in some modest clamping force normally neglected in design. A substantial amount of tightening is needed only for fixed-base columns which rely on clamping forces for moment transfer or for buildings where lateral drift is tightly controlled.

12.6.5 Common anchor bolt locations

For buildings of moderate span and bay sizes, most manufacturers provide two anchor bolts for endwall columns and four for sidewall (frame) columns (Fig. 12.20). Each manufacturer has a standard bolt placement distance from the edge of concrete, usually 4 or 5 in for flush girts (Star Building System uses $5\frac{1}{4}$ in) and larger edge distances such as 12 in for bypass girts. Bolt spacing is also standard (Fig. 12.21). The exact position of anchor bolts at corner columns, and of the columns themselves, also depends on a girt configuration (Fig. 12.22).

Fixed-base columns require a large number of anchor bolts, usually eight, to develop end fixity, as shown on Fig. 12.23. Beware of the designs purporting to develop end fixity with four anchor bolts spaced 3 in apart and located near the middle of a column (Fig. 12.24). This assumption is not realistic! (The ease of assuming nonrealistic conditions brings to mind an old joke about two shipwrecked engineers stranded on an island with one can of beans and no way to open it. Their solution? "Assume a can opener…")

An often-overlooked detail involves anchor bolts at wall bracing clips (Fig. 12.25). These anchors are intended to carry substantial uplift and lateral forces without the benefit of any offsetting column dead loads. It is prudent to provide a reinforced foundation pier, similar to the exterior column piers, at each clip location.

12.6.6 Minimum pier sizes

Do not skimp on column pier sizes. The pier should be large enough not only to accommodate the column base plate, perhaps of yet unknown size, but also to provide ample space for concrete placement around anchor bolts, ties, vertical bars, and formwork. Pier congestion can lead to improper concrete placement and structural failure.[3]

Whenever foundations are designed before the metal building, the contract drawings should indicate the largest acceptable sizes of column base plates.

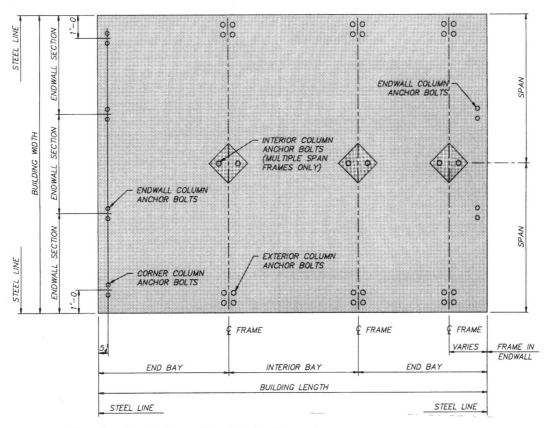

Figure 12.20 Typical anchor bolt layout. (*Star Building Systems.*)

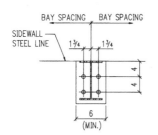

Figure 12.21 Placement of base plate and anchor bolts at exterior sidewall columns. (*Metallic Building Systems.*)

Such restrictions do not indicate paranoia: On a recent project, the manufacturer submitted shop drawings showing a 6-ft-wide column that was to bear on a 2-ft-wide pier.

Similar considerations apply to endwall columns that bear directly on foundation walls. An unfortunate situation of Fig. 12.26 could have been avoided with a proper coordination between the designer and the manufacturer.

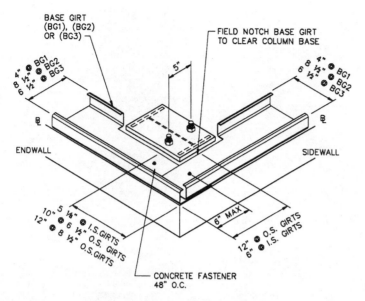

Figure 12.22 Placement of corner column with a base girt. (*Varco-Pruden Buildings.*)

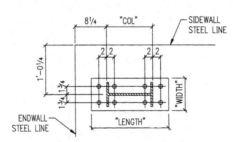

Figure 12.23 Anchor bolt layout for a fixed-base column. (*Metallic Building Systems.*)

12.7 Design of Slabs-on-Grade

A discussion of foundations for pre-engineered buildings would not be complete without at least a brief mention of some slab-on-grade design issues. A good deal of information required to design a slab-on-grade can be found in ACI *Guide for Concrete Floor and Slab Construction*[17] and PCA's *Concrete Floors on Ground.*[18] We focus on only a few critical points.

A typical slab-on-grade consists of compacted subgrade; gravel or crushed stone subbase, usually 6 to 12 in thick; vapor barrier (if needed) covered with sand layer; and the slab itself with joints, reinforcing, and finish (Fig. 12.27).

Proper subgrade preparation is critical to a slab performance, since even a most carefully constructed slab will ultimately fail if placed on a poorly compacted or unsuitable soil. It is quite common to encounter several feet of poor

Figure 12.24 Closely spaced anchor bolts are acceptable for pin-base but not for fixed-base columns.

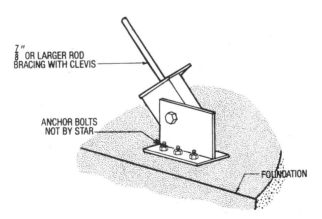

Figure 12.25 Anchor bolts at wall-bracing clip. (*Star Building Systems.*)

Figure 12.26 A column designed without regard to foundation size.

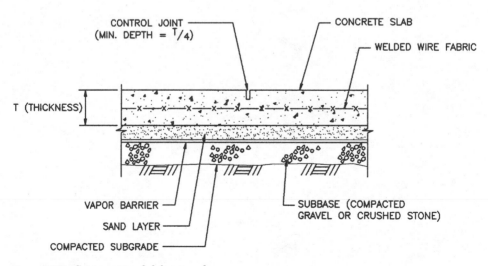

Figure 12.27 Components of slab-on-grade.

soil, or even loose fill, near the surface underlain by a better material. In such situations, geotechnical engineering guidance is indispensable: the soils reports would state whether on-site materials could be compacted or should be removed and replaced with engineered fill, in which case deep foundations in combination with structural floor slab might become an economical alternative.

A subbase helps the slab to span over any poorly compacted spots by spreading concentrated loads over large areas and by providing drainage under the slab. The thicker the subbase, the more effective it is. A subbase may not be required at all if the subgrade consists of easily compactable free-draining granular material.

The pros and cons of vapor barriers have been debated for years. The proponents cite a need to interrupt capillary action of soil and to keep floor finishes dry. The opponents point out that common 6-mil polyethylene vapor barriers may disintegrate within a few years, if not months, and note that the real-life vapor barriers are rarely effective in preventing moisture problems. The polyethylene, however, unintentionally prevents dissipation of water during curing: While surface water evaporates from the top of the slab, water stays trapped at the bottom, resulting in uneven curing rates and, frequently, slab curling. A sand layer on top of the vapor barrier is intended to mitigate the curling problem. Still, many engineers believe that a slab-on-grade located on a proper subbase does not need a vapor barrier unless covered with a moisture-sensitive finish.

Structural design of slabs-on-grade is relatively straightforward, even for concentrated loads. Selection of slab thickness, amount of shrinkage reinforcement, and design approach for concentrated loads are well covered in ACI 302[17] and in Ref. 19. Several items, however, are left to the designer's discretion:

- Support or isolate the slab at the exterior walls? While a commonly encountered continuous isolation joint at the perimeter sounds like a good idea for shrinkage control, it is important to remember that adequate compaction near walls is difficult to achieve. Supporting the slab on the wall and providing wall-to-slab dowels (Fig. 12.14) helps the slab to span over weak spots. The dowels should be field-bent, since compaction near walls is impossible if they simply stick out from the walls horizontally.

- Reinforce the slab or not? If yes, with what? There is no specific code requirement for slab reinforcement. The familiar welded wire fabric (WWF) is intended to minimize the width of shrinkage cracks but not to eliminate them. The same result could be achieved by a close spacing of control and construction joints, by using deformed reinforcement, or by specifying shrinkage compensating (type K) cement. If used, wire mesh or rebars should be properly supported by special bolsters or, better yet, by closely spaced concrete bricks made of the same type of concrete as the slab. The issue of whether to stop welded wire fabric at the slab control joints was debated in Sec. 12.5.2. A common control joint detail is shown in Fig. 12.28.

- How close to space control and construction joints? Clearly, the closer the spacing, the less anticipated shrinkage. Too close a spacing, however, will

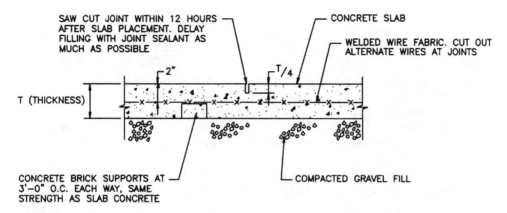

SAW CUT JOINT WITHIN 12 HOURS AFTER SLAB PLACEMENT. DELAY FILLING WITH JOINT SEALANT AS MUCH AS POSSIBLE

CONCRETE SLAB

WELDED WIRE FABRIC. CUT OUT ALTERNATE WIRES AT JOINTS

2"

T/4

T (THICKNESS)

CONCRETE BRICK SUPPORTS AT 3'-0" O.C. EACH WAY, SAME STRENGTH AS SLAB CONCRETE

COMPACTED GRAVEL FILL

Figure 12.28 Typical slab control joint.

increase a cost of the joints and their future maintenance. As a practical rule, the joints are spaced from 15 to 25 ft apart in each direction, hopefully coinciding with the column layout. Depending on a joint spacing, the required amount of shrinkage steel can be determined from the Drag Formula in ACI 302.[17] While in the past slabs were commonly placed in a checkerboard fashion, present-day practice is to use the long-strip method, whereby the slab is cast in alternate strips 20 to 25 ft wide and later divided into squares by control joints. One lesson the author has learned in this regard is not to place construction, isolation, and control joints parallel to each other: Construction and isolation joints tend to absorb the total amount of slab movement leaving control joints uncracked and thus ineffective.

- What type of construction joints to specify? Of the two basic types of construction joints—keyed and doweled—the doweled joints (Fig. 12.29) seem to result in a better load transfer between adjacent slab sections. Careful installation and alignment of the dowels is critical, as misaligned dowels may impede slab movement and induce cracking. Keyed joints are prone to spalling. Diamond-shape isolation joints placed around the columns reduce cracks in that area.

- What surface finish and tolerances to specify? Slab tolerances are a common source of confusion. Until recently, a slab was considered acceptable if a gap under the 10-ft straightedge did not exceed $\frac{1}{8}$ in. The present-day requirements are much more complex, involving two so-called F numbers: F_F, which measures flatness or waviness of the floor, and F_L, which controls slab levelness. These F numbers are used by ASTM E 1155[20] and have been adopted by ACI 117.[21] A good introduction to the subject is given by Tipping.[22]

Quality slab-on-grade construction does not come cheap. Ruddy[1] observes that slab accounts for 5 to 18 percent of a total building cost. The lower end of

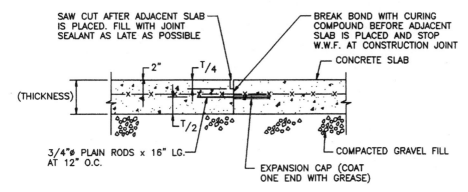

SAW CUT AFTER ADJACENT SLAB
IS PLACED. FILL WITH JOINT
SEALANT AS LATE AS POSSIBLE

BREAK BOND WITH CURING
COMPOUND BEFORE ADJACENT
SLAB IS PLACED AND STOP
W.W.F. AT CONSTRUCTION JOINT

CONCRETE SLAB

(THICKNESS)

2"

T/4

T/2

3/4"∅ PLAIN RODS x 16" LG.
AT 12" O.C.

COMPACTED GRAVEL FILL

EXPANSION CAP (COAT
ONE END WITH GREASE)

Figure 12.29 Typical slab construction joint.

the spectrum applies to office-type occupancies, while the high end relates to manufacturing facilities with special floor toppings. A cost of some high-performance metallic or emery toppings may greatly exceed the cost of the slab itself.

References

1. John L. Ruddy, "Evaluation of Structural Concepts for Buildings: Low-Rise Buildings," BSCE/ASCE Structural Group Lecture Series at MIT, 1985.
2. The BOCA National Building Code, 12th ed., Building Officials and Code Administrators International, Inc., Country Club Hills, Ill., 1993.
3. *Metal Building Systems,* Building Systems Institute, Inc., 2d ed., Cleveland, Ohio, 1990.
4. A. Newman, "Engineering Pre-engineered Buildings," *Civil Engineering,* September 1992, p. 58.
5. *CRSI Design Handbook,* Concrete Reinforcing Steel Institute, Schaumburg, Ill., 1996.
6. Roy E. Hunt, *Geotechnical Engineering Techniques and Practices,* p. 548, McGraw-Hill, New York, 1986.
7. NAVFAC DM-7.2, Foundations and Earth Structures, Department of the Navy, Naval Facilities Engineering Command, Alexandria, Va.
8. ACI Committee 436, Suggested Design Procedures for Combined Footings and Mats, ACI 336.2R-66, American Concrete Institute, Detroit, Mich., 1966.
9. Edwin H. Gaylord, Jr., and Charles N. Gaylord (eds.), *Structural Engineering Handbook,* 2d ed., p. 5-66, McGraw-Hill, New York, 1979.
10. A. Kleinlogel, *Rigid Frame Formulas,* Frederick Ungar Publishing Co., New York, 1964.
11. Mario N. Scacco, "Design Aid: Anchor Bolt Interaction of Shear and Tension Loads," *AISC Engineering Journal,* fourth quarter, 1992.
12. D. W. Lee and J. E. Breen, "Factors Affecting Anchor Bolt Development," *Research Report 88-1F,* Project 3-5-65-88, Cooperative Highway Research Program with Texas Highway Department and U.S. Bureau of Public Road, Center for Highway Research, University of Texas, Austin, August 1966.
13. R. W. Cannon, D. A. Godfrey, and F. L. Moreadith, "Guide to the Design of Anchor Bolts and Other Steel Embedments," *Concrete International,* July 1981.
14. ACI Appendix B—Steel Embedments (1978C), ACI 349-76 Supplement, 1979.
15. John G. Shipp and Edward R. Haninger, "Design of Headed Anchor Bolts," *AISC Engineering Journal,* second quarter, 1983.

16. James M. Fisher, "Structural Details in Industrial Buildings," *AISC Engineering Journal,* third quarter, 1981.
17. ACI Committee 302, *Guide for Concrete Floor and Slab Construction,* ACI 302.1R-89, American Concrete Institute, Detroit, Mich., 1989.
18. *Concrete Floors on Ground,* 2d ed., Portland Cement Association, Skokie, Ill., 1983.
19. Robert A. Packard, *Slab Thickness Design for Industrial Concrete Floors on Grade,* Portland Cement Association, Skokie, Ill., 1976.
20. Standard Test Method for Determining Floor Flatness and Levelness Using the F-Number System, ASTM E 1155-87, ASTM, Philadelphia, Pa., 1987.
21. ACI Committee 117, Standard Specifications and Commentary for Tolerances for Concrete Construction and Materials, ACI 117-90, ACI, Detroit, Mich., 1990.
22. Eldon Tipping, "Bidding and Building to F-number Floor Specs," *Concrete Construction,* January 1992.

13

Some Current
Design Trends

Metal building systems represent one of the youngest and most dynamic sectors of the construction industry. New materials and design applications of metal buildings continue to emerge, expanding the architect's palette of choices. This chapter examines some latest trends in specifying metal buildings, a few winning design solutions, and factors that further increase competitiveness of pre-engineered buildings.

13.1 Facade Systems: Mansards and Canopies

As metal building systems expand their acceptance into commercial, institutional, and community environment, the old bland, utilitarian look of metal-sheathed gable buildings gives way to more interesting and diverse design solutions. Visual interest can be added not only by the wall materials discussed in Chap. 7 but also by various facade treatments ranging from basic canopies to sophisticated fascia panels.

13.1.1 Canopies

A functional and aesthetically pleasing canopy is perhaps the most common facade treatment. The simplest way to build a canopy is to provide a cantilevered extension of the primary frame at the eave level and to continue the roof framing onto the canopy (Fig. 13.1). The eave-line canopy is most appropriate for continuous and wide—up to 10 ft—canopy coverage that extends the full length of the building. For this solution to work, the building exterior must be visually compatible with exposed cantilevered rafter framing, which stays in full view even when soffit panels cover the underside of the canopy's roofing.

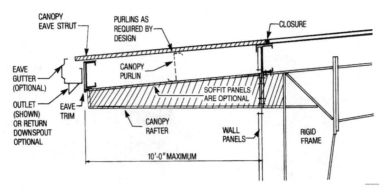

Figure 13.1 Eave-line canopy with exposed rafter. (*Ceco Building Systems.*)

A more refined option is a flush-framing canopy, where all the framing is hidden from view (Fig. 13.2a). At sidewalls, parallel to roof purlins, this canopy is supported by a special cantilevered canopy rafter at each column (Fig. 13.2b), while at the endwalls roof purlins can simply be extended past the wall line (Fig. 13.2c). A flush-framing canopy, while sleek in design, is limited to the maximum width of 3 to 5 ft, or about one-half to one-third that of the eave-line canopy.

For an even more sophisticated treatment, a bullnose canopy can be specified (Fig. 13.3). A bullnose canopy looks and functions best when located some distance below the roof level and attached to a contrasting wall material. Rather than being supported by a roof extension, it is carried by closely spaced frames attached to the building wall, with hat or channel steel members in between. Obviously, the supporting wall must be strong enough and is best made of masonry or concrete, although it is possible to reinforce a metal-panel structure for this purpose. The panels for the bullnose should be curved, as discussed in the next section. The spacing of support frames is controlled by the flexural capacity of hat and channel sections. Structural properties of some representative hat and channel sections produced by MBCI are given in Fig. 13.4. The gage of MBCI sections is rather thin; similar sections made of thicker metal are available from other manufacturers.

13.1.2 Fascias and mansards

Fascia and mansard panels look so natural on pre-engineered buildings that one might forget to specify and detail them separately. A vertical fascia and parapet panel, the most common kind, is commonly supported by its own moment-resisting frame rigidly attached to the primary building framing. The primary frame has to be designed for an additional loading from the fascia. Some common details of this solution are shown in Fig. 13.5.

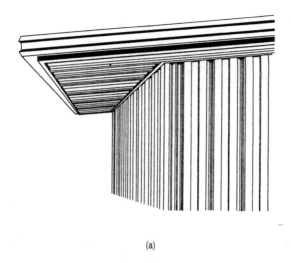

(a)

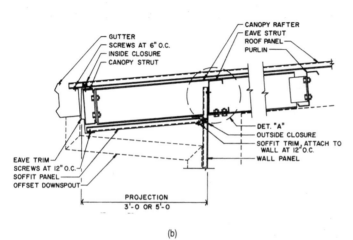

(b)

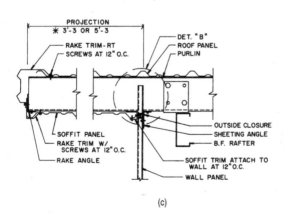

(c)

Figure 13.2 Flush-framing canopy with soffit panel. (*a*) Overall appearance; (*b*) sidewall section; (*c*) endwall section. (*Metallic Building Systems.*)

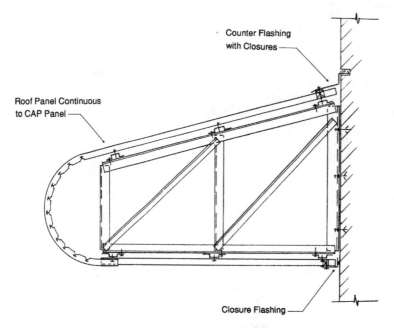

Figure 13.3 Bullnose canopy with soffit. (*Smith Steelite.*)

Mansard-style fascia panels require only some modifications of the vertical panel details (Fig. 13.6), but a completely different type of framing is needed for a so-called double-curve eyebrow panel (Fig. 13.7).

A curved fascia in combination with contrasting wall panels has helped transform what could have been a basic pre-engineered building into a modern-looking office (Fig. 13.8).

For an even more adventurous design, a triple-step curved fascia (Fig. 13.9) can add spice to almost any building.

13.2 Curved Panels

As the illustrations demonstrate, well-proportioned curved panels can make an excellent visual impression. These panels have become extremely popular since 1985 when Curveline, Inc., of Ontario, California, brought into the United States the crimp-curving method of panel bending, first developed and patented in the Netherlands. During crimp-curving, metal panels are being incrementally pushed and pulled into rounded forms in a computer-controlled process.[1] Today, curved panels are specified not only as fascias, mansards, and canopies but also as walkway roofs, decorative column covers, equipment screens, roof transitions, and even as curved formwork for concrete.

Curveline, Inc., remains the industry leader, offering the widest range of products available for curving. The company can form panels 2 to 30 ft long, with a maximum width of 5 ft. Panel depth can range from $\frac{3}{4}$ to 6 in, the thick-

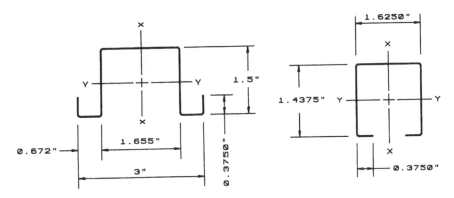

HAT SECTION (HS-1)   CHANNEL SECTION (CS-1)

ALLOWABLE AXIAL LOADS (POUNDS)					
UNBRACED LENGTH (FT)					
	2'	3'	4'	5'	6'
HAT (HS-1)	2400	1590	950	650	480
CHANNEL (CS-1)	3170	2240	1400	1000	780

EFFECTIVE SECTION PROPERTIES								
					TOP IN COMP.		BOTTOM IN COMP.	
PROD. NAME	GAUGE	WT. (PLF)	FY (KSI)	FB (KSI)	IY (IN4)	SY (IN3)	IY (IN4)	SY (IN3)
HAT (HS-1)	22	0.673	33	20	0.064	0.081	0.071	0.094
CHANNEL (CS-1)	18	.798	33	20	0.067	0.079	0.067	0.079

ALLOWABLE UNIFORM LOADS (PLF)										
	STRESS CONTROL					DEFLECTION CONTROL (L/180)				
	3'	4'	5'	6'	7'	3'	4'	5'	6'	7'
HAT (HS-1)	173	97	62	43	32	173	97	62	43	32

NOTES:
1. The effective section modulus is used for allowable loads on a stress basis. The effective moment of inertia is used for allowable loads on a deflection basis.
2. Allowable uniform loads shown for hat section are for bottom in compression. Decrease loads by 15% when top is in compression.
3. Section Properties and Allowable Stresses have been calculated in accordance with the 1986 AISI Specification for the Design of Cold-Formed Steel Structural Members.
4. Steel conforms to ASTM A446-85 Grade A, G-90 Galvanized.
5. The allowable uniform loads are for bending about the Y-Y axis.
6. The allowable loads shown above may be increased by 33% for wind loading.
7. Values shown as allowable loads are based on hats covering three or more equal spans. Multiply the allowable stress values by 0.8 for one and two span conditions. Multiply the allowable deflection values by 0.5 for simple span values.

Figure 13.4 Properties of hat and channel sections. (*MBCI.*)

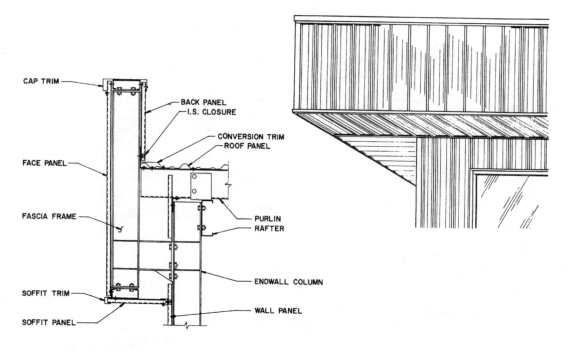

TYPICAL FASCIA SECTION
THRU ENDWALL

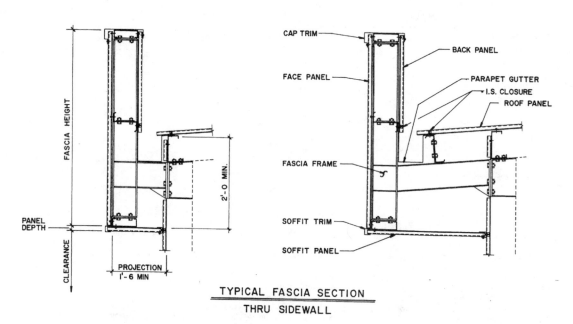

TYPICAL FASCIA SECTION
THRU SIDEWALL

Figure 13.5 Vertical fascias. (*Metallic Building Systems.*)

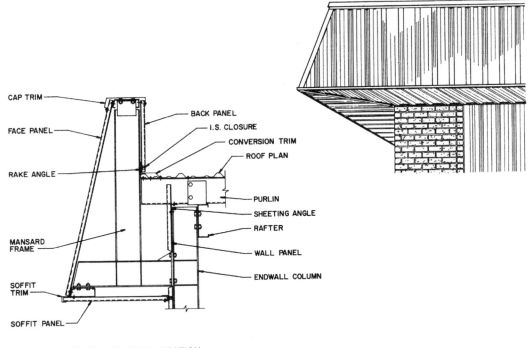

CAP TRIM

FACE PANEL

RAKE ANGLE

MANSARD
FRAME

SOFFIT
TRIM

SOFFIT PANEL

BACK PANEL

I.S. CLOSURE

CONVERSION TRIM

ROOF PLAN

PURLIN

SHEETING ANGLE

RAFTER

WALL PANEL

ENDWALL COLUMN

TYPICAL MANSARD SECTION
THRU ENDWALL

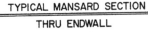

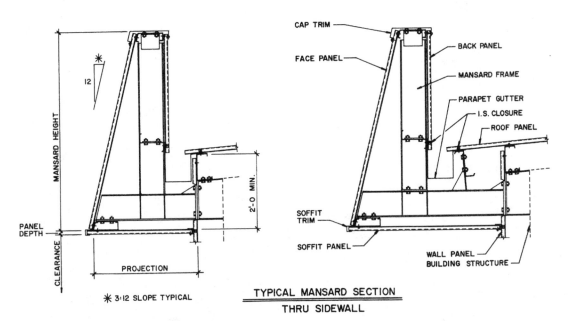

12

MANSARD HEIGHT

PANEL
DEPTH

CLEARANCE

2'-0 MIN.

PROJECTION

✳ 3:12 SLOPE TYPICAL

CAP TRIM

FACE PANEL

SOFFIT
TRIM

SOFFIT PANEL

BACK PANEL

MANSARD FRAME

PARAPET GUTTER

I.S. CLOSURE

ROOF PANEL

WALL PANEL
BUILDING STRUCTURE

TYPICAL MANSARD SECTION
THRU SIDEWALL

Figure 13.6   Mansard panels. (*Metallic Building Systems.*)

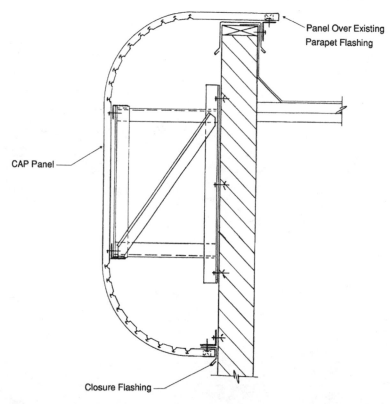

Figure 13.7 Double-curve eyebrow fascia. (*Smith Steelite.*)

Figure 13.8 Curved fascia dresses up an office building. (*Curveline, Inc.*)

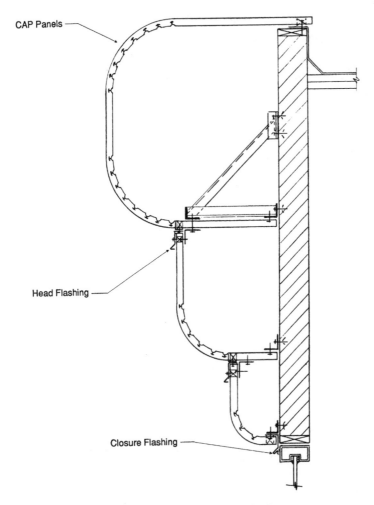

CAP Panels

Head Flashing

Closure Flashing

Figure 13.9 Triple-step curved fascia. (*Smith Steelite.*)

ness from 0.016 to 0.052 in (29 to 18 gage).[2] In addition to a single-curve configuration, Curveline can produce complex and breathtaking multiple and S curves (Fig. 13.10). Exposed fastener panels are most suitable for curving, although some concealed fastener products can also be curved.

The process of crimp-curving approximates true curvature by means of many short chords, a look that some dislike. Where a smoother line is desired, the "chorded" look of crimp-curved panel ribs can be avoided if the panels are turned with their flat parts, rather than the ribs, facing the outside.

Each manufacturer of curved panels has its own standards for the minimum bending radii. For common panel steel conforming to ASTM A 446 grades C, CQ, and D, Curveline requires a minimum *internal* radius of 10.5 in for panels less than 1 in deep, adding to that a 1-ft distance for every inch of the panel

Figure 13.10 Complex curves grace the U.S. Space Camp in Huntsville, Ala. (*Curveline, Inc.*)

depth. For example, a 3-in-deep panel will need at least a 3-ft internal radius. The panels made of thin materials, especially high-strength steel, normally require a bigger bending radius.

The steel composition most favorable for curving, according to Curveline, is ASTM A 446 grade D carbon steel G-90 with a tensile strength of 50,000 lb/in².[3] Panels made of galvanized steel, aluminum, and stainless steel may be curved.

Curved panels are structurally more efficient than the straight ones and can often be made of thinner metal, affording some material savings. For a continuous support, curved girts and purlins conforming to the panel's outline can be produced at the same source.

Wherever curves follow straight panel runs, as in building corners, a separate curved piece may or may not be required, depending on the supplier. According to Curveline, Inc., a separate curved connector piece is usually not needed, and the curve can be built into an end of the straight panel. For both aesthetic and functional reasons, an extra joint is just as well avoided. A notable exception is the mitered corner (Fig. 13.11), which turns out better if shop-fabricated separately.

When factory curving is not practical, field curving is possible. Some companies, such as Berridge Manufacturing of Houston, Tex., offer both roll-forming and curving of the panels on-site. Alternatively, a rounded corner may be obtained without crimp-curving if the panel is bent *parallel* to the ribs, a relatively easy operation.

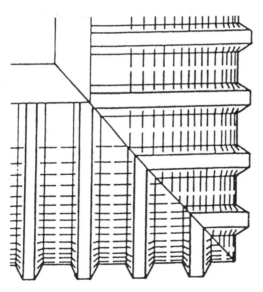

Figure 13.11 Mitered corner. (*Smith Steelite.*)

While curved panels are visually attractive, the panel finish might be severely compromised during curving. Some fabricators that had gotten into the curving business during the 1980s could not overcome the technical difficulties and survive. To this date, some major manufacturers, such as Butler, not only do not offer curved panels themselves but also advise against curving their products by others. For the same reason, many architects avoid specifying curved panels in corrosive climates.

Before specifying crimp-curved panels, designers should contact some local fabricators engaged in this business to inquire about available panel profiles, finishes, bending radii, and product warranties. It is instructive to view some of their past projects, preferably at least several years old, to look for signs of corrosion. During inspection, one should look for any incomplete bending and dimpling of panels, for proper curving of all the trim pieces, and for acceptability of tolerances.

In addition to the firms mentioned above, some other companies involved in production of curved panels include ATAS Aluminum Corp. of Allentown, Pa.; Floline Architectural Systems of St. Louis, Mo.; Petersen Aluminum Corp. of Elk Grove Village, Ill.; Smith Steelite of Moon Township, Pa.; and BHP Steel Building Products USA, Inc., of West Sacramento, Calif.

13.3   Steel-Framed Houses

Always looking for new opportunities, the metal building industry has recently began supplying pre-engineered framing for residential construction—at a spectacular pace. According to AISI, 13,000 steel-framed houses were built in this country in 1993, compared to only 500 built in the two prior years. In 1994, 40,000 steel homes were expected to be built in North America.[4]

Historically, steel has been prohibitively expensive for residential applications, but with wood prices escalating sharply in the early 1990s, steel suddenly became cost-competitive. Apart from the price trends, which might prove transient, steel has some real advantages over wood: It is noncombustible, dimensionally stable, does not warp or rot, and is unaffected by termites. The major disadvantage of steel is its poor thermal properties.

There are three methods of building the house of steel. The first is to simply substitute steel studs and joists for wood, essentially following traditional construction of studs and joists spaced 16 or 24 in on centers. Everything else—roofing, siding, doors, windows—stays the same as in a wood-framed house. This method allows for an easy framing conversion to steel in both standard and custom-designed houses; it is undoubtedly used in most steel-framed houses.

The second method of framing is panelized construction: The structure is built from preassembled steel-stud wall panels and roof trusses. Both studs and trusses are spaced 32 to 68 in on centers, with hat-section subgirts and subpurlins similar to those in Fig. 13.4 spanning in between. Despite the claims of efficiency, this system is rather complex structurally and may require more bracing and anchorage than others.[5] It is wholly unfamiliar to both traditional house builders and pre-engineered building erectors and can introduce a lot of confusion at the job site.

The third method is to construct a true, if small, pre-engineered building, complete with the usual main frames, girts, purlins, rod bracing, and metal roofing. The bay spacing of such pre-engineered structures ranges between 6 and 10 ft. A rigid-frame gable building can be quite appropriate for a large contemporary residence with an open floor plan and cathedral ceilings. The system, however, is unfamiliar to most residential designers and builders. It also differs so much from the traditional construction that to present it as a framing substitution for already-designed projects is difficult. Furthermore, the commonly available commercial-style components of metal building systems such as doors, windows, siding, and roofing might not be in line with the owner's expectations. Since few homeowners dream of a house sheathed in metal siding, either a brick or a wood exterior is desirable. These traditional finishes normally need to be backed by $\frac{3}{4}$-in-thick plywood sheathing, which can span about 4 ft between the metal supports.

The problem of thermal bridging can be solved by applying rigid insulation to the outside of steel studs. Polyisocyanurate insulation offers excellent insulation value (see Chap. 8) and is available in the form of insulating sheathing. Rigid insulation can also be incorporated in the EIFS exteriors, as discussed in Chap. 7. A typical high-quality exterior wall may consist of 6-in steel studs covered with $\frac{3}{4}$-in plywood and 1-in insulating board coated with an EIFS finish. The studs may be filled with 6-in fiberglass insulation covered with a heavy-gage plastic vapor retarder and $\frac{5}{8}$-in drywall.[6]

13.4 Computerization of the Industry

In one word, what has helped to transfer "pre-engineered" designs of old into the modern metal building systems? Computers! Heavy reliance on these machines has allowed metal building manufacturers to discard the old menu of a few predesigned building configurations in favor of unlimited design choices. Indeed, nearly every metal building constructed today is custom-designed for a specific project.

While the architects celebrate the new design freedom, the owners are pleased with the speedy price quotes. Advanced software allows the quotes to be produced in as few as 5 min, a task that used to take days. The builders, in turn, are amazed by the fast delivery schedules: It is not uncommon to compress the delivery time to 5 weeks, a task that only recently required at least 3 months.

The major manufacturers race to develop the most comprehensive and user-friendly software systems that, based on the input data, produce a price quote, design calculations, shop drawings, and even presentation materials. Investment in such premium systems gives the biggest industry players a clear advantage over the small shops. Not surprisingly, the aptly named Butler Advantage System won first place among hundreds of entrants in the manufacturing and distribution category in the Windows World Competition in April

1995, as well as other awards. (Windows World is sponsored by the Microsoft Corp., and by *Fortune* and *Computer World* magazines.) Reportedly, this software is already used by over 1000 Butler builders.

Varco-Pruden Buildings has developed its own Command Computer System which features excellent graphic capabilities and allows order placement 24 hours a day. Some examples of the program's output are reproduced in Chap. 9.

Smaller manufacturers who cannot afford major investment in software development can purchase one of the many off-the-shelf computer programs such as one offered by Loseke Technologies, Inc., of Plano, Tex., or Metal Building Software, Inc., of Fargo, N.Dak.

Computerization allows manufacturers to centralize job costs, accounting systems, inventory control—and to produce more accurate quotes. Furthermore, it permits farsighted manufacturers willing to make the investment to compete in the world of many building codes, languages, and measurement units and to react to market changes faster.

Technological advances are likely to affect the construction side of the industry as well. The Standard Commodity Accounting and Tracking System (SCATS) allows bar-code tracking of every piece of steel on the project. Some Louisiana fabricators of structural steel are already using SCATS to keep track of materials on fast-track projects. With their fabled speed of construction, erectors of metal building systems cannot be far behind.

13.5 Multistory Metal Building Systems

While the vast majority of metal building systems are single-story, the multistory market presents a major growth opportunity. With rising land costs and little available space in built-up areas, multistory pre-engineered buildings are a logical answer to those owners that need more than one level of usable space but still want to capitalize on the advantages offered by the metal-building industry. By some estimates, pre-engineered framing can save about 15 percent of the structural cost in a four-story office building.

Metal building systems utilizing bar joists work well with conventional metal deck and concrete fill floor structure. The systems based on light-gage C and Z sections may face some acceptance problems relating to their fire-rating, deflection, and vibration properties. The designers who wish to specify multistory pre-engineered buildings are wise to inquire first whether local dealers have any experience in this type of construction.

References

1. Marc S. Harriman, "Full Metal Jacket," *Architecture,* March 1992.
2. Steve Moses, "Curved Panels Expand Design Possibilities for Metal Products," *Metal Architecture,* January 1994.
3. *Custom-Curving of Profiled Metal Panels,* Curveline, Inc., Ontario, Calif.
4. Richard Haws, "Steel Framing: The Right Choice in Residential Construction," *Architectural Specifier,* September/October 1994.
5. Juan Tondelli, "Steel Homes Slowly but Steadily Penetrating Residential Market," *Metal Architecture,* December 1993.
6. Sam Milnark, "Steel Makes a Statement on Catawba Island," *Metal Home Digest,* Summer 1994.

14

Reroofing and Renovations of Metal Building Systems

14.1 Introduction

Building reuse and rehabilitation grow ever so popular, and the volume of renovation work now rivals that of new construction. The principles of metal building systems such as cost efficiency and single-source responsibility are applicable not only to new construction but also to building renovation. As the first generations of pre-engineered buildings approach the ends of their useful lives, they can be rehabilitated or partially replaced with new metal building framing. Some components of metal building systems have their place in renovation of buildings constructed conventionally.

A common problem facing the owners of low-rise buildings is leaky roofs. Whenever building renovation is mentioned, reroofing or roof retrofit immediately comes to mind. Accordingly, this chapter focuses first on roof retrofit and then continues to modification and replacement of the existing exterior skin. It concludes with a discussion of some difficult issues surrounding adaptation of existing pre-engineered buildings to new conditions of service.

14.2 Roof Retrofit with Metal Building Systems

14.2.1 The troublesome roofs

No building problem seems to cause more aggravation than a leaking roof. Dealing with the occupants pointing to a leak—a problem that is easy to spot but hard to fix—is among the most frustrating duties of the building owner. After a few rounds of thankless repairs, a total reroofing seems to be the only solution left. Not surprisingly, reroofing work makes up two-thirds of all roofing projects in this country.

Why do roofs fail so fast?

The roof is the hardest-working part of the building, protecting it from the blazing sun in the summer, snow in the winter, and rain year-round. Every windstorm attacks the roof by first trying to literally lift it off the building and then slammming it back. Ultraviolet radiation shortens the lives of unprotected single-ply membranes, slowly robbing them of elasticity and strength; it causes damage to other roofing types as well.

Most conventional low-rise buildings have flat or nearly flat roofs, a solution that was popular until only a few years ago. A minimum pitch of $\frac{1}{8}$:12 was considered adequate for drainage; it probably was—in theory. In real life, however, building foundations settle a bit unevenly, roof beam elevations vary slightly owing to fabrication and installation tolerances, and beams and decking deflect under load. Also, point loading from HVAC equipment, suspended piping, lights, and such causes some roof structural members to deflect more than others. All these factors may result in the actual roof profile being far from the assumed—with some areas of the roof having no slope at all—and lead to an accumulation of ponded water.

Roofing not designed to be submerged for prolonged periods of time, like some built-up asphalt-based products, may slowly start to disintegrate and eventually leak. Other factors leading to roofing failures include local damage from careless foot traffic or equipment maintenance, clogged roof drains—again resulting in ponding—and poorly protected roof penetrations.

The deterioration often starts at the flashing locations, expansion joints, and improperly fastened gravel guards. Regardless of the origin, roof leakage may result in saturation and ruining of fiberglass insulation, staining of finishes, and corrosion of roof decking. If not addressed promptly, damage can progress to the point of making the roof unrepairable, leaving tear-down and replacement as the only solution.

14.2.2   Reroofing options

One popular choice for a reroofing material is the single-ply membrane, especially of the lightweight fully adhered or mechanically fastened varieties. This material is not without drawbacks. To cover an old tar-and-gravel roof with a single-ply membrane, all gravel usually has to be removed. This messy operation, if not handled properly, may result in a badly gouged roof that needs to be overlaid with protection boards or even torn off completely. The roof slope, if previously inadequate, can be changed only with expensive tapered insulation. And, as already mentioned, in sunny locales solar radiation causes the unprotected membranes to fail rather quickly, ruling this system out.

Another increasingly popular option is reroofing with metal. This solution offers numerous advantages. As discussed in Chap. 6, metal roofing comes in a variety of finishes including polyvinylidene-based coatings that are extremely durable and ultraviolet-resistant. Standing-seam roofing with sliding clips can easily handle thermal expansion and contraction. With minimum slopes as low as $\frac{1}{4}$:12 for structural panels and 3:12 for architectural roofing, water can drain faster than in nearly flat roofs. The required slope can be accomplished by erecting a light-gage framework on the old roof.

The total weight of metal roofing and the new framework usually does not exceed 2 to 4 lb/ft^2, placing this system among the lightest available. Quite often this small additional load can be safely taken by the existing roof structure, while the heavier systems such as built-up roofing would overstress it. If the existing roof structure has no excess capacity at all, a system of beams or trusses spanning between the new stub columns on top of the existing building columns may be erected.

Some experienced architects believe that properly designed and constructed metal roofs will last 40 years.[1] While metal roofs may initially cost more than the competing systems, their exceptional durability combined with ease of maintenance often make metal a winner in life-cycle cost comparisons. Being recyclable, metal roofing wins on an environmental scorecard, too.

Metal roofing is very useful in circumstances requiring a replacement of the existing slate or tile roof supported by an aging, and undersized, roof structure. Such roofs can benefit from a metal Bermuda roofing with the slate, shake, or tile profile, or from a PVDF-covered metal shingle product designed to closely resemble the traditional materials.

14.2.3   Tear off or re-cover?

A decision on preserving the existing roofing versus removing it often hinges on a level of moisture in the existing roof system. That the previous leaks caused *some* water to get into the roof insulation is clear; the question is only, how much water. In addition to the already mentioned problems of a diminished insulation performance and corrosion of the existing decking, entrapped moisture may cause offensive smell and growth of mold and mildew, resulting in serious indoor-air quality problems. Also, retrofit fasteners which penetrate the moist space may eventually corrode and undermine the integrity of a newly constructed roof system.

The degree of water saturation can be determined by a moisture survey performed by the design professional or by a specialized consulting firm. The latter may give more reliable results, because specialized firms are likely to employ such advanced testing methods as infrared thermography, capacitance, and nuclear back scatter.[4] The survey produces a rough outline of the areas containing wet insulation and determines the degree of saturation, which can be confirmed by taking a few insulation cores. Only then can the magnitude of the problem be rationally assessed.

A common solution to the problem of entrapped moisture is to install several "breather" vents and hope that the moisture escapes prior to the final enclosure. This approach works only for very modest moisture levels, however. If the existing roofing and insulation are totally saturated with water from frequent leaks, venting through a few holes might not be adequate, especially when structural decking or a vapor barrier restricts the downward moisture migration.

Studies indicate that it would take 30 to 100 years for the insulation to dry out in such circumstances, even with the vents installed.[4] A better course of action is to remove the roofing and the wet insulation. Tobiasson[5] notes: "In most cases, wet insulation should be viewed as a cancer that should be removed before reroofing." He points out that every inch of saturated insulation can add up to 5 lb/ft² to the dead load—a significant amount. A moisture survey that indicates numerous areas of wet insulation is to be taken seriously: a complete tear-off might be the only prudent option left.

A survey of the roof structural decking is also extremely helpful. The persistent leaks might have led to a widespread decking corrosion beyond repair. Similarly, a presence of some potentially corrosive roof components could have degraded the deck. Phenolic-foam roof insulation produced in the late 1980s until 1992 is a case in point. It has been reported that this type of insulation, when wet or damaged, can contribute to corrosion of metal deck, sometimes to the point of making it unsafe to walk on. A replacement of roof decking is a serious matter since it opens the inside of the building to the elements and affects the operations.

There are arguments against a complete roof tear-off, the most obvious being high cost. The bill for a disposal of the removed materials, perhaps containing hazardous waste such as asbestos roofing felts, could also be significant.

The pros and cons of the two approaches require careful consideration. Curiously, the 2 billion ft² of reroofing work performed annually in this country are evenly divided between the tear-off and re-covering.[4] Of course, when the local building code prohibits the addition of another roofing layer, the decision on tear-off versus reroof comes easily.

14.2.4 The issue of design responsibility

Who determines whether an existing roof is structurally capable of carrying the extra load, however modest, from reroofing? As we discussed earlier, the manufacturers of metal building systems are unwilling to get involved beyond the design of metal components; they normally disclaim any responsibility for evaluation of the existing roof structure and its capacity to support additional loads. Hire a local structural engineer to analyze the existing roof structure, they suggest.

The problem is, the engineer can readily check only the average *uniform load* capacity of the roof, a computation usable only if the new roofing is simply laid on top of the existing. Any change in slope, however, requires a new ("retrofit") framework supported by some discrete columns that will transmit concentrated, not uniform, loads to the existing structure. At the evaluation stage, the engineer often has no way of knowing which manufacturer will be selected to do the work and what the column spacing will be.

If the manufacturer is already on-board, so much the better. If not, engineers can select one of the popular systems, such as the one described below, base their analysis on that system, and require the contractor to adhere to it. Or, they can assume the worst-case scenario and use a rather expensive approach of requiring the new trusswork to bear only on the existing columns, bypassing the existing roof framing altogether. Alternatively, they might require that the new supports be spaced so closely as to approximate a uniform load—not the most cost-effective solution, either.

To make matters even more complicated, any significant change in the roof slope will increase the vertical projected area of the roof and result in a larger design wind loading on the building. Now, the whole building's lateral load-resisting system may have to be rechecked, involving the engineer even deeper into the project.

To be ready for such complications and be able to make educated design decisions while preparing the construction documents, specifying design professionals are wise to learn about the available types of support framing for metal reroofing.

14.2.5 Structural framework for slope changes

The proposed reroofing framework has to provide the same level of strength and rigidity as any other metal roof structure. In practice it means that the spacing of the new ("retrofit") purlins is probably limited to 5 ft or so (Fig. 14.1). A closer, perhaps half as wide, spacing is needed at the "salient corner" areas near the eaves, rake, and ridge (for certain roof slopes), to resist the increased wind loading there. Similarly, purlin spacing is reduced in the areas of a potential snow drift accumulation.

Whenever the existing roof structure stays in place, the new framework resists not only the wind and snow loads on the new roof but also the wind loads on the new gabled endwalls. Thus two kinds of bracing are required for stability: vertical, between the framing uprights, and horizontal, in the plane of the new roof, to act as a diaphragm. The diaphragm action can be provided by rod or angle bracing, by steel deck, or by certain types of through-fastened roofing (but not standing-seam roofing).

Another important issue to consider is lateral bracing of the new purlins. When closely spaced and cross-braced, the framework verticals provide the necessary bracing, but this may not be the case when the supports are far apart. As was pointed out in Chap. 5, the manufacturers' design practices for lateral support of purlins vary widely and range from conservative to ignorant. To assure a uniformity of design assumptions among the bidders, the owner's requirements relating to the acceptable diaphragm construction and purlin bracing methods should be spelled out in the contract documents.

To avoid a blow-off of the new metal roofing which in a sense acts as a giant sail erected on top of the existing building, proper anchorage into the existing structure is critical. The anchorage details should be designed by the metal system manufacturer and carefully checked by the engineer of record. The details should be custom designed for the actual existing roof structure, instead of showing the new fasteners terminating in a mass of concrete, an easy but useless "solution" submitted all too often.

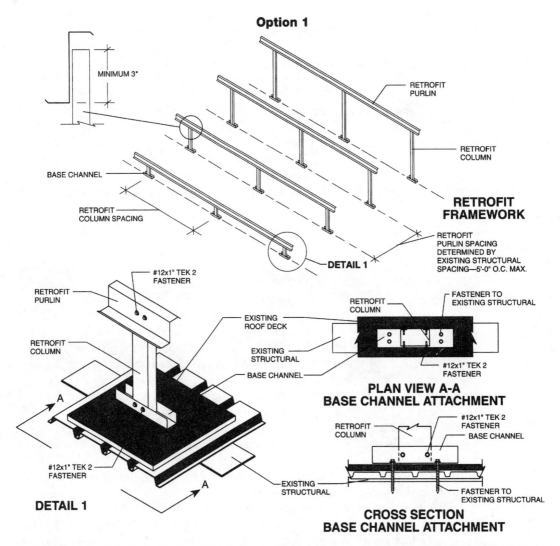

Option 1

MINIMUM 3"

RETROFIT PURLIN

RETROFIT COLUMN

BASE CHANNEL

RETROFIT COLUMN SPACING

RETROFIT FRAMEWORK

DETAIL 1

RETROFIT PURLIN SPACING DETERMINED BY EXISTING STRUCTURAL SPACING—5'-0" O.C. MAX.

#12x1" TEK 2 FASTENER

RETROFIT PURLIN

RETROFIT COLUMN

#12x1" TEK 2 FASTENER

EXISTING ROOF DECK

EXISTING STRUCTURAL

BASE CHANNEL

A

A

DETAIL 1

RETROFIT COLUMN

FASTENER TO EXISTING STRUCTURAL

#12x1" TEK 2 FASTENER

PLAN VIEW A-A BASE CHANNEL ATTACHMENT

#12x1" TEK 2 FASTENER

BASE CHANNEL

RETROFIT COLUMN

EXISTING STRUCTURAL

FASTENER TO EXISTING STRUCTURAL

CROSS SECTION BASE CHANNEL ATTACHMENT

Figure 14.1 Option 1: Retrofit framing with columns located over existing roof purlins spaced not over 5 ft 0 in on centers (*MBCI.*)

If the existing roof structural decking—not just the roofing—needs to be removed for an easier attachment to the existing framing, or because of excessive corrosion, one should remember to replace it with a new decking or horizontal bracing to provide lateral support for the existing roof beams and to restore the existing roof diaphragm.

The details of roofing for the new work, such as clip design, endlap fastening, placement of stepped expansion joints, and the like, remain the same as for new roofing (see Chap. 6). The new "attic space" needs to be carefully assessed in terms of code requirements for fire safety, ventilation, and egress.

14.2.6   Determination of new support locations

The locations of new roofing supports are determined by the type and spacing of the existing roof structural members. The new supports are best located directly above the existing roof purlins, whether the purlins are made of steel, concrete, or wood. This approach avoids bearing on the existing roof decking, which could be corroded or decayed by water leakage, even if theoretically adequate.

The desired roof slope is accomplished by varying the height of the new columns. All the variable-height columns must be precut to the exact size unless a license to use adjustable supports is first obtained from the patent holder. Columns of adjustable height—no matter how achieved—are covered by several patents held by Re-Roof America Company of Tulsa, Okla., which enforces its rights.

A grid of purlins and columns spaced not wider than 5 ft apart in both directions, and perhaps closer in some areas, is ideal. This is possible only when the existing roof purlins are also spaced 5 ft apart or less (Fig. 14.1).

The actual spacing of the new support columns along the purlin length may depend more on the type and span of the existing purlins than on the limitations imposed by the new framing. For example, if the existing roof is framed with bar joists and steel deck, the new posts should be located directly above the panel points of the joists and be spaced at one-quarter points of their span, or even closer, to approximate a uniform load. The reason: because of their proprietary design, the only structural information normally available about the joists is their uniform-load capacity; unlike hot-rolled beams, bar joists cannot be readily checked for concentrated loads. Keep this nuance in mind when a bidder starts to argue that there is no need for such a close support spacing—the bidder is not responsible for evaluation of the existing roof's strength.

When bearing a support column directly on the existing corrugated steel deck, remember to check the buckling strength of the deck ribs. A bearing plate or channel is frequently needed to spread the column load over several corrugations as shown in Fig. 14.1.

In a more general—and difficult—situation, the existing roof purlins are not spaced 5 ft apart. In this case, another design approach is called for, whereby a grid of base support members, usually C and Z purlins, is provided at 5 ft 0 in on centers perpendicular to the existing roof purlins. The new retrofit columns bear on these members and support the retrofit purlins (Fig. 14.2). Note that the bearing plates are provided under the base support members and that the retrofit columns are located in a checkerboard fashion. Lateral stability of the base support members can be achieved by adequate attachment to the existing deck and by proper bracing or blocking between the points of attachment.

14.2.7   Lateral bracing requirements

As was already mentioned, bracing requirements for light-gage C and Z purlins are surrounded by some controversy and misunderstanding. We recommend a

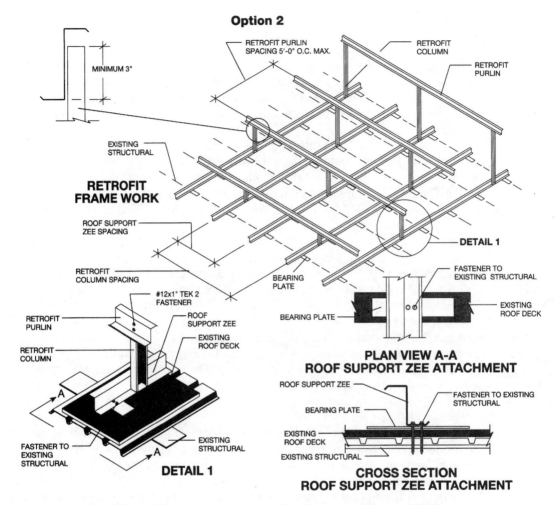

Figure 14.2 Option 2: Retrofit framing over existing roof purlins—a general case. (*MBCI.*)

closely spaced bracing for these members, perhaps 5 to 6 ft apart. The retrofit columns may be assumed to brace both the top and the base purlins, but at least some strapping or angle bracing is still needed above and between the columns. Some manufacturers seem to agree and provide such bracing at both flanges of the retrofit purlins at a maximum of 5 ft on centers if the columns are spaced over 5 ft on center and standing-seam roofing is present.[2] The strapping locations are illustrated in Fig. 14.3. Consistent with our position expressed in Chap. 5, we would prefer to see the light-gage angles or other members capable of acting in compression instead of the tension-only strapping.

In addition to purlin bracing, cross bracing for lateral load resistance of the whole assembly is also normally needed. The primary purpose of the cross bracing is to provide lateral stability to the new framework and to transfer wind

Strapping Requirements

A. 2"x24 gage strapping is to be attached to the top of the purlins, running continuous from eave to ridge, at each column line. If strapping is to be spliced, splice must occur over a purlin. Use four fasteners to attach the splice to the purlin.

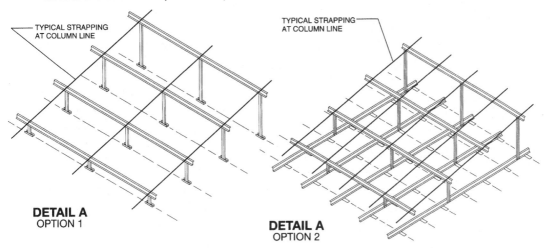

B. When a MBCI standing seam panel is used, purlin stabilization strapping is required at the top and bottom flanges of the retrofit purlin at a maximum of 5'-0" o.c. between column lines, if column spacing exceeds 5'-0" o.c.

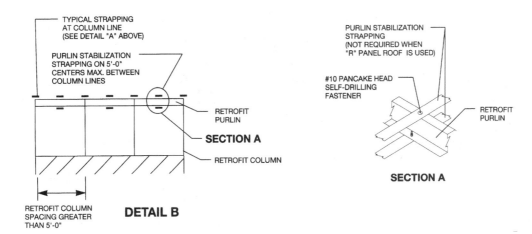

Figure 14.3 Bracing of retrofit purlins. (*MBCI.*)

loads into the existing building structure. The exact configuration and spacing should be left for the manufacturer to determine, but at least *some* bracing should be provided at regular intervals. Typical cross-bracing requirements are illustrated in Fig. 14.4; some typical details are shown in Fig. 14.5.

14.2.8   Reroofing over existing metal roof

For an old metal roof deteriorated beyond repair, reroofing may be considered. Whenever the existing slope is sufficient, a new buildup framework is not needed; the new roofing can be installed directly over the old.

A common situation involves standing-seam metal panels plus fiberglass insulation installed on top of an old through-fastened roof. Here, the only new structural framing consists of light-gage hat channels—sometimes even pressure-treated wood 2 by 4s—located directly above the existing purlins (Fig. 14.6). For added insulation value, thermal spacer blocks may be installed over the hat channels where the insulation is compressed.

A weak point of this design is the existing metal roofing: if it is corroded or structurally deficient for the new loading—two very common scenarios—it can hardly serve as a proper support for the structure. A direct attachment to the

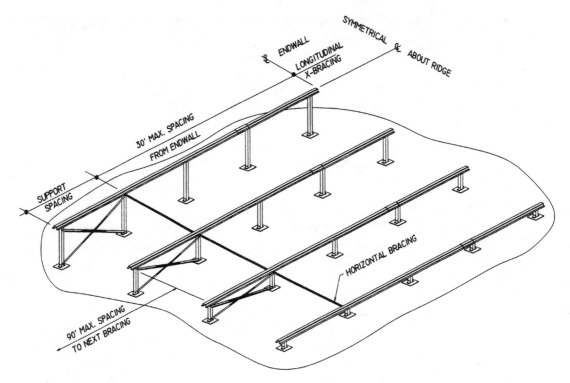

Figure 14.4 Typical longitudinal bracing layout for slope buildup. (Bracing in the transverse direction is not shown for clarity.) (*Butler Manufacturing Co.*)

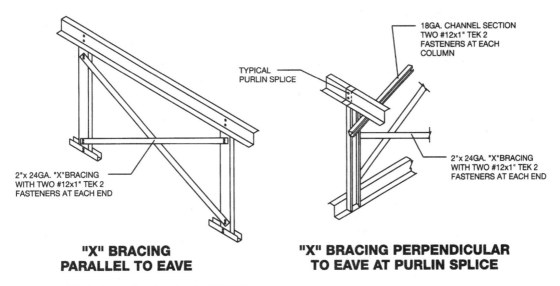

**"X" BRACING
PARALLEL TO EAVE**

**"X" BRACING PERPENDICULAR
TO EAVE AT PURLIN SPLICE**

Figure 14.5 Typical cross-bracing details. (*MBCI.*)

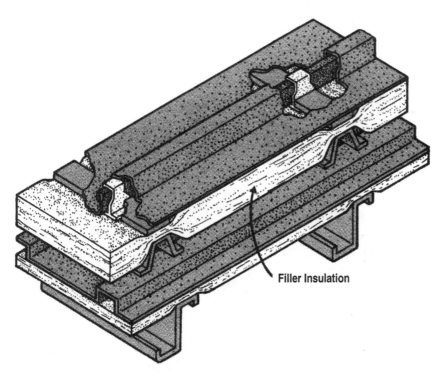

Figure 14.6 Reroofing over an existing metal roof with adequate slope. (*Courtesy of NAIMA Metal Building Committee.*)

purlins via some filler material that fits within the roofing profile is then the only solution.

One recently introduced product that makes this attachment possible is Roof Hugger,* basically a small-height Z purlin notched out at the bottom flange and the web to fit over the existing roofing corrugations (Fig. 14.7). According to the company, this product has been proved in numerous applications and is becoming extremely popular.

14.2.9   Design details

Each manufacturer involved in reroofing applications has developed its own details for various conditions, such as shown above and in Fig. 14.8. (The details from the MBCI catalog[2] are used for consistency's sake.) The details reproduced here cover only the condition when the existing roof purlins are perpendicular to the new roof slope; slightly different details are used if the existing framing runs parallel to the new slope. Still, the general concepts discussed in this chapter should be applicable to any manufacturer and to any roof condition.

One design detail that should not be forgotten deals with a treatment of additional insulation which might be needed to meet the code-mandated U values.

---

*Roof Hugger is a registered trademark of Roof Hugger, Inc.

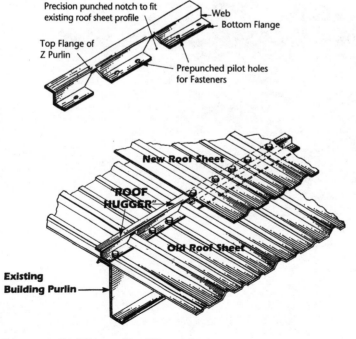

Figure 14.7   Roof Hugger. (*Roof Hugger, Inc.*)

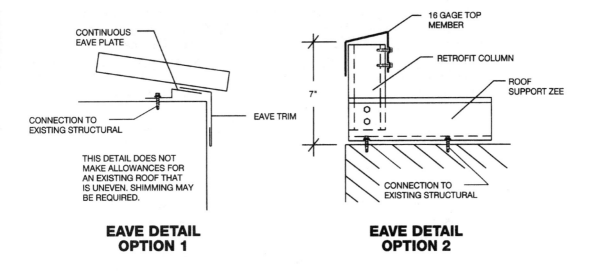

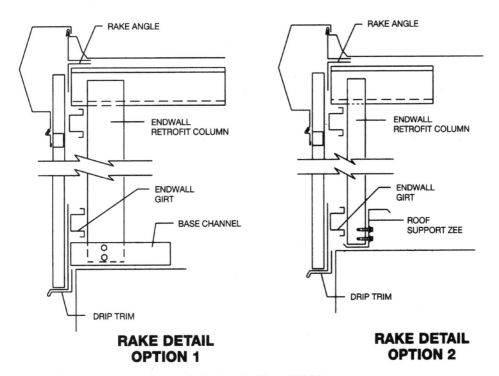

Figure 14.8 Typical eave and rake details for slope buildup. (*MBCI.*)

Quite often, rigid foam or fiberglass blanket insulation is laid on top of the existing roof between the new supports. The issue facing the specifiers is whether to provide a vapor retarder on such insulation. If the existing roofing system has insulation with a vapor retarder of its own, or if the roofing is left essentially intact, the new vapor retarder is probably not needed. If, on the other hand, the existing roofing has been cut for venting purposes, it might be.

A similar situation occurs when a new rigid insulation is mechanically attached to an existing roof. A multitude of the resulting holes may compromise the vapor-retarding qualities of the old membrane, although we don't quite know whether this will in fact happen.[3]

Another common problem is treatment of existing rooftop HVAC components. With the new slope buildup framing, all existing roof vent pipes can be extended. Similarly, exhaust fans can be relocated to the new roof surface and be supported by proper curbs. But what to do about major equipment such as chillers, air conditioners, or cooling towers? The new lightweight retrofit framing is clearly not strong enough to support them. It is possible, of course, to extend structural framing to a higher elevation, but the cost might be prohibitive. It is also possible to treat the space between the new and existing roofs as a sort of mechanical penthouse and fill the new gable walls with louvers or large vents. In that case, the HVAC equipment could stay on the existing roof and be completely or partially enveloped by the retrofit roofing.

It is wise to specify that the existing roofing being penetrated by retrofit columns or connections be patched to provide an interim leak protection.

14.2.10 What the manufacturer needs to know for metal reroofing

The contract documents for metal reroofing should include at least the following information:

- The governing building code and edition; desired UL and insurance ratings
- Design live, snow, wind, and seismic loads and load combinations
- Existing building dimensions and construction details (some original drawings could be attached)
- Proposed roof plan, slope, and configuration
- Desired type of roofing including profile and finish
- Structural requirements for the new support framing such as column spacing, bracing, and roof diaphragm construction
- Provisions dealing with partial or complete removal of the existing roofing, if appropriate, or with roof venting

It is prudent to require a submittal of structural design calculations and detailed shop drawings accompanied by the certification by a professional engineer that the new roof meets the contract requirements.

14.2.11   Rehabilitation of existing metal roofing

As with any other roofing type, metal roofing has limited service life which can be shortened by severe weather exposure or improper installation. A failure of metal roofing may manifest itself in standing seams that open up, fasteners that back out, or rust that spreads from the panel ends to the rest of the roof. All of these lead to leaks. Sometimes, the finish fails first and the roof simply *looks* bad.

The first step in addressing the problem is a roof survey. If the existing roof slope, insulation, and other parameters are satisfactory and the trouble exists only in the deteriorated finish, a complete reroofing job might not be needed. A simple recoating might be enough to dramatically improve the building's image. The most troublesome parts of the roof—flashing, gutters, panels at eaves, and roof penetrations—need to be closely examined and replaced, if needed, prior to the recoating.

Obviously, rusted-through metal cannot be helped by a new finish. The supporting roof framing and fasteners are also vulnerable and should be examined for signs of corrosion. The most critical areas in this regard are the edges of the holes in the purlins made by the roofing fasteners, where the purlins' protective coating is interrupted and the bare metal is exposed. The only protection this area receives is from the fastener's compressible gasket, if any is present, but the gasket's performance depends greatly on the skill of the installer. Widespread corrosion around the fasteners could dramatically weaken the roof's anchorage and require installation of new fasteners, bringing cost-effectiveness of the whole recoating project into question.

The actual rehabilitation process, similar to a repainting job, consists of complete or partial stripping of the existing finish (or sometimes just a power wash), priming, and recoating with a new material compatible with the substrate. The field is crowded with recoating products designed specifically for aging metal roofs. Many such products are based on elastomeric membranes that cover the roof without any joints; some utilize seamless foam systems that consist of sprayed-on polyurethane foam insulation covered with a sprayed-on silicone membrane. Prudent specifiers thoroughly investigate the actual performance of these products prior to their selection.

14.3   Exterior Wall Replacement and Repair

Metal-clad walls age much in the same way as metal roofs do: the finishes deteriorate, seams open up, fasteners loosen and corrode. Sooner or later, the siding cries for attention. Fortunately, the walls are normally easier to upgrade than the roofs.

The simplest kind of wall rehabilitation is, of course, repainting, a relatively straightforward and familiar process. A more complicated situation arises when the walls are deteriorated beyond repair and need to be replaced with a product that works with the existing girt system.

Structural evaluation of the existing girts should be the first order of business. Are the girts properly sized? Is there excessive corrosion? What about lateral bracing? Connections? These questions must be satisfactorily answered before any siding replacement takes place. If the girts are spaced too widely or are deficient in some respect, it might be more economical to add the new ones and use standard siding than to specify extra-strong replacement panels able to span long distances.

Most metal wall replacement projects include new windows, doors, flashing, and perhaps even framing around the openings to provide a coordinated exterior wall system. A special, if uncommon, situation arises when the existing metal panel walls need to be replaced with "hard" walls. In this case the existing building is probably not rigid enough to laterally support such walls and possibly has to be replaced. To keep the operations uninterrupted in a facility of moderate size, it might be possible to build the new exterior "hard" walls and the new roof structure outside of the existing building envelope and later remove the old building piece by piece.

14.4 Expansion of Existing Metal Buildings

One of the often-stated advantages of metal building systems is the ease of expansion. True, it is relatively easy to extend a pre-engineered building by adding several more bays of matching framing to its endwall and cutting in a door in the wall. The complications begin when anything more than that is attempted.

A case in point concerns removal of the old endwall framing to unify the new and the existing spaces. As was discussed in Chap. 3, this task is easy only if the building had been programmed for expansion and a moment-resisting frame installed in each endwall. Otherwise, the elimination of an old endwall requires some engineering gymnastics such as temporarily shoring the purlins and girts that bear on the endwall framing, removing the latter, and erecting a new clear-spanning moment frame in its place—hardly an easy chore.

Expansion alongside the existing building is even more treacherous. In northern regions, two gable buildings sharing a common wall create a valley likely to be filled with drifted snow (Fig. 14.9). The resulting design roof snow

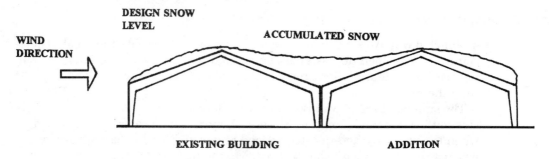

Figure 14.9 Building addition alongside an existing structure can result in its overstress.

load will probably exceed that of the original building. While the addition can certainly be designed for the larger loading, the existing building could be in serious danger. In fact, the losses from failures attributable to this very condition are measured in hundreds of millions of dollars.[5] Unless a costly structural upgrade is contemplated, it is better to avoid building expansion in this manner.

A similar problem arises whenever the addition has a higher roof elevation than the original building. Again, the snow drifted onto the lower existing roof can result in overstress and failure. In the next section, we look at this rather common situation again from a different perspective.

14.5 Changes in Loading Conditions

Quite often, changes in ownership, occupancy, manufacturing process, or mechanical systems entail some changes in structural loading as well. Evaluation of such changes in conventionally designed buildings is a relatively straightforward task for structural engineers. Pre-engineered buildings, however, need to be "re-engineered," or at least evaluated for the new conditions, which is not straightforward at all given the proprietary nature of the framing. A few common examples:

- Change in overhead crane loading or layout. As the next chapter explains, the cranes exert concentrated loads on the metal building framing. Top-running overhead cranes are usually supported by the building columns, but monorails are commonly suspended from the frame rafters. Any increase in the monorail capacity, or relocation of the crane runway, will impose new loading on the pre-engineered frames for which the frames were not designed.

- A new overhead door is needed at the bay containing an existing cross bracing. Can the bracing be moved to another bay?

- A new, higher, building or an addition is going to be erected next to "our" structure. As a result, drifted snow will likely accumulate on the existing roof and perhaps overstress it. Can the building roof be strengthened?

- As a result of the mechanical system upgrade, new heavy rooftop-mounted HVAC equipment is proposed. Is the pre-engineered roof framing strong enough to support it?

- A new state-of-the-art process equipment can fit within the existing building only if one of the main-frame rafters "loses" a few inches in depth. Is this possible?

In all of those cases, an intelligent answer can only be given if the metal framing sizes are known and the building can be readily analyzed. Step one in this endeavor is a search for the original building structural drawings and calculations prepared by the metal building manufacturer. These could be in the owner's file or at the city's building department or, if an architect was involved, in the architect's files. Alas, if anything is found at all, it is likely to consist only of the erection plans without any information on the member sizes. Recall that neither the fabrication drawings nor the calculations are normally provided by the manufacturer unless required by contract. Still, the plans can yield at least the building manufacturer's name and job number. Armed with this information, the owner's engineers can move to step two—contacting the original building manufacturer.

If still in business, manufacturers may have the coveted design files which contain the member sizes. In any case, they can help identify the design assumptions, grade of steel, and similar information needed for the analysis.

If the document search proves fruitless, it it usually possible to field measure the framing sizes for a new analysis. Such analysis may seem daunting to some structural engineers unfamiliar with the pre-engineered building design. In this case, step three is to seek assistance from a friendly metal-building manufacturer who could be in a better position to reanalyze the building for the new loading conditions. However rare, the need for such assistance is a good enough reason to keep working relationships with a few manufacturers who could be called upon to help.

References

1. James E. Grimes, "Metal Reroof Systems Deserve a Closer Look by Specifiers," *Metal Architecture,* November 1993.
2. *NuRoof Design/Installation Information,* MBCI Publication, Oct. 7, 1993.
3. Rene Dupuis, "How to Prepare Comprehensive Re-roofing Specifications," *1989 Handbook of Commercial Roofing Systems,* Edgell Communications, Inc., Cleveland, Ohio, 1989.
4. Maureen Eaton, "Re-covering Roofs Requires Careful Evaluation," *Building Design and Construction,* December 1994.
5. Wayne Tobiasson, "Some Thoughts on Snowloads," *Structure,* Winter 1995.

Specifying
Crane Buildings

15.1 Introduction

Two out of every five metal building systems are constructed for manufacturing facilities where cranes are frequently needed for material handling. A building crane is a complex structural system which consists of the actual crane with trolley and hoist, crane rails with their fastenings, crane runway beams, structural supports, stops, and bumpers. A motorized crane would also include electrical and mechanical components that are not discussed in this chapter. Our discussion is further limited to interior building cranes.

The main focus of this chapter is on proper integration of the crane and the metal building into one coordinated and interconnected system. Any attempts to add a heavy crane to the already designed and constructed pre-engineered building are likely to be fraught with frustration, high costs, and inefficiencies. If the required planning is done beforehand, however, a cost-effective solution is much more likely. We are also interested in a relationship among the three main parties with design responsibilities—architect-engineer, metal building manufacturer, and crane supplier. Occasionally, disputes arise when the contract documents do not clearly delineate their respective roles in the project.

15.2 Building Cranes: Types and Service Classifications

Several types of cranes are suitable for industrial metal building systems, the most common being bridge cranes (either top-running or underhung), monorail, and jib cranes. Occasionally, stacker and gantry cranes may be required for unique warehousing and manufacturing needs. Jib, monorail, and bridge cranes are examined here in this sequence—in order of increasing structural demands imposed on a pre-engineered structure. Constraints of space prevent us from discussing gantry and stacker cranes, as well as conveyors and similar material handling systems.

Within each type, the cranes are classified by the frequency and severity of use. Each crane must conform to one of six service classifications established by the Crane Manufacturers Association of America (CMAA). The six classes are: A (standby or infrequent service), B (light), C (moderate), D (heavy), E (severe), and F (continuous severe).

Guidance for assigning a service classification is contained in CMAA standards 70[1] and 74[2] and in the MBMA *Manual*.[3] The *Manual*'s Design Practices apply only to the cranes with service classifications A to D. Information on cranes with service classification E or F, including design loads and impact factors, is given in *AISE Technical Report* 13.[4]

Another way to classify the cranes is by kind of movement—hand-geared or electric. Hand-geared cranes are physically pulled along the rail by the operator and are less expensive, but slower, than electric cranes. Hand-geared cranes act with less impact on the structure than their faster-running electric cousins. The operator controlling an electrically powered crane can be either standing on the floor using a suspended pendant pushbutton station or sitting in a cab located on the moving bridge.

15.3 Jib Cranes

Jib cranes require relatively little planning from the pre-engineered building designer. Floor-mounted jib cranes (also known as pillar cranes) do not depend on the building superstructure for support and bear on their own foundations (Fig. 15.1). Column-mounted jib cranes, on the other hand, are either supported from or braced back to the metal building columns and thus impose certain requirements on strength and stiffness of the structure. For example, Ref. 5 recommends that building columns supporting jib cranes be rigid enough so that the relative vertical deflection at the end of the boom is limited to the boom length divided by 225. Floor-mounted jib cranes can rotate a full 360°, while column-mounted cranes are usually limited to a 200° boom rotation.

A jib crane picks up the load by a trolley that travels on the bottom flange of the boom and carries a chain hoist. The hoist can be either electric or manually operated. Upon lifting the load, the boom rotates around the crane's stationary column and lowers the object to the desired location. These two operations—travel of the trolley and rotation of the jib—are frequently performed manually.

The length of the jib crane's boom varies from 8 to 20 ft. The lifting capacity ranges from $\frac{1}{4}$ to 5 tons,[3] with $\frac{1}{2}$- and 1-ton jib cranes being the most popular.

Common applications of jib cranes include machinery servicing operations, assembly lines, steam hammers, and loading docks. Sometimes, a pair of jib cranes and a monorail in combination with forklifts is sufficient to transport cargo from a loading dock to the area serviced by overhead or stacker cranes. Inexpensive jib cranes can relieve main overhead cranes of much minor work that would tie them up for a long time.

Figure 15.1 Floor-mounted jib crane. (*American Crane and Equipment Corp.*)

Manufacturers of floor-mounted jib cranes normally supply the suggested foundation sizes for their equipment, but the foundation design is still the specifying engineer's responsibility.

Whenever floor-mounted jib crane foundations are added to an existing metal building, care should be taken not to interrupt any floor ties or hairpins which could be located in the slab. Otherwise the lateral-load-resisting system of the building could be damaged. Obviously, an addition of the column-mounted crane needs to be approached even more carefully, because the existing building columns will probably need strengthening to resist the newly imposed loads.

15.4 Monorails

15.4.1 The monorail system

The monorail crane is a familiar sight in many industrial plants, maintenance shops, and storage facilities. Monorails are cost-effective for applications requiring material transfer over predetermined routes without any side-to-side detours; their range of travel can be expanded with the help of switches and turntables. The monorail crane is essentially a hoist carried by trolley, the wheels of which ride on the bottom flange of a single runway beam. Monorails can be used to move the loads from 1 to 10 tons and can be either hand-geared or electric.

Monorail runway beams have been traditionally made of standard wide-flange sections that could accept straight-tread wheels or from I beams supporting tapered-tread wheels. (The straight-tread wheel is essentially a short cylinder; the tapered-tread one is a short truncated cone.) Today, these standard beam sections are being increasingly displaced by proprietary built-up runway beam products with unequal flange configuration. Figure 15.2 illustrates one of these hard-alloy-steel inverted T products offered by crane manufacturers; it also shows the loads exerted on the runway by the hoist.

Some advantages of the proprietary tracks over rolled beams include better wear resistance, easier rolling, longer service life, and weight savings. The tracks are specially engineered to overcome such common problems of the standard shapes as excessive local flange bending due to wheel loading. The advantages of the proprietary products make their use worthwhile for most monorail applications.

15.4.2 Loads acting on monorail runway beams

The vertical load V indicated in Fig. 15.2 includes the lifted weight and the weight of the hoist and trolley. It also includes impact that the AISC specification[6] prescribes to be taken as 10 percent of the maximum wheel load for pendant-operated cranes and 25 percent of the wheel load for cab-operated cranes. The side thrust S is specified as 20 percent of the lifted load including the crane trolley; this lateral force is equally divided among all the crane wheels. According to AISC, the longitudinal force L caused by trolley deceleration is to be taken as 10 percent of the maximum wheel loads.

The load percentages of the AISC specification apply only if not otherwise specified by other referenced standards. As will be demonstrated later for overhead cranes, various design authorities impose higher minimum requirements

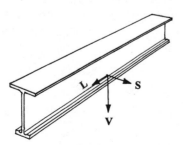

Figure 15.2 Loads acting on monorail runway beam (proprietary track shown).

for impact and lateral loads acting on runway beams. For example, ANSI MH 27.1[8] specifies the impact allowance for electric-powered hoists as ½ percent of the rated load for each foot per minute of hoisting speed, with a minimum allowance of 15 percent and a maximum of 50 percent. For bucket and magnet applications, it requires the impact allowance to be 50 percent of the rated load. Each of these forces must be resisted by suspension supports and lateral bracing.

15.4.3 Suspension and bracing systems

Depending on the available vertical clearance to the frame, suspension structure can be a simple bracket welded to the underside of the rafter and bolted to the runway beam (Fig. 15.3a), or a longer steel section with diagonal angle bracing to resist side thrust (Fig. 15.3b). Both these brackets represent a *rigid suspension* system. Another system, *flexible suspension,* utilizes hanger rods instead of the brackets to resist the vertical reaction V. Flexible suspension results in lower crane loads and reduced wear.[3,7] Monorail beams supported by hanger rods always require antisway lateral bracing for stability. Any suspension system should be vertically adjustable to bring the monorail beams to a true horizontal position prior to installation of the lateral bracing.

The side thrust S has been traditionally resisted by a shop-welded channel laid flat on top of an I beam (Fig. 15.3). The proprietary track, with its wide top flange, makes the channel unnecessary. The side thrust can also be resisted by

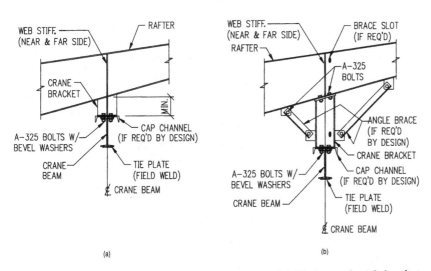

Figure 15.3 Monorail supports with rigid suspension. (a) Minimum- length bracket; (b) long bracket. (*Metallic Building Systems.*)

intermittent lateral bracing of the girder's top flange; such bracing must be designed not to interfere with vertical deflection of the monorail beam. Lateral bracing of this sort might be impractical in pre-engineered buildings with cold-formed purlins: The purlins have little lateral stability of their own and cannot accept bracing loads.

The longitudinal runway force L may be resisted by a diagonal angle brace located in the plane of the monorail beam at approximately 100-ft intervals and at all runway turns.[5]

The locations and conceptual details of all vertical and lateral supports should be indicated in the contract documents. The vertical supports should ideally occur at each main frame or at 20- to 25-ft intervals if the monorail runs alongside the frame.

15.4.4 Design considerations for runway beams

ANSI MH 27.1 requires that the allowable stress in the lower (tension) flange of monorail runway beams be limited to 20 percent of the ultimate steel strength. It also specifies the deflection criterion for monorail runway beams with spans of 46 ft or less as the length between vertical supports divided by 450.

Monorail beams must be carefully spliced to allow for smooth wheel movement between the individual beams. The best splice detail involves full-penetration welding of the bottom flange in combination with bolted shear plates in the web. Some pre-engineered manufacturers prefer to use field-welded tie plates and locate the splices under each hanger (Fig. 15.4).

A special case arises when two parallel monorail beams must transition into a single perpendicular beam. This problem can be solved with monorail switches.

Who supplies the runway beam and its supports? The metal building manufacturer already provides the suspension supports and bracing and could also provide the runway beam if specifically required to do so by contract (normally, crane work is excluded from the manufacturer's scope of work). The runway beam supplied by a building manufacturer is likely to be of standard structural shape. Whenever proprietary runway beams or switches are specified, they should be furnished and installed by the monorail supplier.

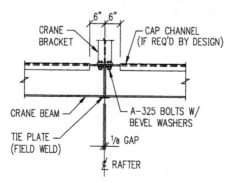

Figure 15.4 Runway beam splice at hanger locations. (*Metallic Building Systems.*)

15.4.5 Special requirements for supporting frame rafters

Whichever suspension system is selected, the weight of the loaded monorail needs to be transferred into the supporting frame rafter. A suspended load that attempts to tear the bottom flange of the rafter away from its web must be resisted by welds between the flange and the web; the web tearing is resisted by stiffeners (Fig. 15.3).

Pre-engineered building manufacturers usually provide single-side web-to-flange welds in their primary frames. Such welds may be inadequate to resist high localized suspended loads. The built-up rafters welded on only one side and subjected to cyclical suspended loads may suffer from fatigue problems caused by a notch which is sometimes produced by one-sided welding.[11] For this reason, the areas around the hangers may have to be reinforced with double-sided welds, to act in combination with web stiffeners in resisting the suspended loads. Occasionally a large part of the frame may have to be reinforced with double welds. This nuance is just one more reason to coordinate the design of metal-building and overhead-crane systems.

Loads acting on the metal building system from monorails, or from any other crane supported by the building structure, should be entered into the loading combinations discussed in Chap. 3.

15.5 Underhung Bridge Cranes

15.5.1 System description

An underhung bridge crane, as the name implies, features a hoist trolley which moves along the crane bridge being "hung" from the runway beams. The crane bridge is usually a single, and occasionally double, girder supported by two end trucks with wheels running on the bottom flanges of the runway beams (Fig. 15.5). The runway beams, in turn, are suspended from the building frame rafters or trusses. (In the latter case, supports are at the truss panel points.) A minimum clearance of 2 in is required by ANSI MH 27.1 between the underhung crane and any lateral or overhead obstruction.

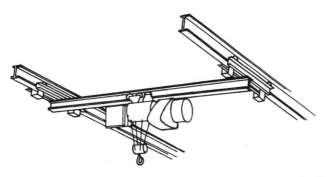

Figure 15.5 Underhung single-girder crane. (*FKI Industries, Inc.*)

Underhung bridge cranes have relatively modest lifting capacities—from 1 to 10 tons—and are usually confined to spans of 20 to 50 ft. Both hand-geared and electric-powered underhung cranes are available. The electric cranes are usually controlled by a pendant pushbutton station, although cab-operated and automatically controlled underhung cranes exist, too.

The chief advantage of the underhung design lies in the fact that the crane span need not extend all the way between the building columns. Thus underhung cranes are especially appropriate when only a part of the building aisle needs crane service and when the building has a large clear span. The underhung design allows the trolley to travel beyond the centerlines of the runway beams and permits load transfer between the adjacent crane aisles or between several parallel underhung cranes in one aisle.

15.5.2   Runway beams

Design and construction of runway beams for underhung cranes are similar to those of monorails. Both types of cranes traditionally relied on runways made of I beams (now called S shapes) with thick webs for good lateral-load resistance and tapered flanges helpful for self-alignment of trolley wheels. Recently, proprietary tracks described above have been increasingly gaining in popularity. The runway beam suspension and bracing details for underhung cranes are the same as for monorails, except that only one of the two runways needs to be laterally braced to allow for variations in alignment and crane deflections.[3]

Since both monorails and underhung cranes can use the same design of patented track, they can be interconnected by various interlocking devices and switches. The resulting combined crane coverage can be custom-tailored to the process at hand, yet very cost-efficient.

15.5.3   Design data

An example of minimum clearances required for single-girder underhung cranes is given in Fig. 15.6, reproduced with permission from Ref. 9. Figure 15.6 also includes maximum crane-wheel loads, crane weights, and hoist data. ANSI MH 27.1 provides additional information on some design issues relating to underhung cranes.

As was explained in Sec. 15.4.5, rafters of main frames which support underhung cranes should have two-sided welds connecting the flanges to the web.

Whenever proprietary runway-beam sections, stops, or switches are specified, it is best to let the crane supplier provide all of them for the sake of product compatibility. Standard-section runway beams can be provided by the metal building manufacturer if required by contract.

15.6   Top-Running Bridge Cranes

15.6.1   System description

A typical top-running bridge crane is supported on building columns and provides crane coverage for most of the aisle—the space between the columns.

Single girder underhung crane

As the name implies, this crane runs on the bottom flanges of a rail suspended from the roof joists.

This style is most economical for clear-span gabled buildings with continuous rigid frames that feature tapered wall columns if the crane and its load can be supported from the building trusses.

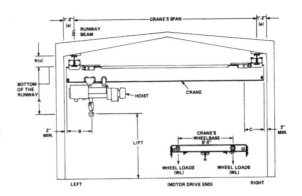

Capacity in Tons	Crane Max. Span in feet	Bottom of Running Beam to Upper Hook Position in inches A	Min. Hook Approach in inches B	Min. Hook Approach in inches C	Crane Product #	Crane Wt.	Wheel Load Per Pair in lbs. WL (b)
1	20	36-1/2	20	21	5260020	1,960#	1,880#
	30	42-1/2	20	21	5260040	3,040#	2,180#
	40	44-1/2	20	21	5260070	4,700#	2,640#
	50	48-1/2	20	21	5260090	6,240#	3,060#
2	20	39-1/4	21	20	5260110	2,180#	3,110#
	30	39-1/4	21	20	5260140	3,000#	3,330#
	40	44-1/4	21	20	5260160	4,700#	3,800#
	50	48-1/4	21	20	5260180	6,240#	4,230#
3	20	45-3/8	41	25	5260210	2,450#	4,720#
	30	42-3/8	45	25	5260230	3,000#	4,870#
	40	47-3/8	49	25	5260250	5,000#	5,420#
	50	51-3/8	56	25	5260270	6,630#	5,870#
5	20	47-3/8	40	25	5260380	2,950#	7,200#
	30	45-3/8	44	25	5260400	3,810#	7,440#
	40	47-3/8	48	25	5260420	5,030#	7,770#
	50	51-3/8	56	25	5260450	7,220#	8,370#
7-1/2	20	50-7/8	35	26	5260580	3,660#	10,750#
	30	52-7/8	38	26	5260600	4,560#	11,000#
	40	56-7/8	42	26	5260630	7,010#	11,670#
	50	57-7/8	50	26	5260650	9,160#	12,260#
10	20	50-7/8	35	26	5260680	3,390#	13,550#
	30	52-7/8	38	26	5260710	4,730#	13,920#
	40	56-7/8	42	26	5260740	6,890#	14,510#
	50	57-3/8	50	26	5260760	9,930#	15,350#

(a)   This dimension includes OSHA minimum lateral clearance of 2 inches.

(b)   Wheel load includes allowance of 15% impact with a maximum hoist speed of 30 FPM standard industrial service. Refer to Acco Structural Beam Guide for other requirements.

(c)   This dimension represents the height of the runway beam. Step 4 of this planning guide will determine this dimension.

Hoists for Single Girder Cranes

Capacity in Tons	Hoist Product #	Bridge Speed (FPM)	Hoist Speed (FPM)	Trolley Speed (FPM)	Hoist Lift in Feet	Hoist Wt.
1	2214600	70	16	65	20	350#
2	2215180	70	15	65	20	380#
3	3250360	70	15	65	23	1,080#
5	3250420	70	15	65	23	1,160#
7-1/2	3373950	70	15	65	25	2,030#
10	3374010	70	12	65	25	2,030#

NOTE: Hoists are single speed with single speed trolley. The 1 and 2 ton hoists are single reeved units. Hook approaches B & C are approximate. Hook moves lateral from high to low hook position. If necessary contact Acco representative for actual dimensions. The 3, 5, 7-1/2 and 10 ton hoists are double reeved units.

Figure 15.6   Dimensional and loading data for single-girder underhung cranes. (*FKI Industries, Inc.*)

Crane movement is effected by two end trucks riding on top of rails supported by runway beams. The speed of travel is generally higher than that of underhung cranes.

The crane bridge can consist of a single girder that carries a trolley hoist traveling on its bottom flange monorail-style (Fig. 15.7a), or of a double girder that supports a top-running trolley (Fig. 15.7b). Double-girder cranes can lift heavier loads and can accommodate greater lifting heights than single-girder cranes; even greater loads can be lifted by box-girder cranes.

Top-running bridge cranes are mostly electrically operated, except for some single-girder models, and are normally controlled by pushbutton pendants. Some heavy box-girder cranes feature operator-controlled attached cabs.

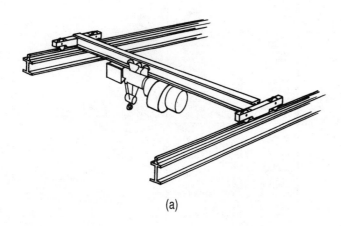

(a)

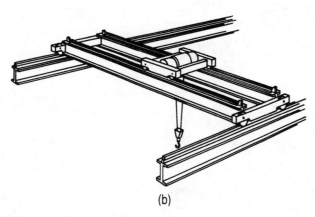

(b)

Figure 15.7 Top-running bridge cranes. (a) Single-girder; (b) double-girder. (*FKI Industries, Inc.*)

Single-girder top-running cranes are limited in lifting capacities from 1 to 10 tons and to spans from 20 to 60 ft. They represent a good choice for budget-conscious owners whose modest material handling needs can be satisfied with the crane of a low service classification. A sample of dimensional and loading data for single-girder cranes is shown in Fig. 15.8.

Double-girder pendant-operated cranes can lift from 5 to 25 tons, can be designed to higher service classifications, and can span from 30 to 100 ft. The lifting requirements in excess of 25 tons normally call for a box-girder bridge. Some cab-operated box-girder models can lift 250 tons and span up to 100 ft. Double-girder bridge cranes with top-running trolleys offer the largest lifting distances of all the crane types discussed here, and models with low-headroom profile are available for those cases where a few extra inches of headroom are worth the inevitable sacrifices in span and lifting capacity.

An example of dimensional and loading requirements for double-girder cranes is given in Fig. 15.9. (As with all such tables reproduced in this chapter, the cranes produced by other manufacturers do not necessarily conform to these data.) For all top-running cranes, a minimum overhead clearance of 3 in and a minimum side clearance of 2 in are required by CMAA 70 and CMAA 74.

15.6.2 Forces acting on top-running crane runways

Runway beams supporting top-running cranes are generally made of standard structural sections; a combination of wide-flange beam and capping channel is typical. Because wheels of these cranes travel on top of the rails and not on the actual beams, hardened proprietary tracks are not necessary.

The crane wheels exert the same three kinds of reactions on the runway beam as described in Sec. 15.4.2, but applied at the top flange (Fig. 15.10). The AISC Specification assigns equal factors for vertical impact, lateral, and longitudinal forces for all cranes, regardless of capacity. Other sources recommend a sliding-scale approach, designing heavier-duty cranes for proportionally larger loads. For example, Weaver[13] suggests adjusting the load values according to the CMAA service classification of the crane. The cranes of CMAA class A would be designed for a vertical impact of 10 percent, lateral load of 10 percent, and longitudinal load of 5 percent of the wheel loads, while the cranes of class F would be designed for 25 to 50 percent, 15 to 20 percent, and 20 to 30 percent, respectively. For the majority of CMAA service classifications, this approach results in higher design loading than required by AISC.

CMAA 70 and CMAA 74 prescribe taking the impact factor (hoist load factor) as 0.5 percent of the hoisting speed in feet per minute, but not less than 15 percent and not more than 50 percent. For bucket and magnet cranes, the impact factor is to be taken as 50 percent of the hoist's capacity. In addition to this lifted-load impact factor, CMAA 70 and 74 also require that impact factor be assigned to the dead load of the crane, trolley, and its associated equipment. The dead-load factor is specified as 1.1 for cranes with travel speeds of up to 200

Single girder top running crane

Top running cranes operate on runway rails and are excellent installations where the runway rails can be supported from building columns.

The main advantage of top running cranes is that it can be added to existing structures with relative ease through the use of additional columns to support the runway rail.

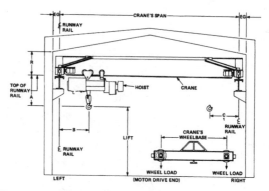

Capacity in Tons	Crane Max. Span in feet	Top of Rail to Upper Hook Position in inches A	Min. Hook Approach in inches B	Min. Hook Approach in inches C	Min. Side Clearance of Crane in inches EG (a)	Min. Overhead Clearance of Crane in inches R (a)	Wheel Base in feet WB	Crane Product Number	Crane Wt. in lbs.	Wheel Load Per Pair in lbs. WL (b)
1	20	24-5/8	30	31	7	14	5'-0"	5360020	1,880#	1,860#
	30	23-5/8	30	31	7	21	5'-0"	5360040	2,930#	2,150#
	40	25-5/8	30	31	7	21-1/4	5'-0"	5360070	4,560#	2,600#
	50	25-5/8	29	32	7-3/4	25-1/4	8'-4"	5360100	7,130#	3,300#
	60	28-3/8	29	32	7-3/4	28-1/2	8'-4"	5360120	10,300#	4,180#
2	20	25-3/8	31	38	7	16	5'-0"	5360140	2,090#	3,080#
	30	25-3/8	31	38	7	16-1/4	5'-0"	5360170	2,970#	3,330#
	40	25-3/8	31	38	7	21-1/4	5'-0"	5360190	4,560#	3,760#
	50	25-3/8	32	40	7-3/4	25-1/4	8'-4"	5360220	7,430#	4,550#
	60	28-1/8	32	40	7-3/4	28-1/2	8'-4"	5360240	10,300#	5,340#
3	20	26-1/2	41	33	7	21	5'-0"	5360270	2,410#	4,710#
	30	28-1/2	45	33	7	16-1/4	5'-0"	5360290	2,970#	4,860#
	40	28-1/2	49	33	7	21-1/4	5'-0"	5360310	4,560#	5,300#
	50	28-1/2	57	33	7-3/4	25-1/4	8'-4"	5360340	7,430#	6,090#
	60	31-1/2	61	33	7-3/4	28-1/2	8'-4"	5360360	11,370#	7,170#
5	20	28-1/8	41	33	7	21	5'-0"	5360500	2,600#	7,100#
	30	26-1/2	45	33	7	21-1/4	5'-0"	5360520	3,570#	7,370#
	40	28-1/8	49	33	7	21-1/4	5'-0"	5360540	4,790#	7,710#
	50	28-1/8	57	33	7-3/4	25-1/4	8'-4"	5360580	8,150#	8,630#
	60	31-1/2	61	33	7-3/4	28-1/2	8'-4"	5360600	11,370#	9,510#
7-1/2	20	34-1/8	35	33	7-1/2	18-3/4	5'-0"	5360760	3,120#	10,600#
	30	31-5/8	39	33	7-1/2	23-1/4	5'-0"	5360780	4,030#	10,850#
	40	31-5/8	43	33	7-1/2	27-1/2	5'-0"	5360810	6,150#	11,430#
	50	32-1/8	51	33	8-1/2	27-1/2	8'-4"	5360840	9,990#	12,490#
	60	34-5/8	55	33	8-1/2	30-1/2	8'-4"	5360860	12,350#	13,140#
10	20	34-1/8	35	33	7-1/2	18-3/4	5'-0"	5360890	2,790#	13,380#
	30	31-5/8	39	33	7-1/2	23-1/4	5'-0"	5360920	4,130#	13,750#
	40	31-5/8	43	33	7-1/2	27-1/2	5'-0"	5360940	6,170#	14,310#
	50	31-5/8	51	33	8-1/2	30-1/2	8'-4"	5360970	9,770#	15,300#
	60	34-5/8	55	33	8-1/2	30-1/2	8'-4"	5360990	12,350#	16,010#

(a) This dimension includes OSHA minimum 2 inch lateral clearance and 3 inch vertical clearance.
(b) Wheel load includes allowance of 15% impact with a maximum hoist speed of 30 FPM standard industrial service. Refer to Acco Structural Beam Guide for other requirements.

Hoists for Single Girder Cranes

Capacity in Tons	Hoist Product #	Bridge Speed (FPM)	Hoist Speed (FPM)	Trolley Speed (FPM)	Hoist Lift in Feet	Hoist Wt.
1	2214600	70	16	65	20	350#
2	2215180	70	15	65	20	380#
3	3250360	70	15	65	23	1,080#
5	3250420	70	15	65	23	1,160#
7-1/2	3373950	70	15	65	25	2,030#
10	3374010	70	12	65	25	2,030#

NOTE: Hoists are single speed with single speed trolley. The 1 and 2 ton hoists are single reeved units. Hook approaches B & C are approximate. Hook moves lateral from high to low hook position. If necessary contact Acco representative for actual dimensions. The 3, 5, 7-1/2 and 10 ton hoists are double reeved units.

Figure 15.8 Dimensional and loading data for single-girder top-running cranes. (*FKI Industries, Inc.*)

Double girder top running crane

The double girder crane has two main girders which support a hoist mounted on a top running trolley.

Selection of a double girder crane will allow heavier capacity, longer spans and maximum height of vertical load travel.

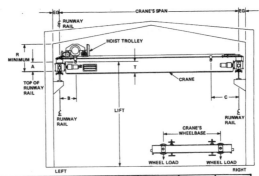

Capacity in Tons	Crane Span in feet	Top of Rail to Upper Hook Position in inches A	Min. Hook Approach in inches B	Min. Hook Approach in inches C	Min. Side Clearance of Crane in inches EG (a)	Min. Overhead Clearance in inches R (a)	Upper Hook Position to Bottom of Crane Girder in T inches	Wheel Base WB	Crane Product Number	Crane Wt. in lbs.	Wheel Loads in lbs. WL (b)
5	30	7-1/2	27	25	7-3/4	42	9-1/2	8'-4"	5430130	8,350#	8,940#
	40	7-1/2	27	25	7-3/4	42	11-1/2	8'-4"	5430150	11,350#	9,690#
	50	7-1/2	27	25	7-3/4	42	17	8'-4"	5430190	15,070#	10,700#
	60	9-1/2	26	24	8-1/2	44	23	9'-4"	5430230	19,650#	11,800#
	70	15-3/4	25	23	9-3/8	50-1/2	26	12'-4"	5431060	26,860#	13,600#
	80	21-1/2	25	23	9-3/8	56	33-1/2	13'-0"	5530170	28,520#	14,000#
	90	26-1/2	25	23	9-3/8	61	38-1/2	16'-0"	5530230	36,840#	16,100#
	100	26-1/2	25	23	9-3/8	61	44-1/2	16'-0"	5530270	41,430#	17,300#
7-1/2	30	4	26	24	8-1/2	44	4	8'-4"	5430270	8,040#	11,900#
	40	4	26	24	8-1/2	44	11-1/2	8'-4"	5430330	13,170#	13,200#
	50	4	26	24	8-1/2	44	12	8'-4"	5430370	16,820#	14,100#
	60	4	26	24	8-1/2	44	17-1/2	9'-4"	5430410	21,690#	15,300#
	70	10-1/2	25	23	9-3/8	50-1/2	23	12'-4"	5431120	29,590#	17,300#
	80	16	25	23	9-3/8	56	28	13'-0"	5530450	29,230#	17,200#
	90	21	25	23	9-3/8	61	33	16'-0"	5530510	37,760#	19,300#
	100	21	25	23	9-3/8	61	39	16'-0"	5530550	42,410#	20,500#
10	30	5-1/2	28	26	8-1/2	47	7-1/2	8'-4"	5430470	10,120#	15,400#
	40	5-1/2	28	26	8-1/2	47	13-1/2	8'-4"	5430510	13,120#	16,200#
	50	5-1/2	28	26	8-1/2	47	14	8'-4"	5430550	18,250#	17,500#
	60	5-1/2	28	26	8-1/2	47	23	9'-4"	5430590	21,930#	18,400#
	70	12	27	25	9-3/8	53-1/2	26	12'-4"	5431180	31,570#	20,800#
	80	22-1/2	27	25	9-3/8	64	34-1/2	13'-0"	5530730	29,880#	20,400#
	90	22-1/2	27	25	9-3/8	64	40-1/2	16'-0"	5530790	38,730#	22,600#
	100	22-1/2	27	25	9-3/8	64	41	16'-0"	5530830	45,410#	24,300#
15	30	8-1/4	28	27	9-3/8	56	9-1/2	9'-4"	5430630	12,050#	22,100#
	40	8-1/4	28	27	9-3/8	56	10	9'-4"	5430670	15,670#	23,000#
	50	8-1/4	28	27	9-3/8	56	18-1/2	9'-4"	5430730	21,390#	24,400#
	60	8-1/4	28	27	9-3/8	56	21	9'-4"	5430750	25,370#	25,400#
	70	18-1/2	28	27	9-3/8	66-1/2	30-1/2	13'-0"	5530970	28,240#	26,100#
	80	18-1/2	28	27	9-3/8	66-1/2	36-1/2	13'-0"	5531010	32,220#	27,100#
	90	18-1/2	28	27	9-3/8	66-1/2	37	16'-0"	5531070	42,390#	29,600#
	100	24-3/4	28	27	9-3/8	72-1/2	43	16'-0"	5531110	47,470#	30,900#
20	30	4-1/2	28	27	9-3/8	56	6	9'-4"	5430800	12,570#	28,000#
	40	4-1/2	28	27	9-3/8	56	9	9'-4"	5430890	19,360#	29,700#
	50	4-1/2	28	27.	9-3/8	56	15	9'-4"	5430950	22,000#	30,400#
	60	4-1/2	28	27	9-3/8	56	17-1/2	9'-4"	5430980	27,150#	31,700#
	70	15	28	27	9-3/8	66-1/2	27	13'-0"	5531310	31,110#	32,700#
	80	15	28	27	9-3/8	66-1/2	33	13'-0"	5531370	34,610#	33,600#
	90	21-1/4	28	27	9-3/8	73	39	16'-0"	5531460	44,640#	36,100#
	100	21-1/4	28	27	9-3/8	73	39-1/2	16'-0"	5531520	51,800#	37,900#
25	50	– 4	44	46	9-3/8	63	9	9'-4"	5431440	22,780#	38,000#
	60	6-1/4	44	46	9-3/8	73-1/2	18-1/2	13'-0"	5531610	23,340#	38,200#
	70	6-1/4	44	46	9-3/8	73-1/2	24-1/2	13'-0"	5531670	30,960#	40,100#
	80	6-1/4	44	46	9-3/8	73-1/2	24-1/2	13'-0"	5531710	35,710#	41,300#
	90	13	44	46	9-3/8	80	31	16'-0"	5531770	48,090#	44,400#
	100	13	44	46	9-3/8	80	36-1/2	16'-0"	5531810	56,000#	46,400#

Hoists for Double Girder Cranes

Capacity (Tons)	Hoist Product #	Bridge Speed (FPM)	Hoist Speed (FPM)	Trolley Speed (FPM)	Hoist Lift	Hoist Wt.
5	3370610	70	20	70	27'	2200#
7-1/2	3371090	80	15	70	23'	2400#
10	3471300	80	15	70	24'	2700#
15	3570890	80	20	70	32'	3500#
20	3571390	80	15	70	27'	3700#
25	3670050	80	10	60	29'	7100#

(a) This dimension includes OSHA minimum 2 inch lateral clearance and 3 inch vertical clearance.

(b) Wheel load includes allowance of 15% impact with a maximum hoist speed of 30 FPM standard industrial service. Refer to Acco Structural Beam Guide for other requirements.

Note 3 & 30 Ton are also available. Contact your local Acco Representative.

NOTE: Hoists are all single reeved units with single speed hoist and trolley except those product numbers beginning with 36 which are double reeved with variable speed hoist and single speed trolley.

Figure 15.9 Dimensional and loading data for double-girder top-running cranes. (*FKI Industries, Inc.*)

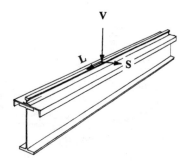

Figure 15.10 Forces acting on runway beam for top-running crane.

ft/min and as 1.2 for faster cranes. CMAA 70 and 74 also include other crane loads and load combinations for which the crane supports should be designed.

Alternatively, some engineers assume an impact factor of 0.25 for preliminary design of most overhead cranes, as suggested in Ref. 16, for example.

The support systems and bracing capable of resisting large loads exerted by top-running cranes are much more complex than those for support of monorails or underhung cranes. These support systems are explained in separate sections below. For the sake of simplicity, our discussion is limited to the buildings housing a single crane. Additional considerations for buildings with multiple cranes are covered in MBMA Design Practices.[3]

15.6.3 Structural design of crane runway beams

Structural design of runway beams for combined loads is well treated in many engineering handbooks as well as in *AISC Design Guide* 7,[5] so only a general procedure will be outlined here.

The first design step is to determine whether fatigue controls the design. Fatigue cracking is blamed for perhaps nine out of ten crane girder problems.[17] Given the anticipated number of loading cycles supplied by the owner and a life span of the building—50 years may be assumed as a default value—one follows the procedure of AISC Specification Appendix K to determine the allowable stress range. Crane girders of CMAA classes D, E, and F are often controlled by fatigue, meaning that the allowable bending stress in those members is reduced from 24 to, perhaps, 16 kips/in^2.[16]

If fatigue does not come into play, the allowable combined beam stress can be found in a conventional way, as a function of the beam properties and its unbraced length. In the absence of any additional lateral bracing, the unbraced length of a simply supported runway beam equals the column spacing (bay size). For bay sizes found in most pre-engineered buildings—20 to 30 ft—the allowable combined stress often ends up being equal to $0.6 F_y$.

The second step involves a computation of the required stiffness (moment of inertia) of the runway beam based on the allowable deflection criteria. Those readers who followed our discussion in Chap. 11 might remember that there is no consensus among engineers on the deflection criteria in general; deflections of crane runways are no exception. One source of information is *AISC Design Guide* 7, which recommends the following design criteria:

- For CMAA classes A through D, vertical deflection of runway beams under wheel loading is limited to $L/600$.

- For CMAA classes E and F, the maximum vertical deflection is limited to $L/1000$.

- Maximum *lateral* deflection of runway beams is limited to $L/400$ for all crane classifications.

Other sources suggest somewhat more restrictive criteria for vertical deflections, such as $L/1000$ for CMAA classes A, B, and C and $L/1200$ for CMAA classes D, E, and F. The lateral-deflection criterion of $L/400$ seems to be universally accepted. The impact factors used for stress analysis need not be included in deflection calculations.

The third step is a determination of the maximum bending moments from horizontal and vertical moving loads. Horizontal loads may be assumed to be resisted only by the top flange of the runway beam and the cap channel, if any is present.

Next, a trial section is selected which keeps both the deflection criteria and the combined stresses within the limits established before. Beam tables of AISC *Manual*,[14] *AISC Design Guide* 7, or Ref. 10 can be useful for this task.

The last step usually involves checking the selected section for sidesway web buckling in accordance with Sec. K1.5 of the AISC Specification.

AISC Design Guide 7 points out that there is yet another step that is commonly overlooked: a calculation of the local longitudinal bending stress in the top flange of the runway girder caused by moving wheels of the crane. This additional contributor to the total bending stress in the top flange can increase it by 1 to 4 kips/in^2.

In most cases, the design procedure outlined above results in a selection of the wide-flange beam–capping channel combination. For the heaviest cranes or for long runway spans, built-up steel girders with built-up cap channels, or even with top-flange horizontal trusses, could be required. For light cranes and relatively small bay sizes, it might be possible to select a single heavy wide-flange beam without a top channel. The increased beam weight might be more than offset by savings in labor required to weld the channel. A rule of thumb is that to be economical, a wide-flange–channel combination has to be at least 20 lb/ft lighter than a single wide-flange beam. If a single beam is used, its flange should be wide enough to allow for rail fastening hardware.

One problem with capping channels involves tolerances: Since neither the channel nor the wide-flange beam are perfectly straight, there are likely to be small gaps between the two. As the crane wheels pass over the gaps, some distress in the connecting welds or in the channel itself may occur. For this reason, capping channels or capping plates should be avoided in girders used for crane classifications E and F.[17]

Crane runway beams can be of simple-span or continuous design. Continuous beams will deflect less under load and require lighter, and therefore less expensive, sections. Continuous members, however, are susceptible to damage from unequal settlement of the supports and to buildup of thermal stresses. Simple-span runways not only are virtually unaffected by such problems but are also easier to design, erect, and replace if needed. We recommend that all runway beams be designed as simple-span members.

15.6.4   Supports for runway beams

The easiest, and perhaps the most common, method of supporting runway beams for top-running cranes is by brackets shop welded to rigid-frame columns (Fig. 15.11). The bracket supports are most appropriate for relatively light cranes, up to a 20-ton capacity; for heavier cranes, the eccentrically loaded building columns become uneconomical. Also, every slight impact on the runway is transmitted into the metal building structure, possibly causing vibrations and annoying the occupants.

In this system, building columns are typically reinforced with web stiffeners at the points of bracket attachment. In addition, continuous double-side welding of frame web to flanges is recommended to improve fatigue behavior of the frame. The bottom flange of the runway beam is attached to the bracket with high-strength bolts.

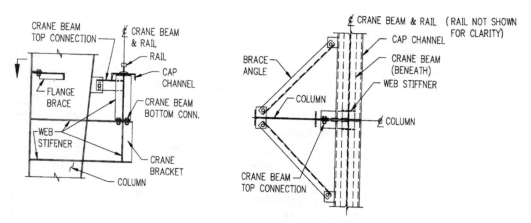

Figure 15.11   Bracket-supported runway beam for top-running crane. (*Metallic Building Systems.*)

The top flange is laterally braced by a short piece of angle or bar stock which transfers horizontal reactions into the building column. The top connection should allow for a slight vertical movement of the runway beam ends under load. This slight movement can be accommodated by slotted holes receiving the connecting bolts (Fig. 15.11) or by flexibility of the connector piece.

A second method of runway support utilizes stepped building columns, a solution common in old mill buildings. Stepped columns are appropriate for heavy-duty cranes and for the buildings with large eave heights that can benefit from a substantial stiffness of such columns. With this design, as with the previous one, crane vibrations are likely to be transferred to the rest of the building and be felt by the occupants.

Runways of top-running cranes with capacities exceeding 20 tons can be economically supported by a third method—separate crane columns. The separate columns are positioned directly under the runway beams and receive only vertical loading, while the building frame resists only lateral loading from the crane. A separate set of small columns may be cost-effective even for crane capacities less than 20 tons but with spans exceeding 50 ft.[3] Lateral reactions are transferred to the building frame by bracing between the two sets of columns which also acts as a lateral support for the top flange of the runway beam (Fig. 15.12). The runway-supporting columns are normally oriented with their webs perpendicular to those of main frames. Some engineers prefer to design these columns with fixed bases to decrease the column drift.

15.6.5 Bracing against lateral and longitudinal runway forces

Of the three forces that act on a runway girder shown in Fig. 15.10, the vertical reaction V is taken by the supporting bracket or column. The side thrust S is resisted by a cap channel or the girder's top flange and transferred into the building column via the top connection described above. For heavy cranes, a horizontal truss or a built-up member may be needed in lieu of the cap channel.

In order to design a bracing system to resist the side-thrust forces, one must make an assumption on whether or not to divide these forces between the runways at the opposite sides. The answer depends on a shape of the crane's wheels. Most medium-duty cranes have double-rimmed wheels which "grip" the rail well and assure a good lateral guidance. Whenever these wheels are used, the side thrust can be equally divided between both runway girders. Some long-span bridge cranes have double-rimmed wheels only on one side, with rimless rollers on the other side. This design is intended to prevent wheel binding because of changes in the bridge length due to temperature fluctuations. Whenever such wheels are used, the entire side thrust should be resisted by only one runway girder.[16]

The longitudinal force L presents some difficulties. Caused mostly by the crane braking or accelerating, as well as by its impact on the runway stops, the longitudinal force subjects the building columns to a racking action.

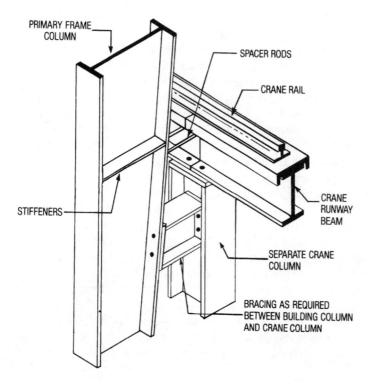

Figure 15.12 Crane runway beam supported by separate columns. (*Ceco Building Systems.*)

Unfortunately, most crane buildings are poorly braced in the runway direction. In fact, as Mueller[15] observes, "The most frequently heard complaint about steel mill buildings in the past has been too much sidesway, too much vibration, too much movement of the structure." This sentiment is echoed by many crane manufacturers supplying their products for metal building systems.

The only certain way to minimize longitudinal movement of the structure is to provide cross bracing made of angle sections under the runway beams to transfer the longitudinal forces from the runway beams to foundations. For multibay runways, Mueller (as well as the MBMA *Manual*) recommends that the bracing bents be located in the middle bay rather than at the ends of the runway. He strongly discourages the use of knee braces to stiffen the runway beams.

Proper splicing of runway girders is extremely important. As Mueller demonstrates, a simple splice utilizing web plates can lead to failure of the girder web, because the web plate restricts rotation of simply supported girder ends. Instead, he recommends that the girder's bottom flange be bolted to a cap plate of the crane column with the bolts designed to transfer the longitudinal forces from the runway to the cross bracing below.

As already noted, the lateral tie-back detail should also allow for girder-end rotation. Instead of rigid diaphragms favored decades ago, today's thinking is to use a bent plate, with web horizontal, welded between the top flange of the runway girder and the building column. Note that the commonly used spacer rods illustrated in Fig. 15.12 are attached to the girder web, a questionable solution. A somewhat better connection shown in Fig. 15.11 is made to the girder's web stiffeners; still, we would prefer the tieback angles to be welded instead to the cap channel, the same way the column flange braces are welded in this picture.

A premium method of lateral bracing for runway girders involves a proprietary tieback linkage, such as that illustrated in Fig. 15.13. The linkage allows for rotation and movement of girder ends while providing the desired resistance to side thrust.

The details described above are intended for light- and medium-capacity cranes commonly found in pre-engineered buildings. For heavy cranes, much more substantial details are needed, some of which are illustrated in *AISC Design Guide* 7. Heavy mill cranes may also require additional bracing not discussed here. For example, AISE 13 calls for the *bottom* flanges of the crane girders longer than 36 ft to be laterally braced.

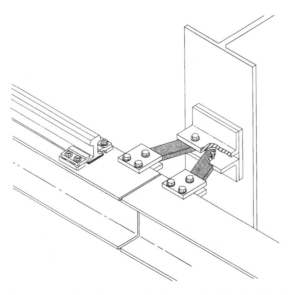

Figure 15.13 Proprietary rail clips and tieback linkage. (*Gantrex Corporation.*)

15.6.6   Specifying crane rails and rail attachments

Crane rails are on the borderline between the overhead crane and the metal building system. While the rails are normally selected and provided by the crane supplier, the methods of their attachment, splicing, and layout are in the specifier's domain. Crane rails, if installed incorrectly, can lead to a host of crane performance problems, such as premature wear-out of various crane components, high levels of noise and vibration, and even failures of runway girders.

The rails should be laid out in such a manner that on both runway girders their splices, as well as splices in the girders and in the cap channels, are staggered by at least 1 ft relative to each other. The crane supplier should verify that the crane's wheelbase does not happen to equal the amount of stagger. All crane rails should be specified as being "for crane service."

Rail splices can be made by bolting or welding. Successful welded splices require special end preparation and very close control of the operation; both could be difficult to achieve at the job site. Bolted splices offer a more practical alternative. The best splice detail is a tight joint between the rail sections instead of the commonly used joint with a small ($\frac{1}{16}$- to $\frac{1}{8}$-in) gap.[14] This detail requires mill finish of rail ends and a particular rail drilling pattern. The adjacent rail sections are fastened with high-strength bolts through the rails and two connecting bars which match the rail profile.

Perhaps the biggest challenge confronting the specifiers of crane rails is how to attach the rails to the girders. The ideal connection should be able to resist the side thrust from the crane wheels without constraining the rail's capacity to expand and contract with temperature changes independent of the girder. (Hence, the rails should never be welded to the runway.)

Rails can be attached to crane girders in three different ways. The first one is by hook bolts which used to be popular but now are not recommended except for the lightest of cranes. Hook bolts are not very good in resisting side-thrust forces, because the bolts tend to stretch and the nuts loosen under load. For this reason, hook bolts require frequent inspection and adjustment to maintain rail alignment.[12] Hook bolts are dimensionally suitable for capless crane girders with narrow flanges that leave no room for any other attachment method. Hook bolts are supplied in pairs and are usually spaced 2 ft on centers.

Rail clamps perform somewhat better than hook bolts and can be used for nearly any crane. The clamp assembly consists of a clamp plate connected by two high-strength bolts through a filler bar to the girder. Rail clamps are spaced not more than 3 ft apart. The clamps can be of tight or floating type. Floating clamps are intended to allow for a thermal expansion of the rail but, unfortunately, may also permit excessive rail movement in the transverse direction, making it difficult to maintain rail alignment. Tight clamps, on the other hand, tend to prevent any rail movement at all and may lead to a buildup of thermal stress in the rail and in the girder.

Hook bolts and rail clamps are illustrated in the AISC *Manual,* which also contains information on the rail properties and splice bolting.

The third method of rail attachment involves proprietary adjustable rail clips offered by various crane manufacturers. The patented clips claim to both allow for a longitudinal expansion of the rails and restrict their transverse movement. Some crane manufacturers also advocate the use of synthetic rubber pads under the rail at the clip locations that reportedly provide for a more uniform rail bearing and for reduced impact, noise, and girder wear. An example of proprietary clips and pads is shown in Fig. 15.13. At this time, adjustable clips represent the best available method of rail attachment.

Regardless of the rail attachment method, periodic inspection and maintenance is required to keep the bolts properly tightened and the rail aligned.

15.6.7 Runway stops and bumpers

Runway stops and bumpers are frequently the last elements considered in the design of a crane system. They should not be the least, however, because a poorly designed crane stop may ruin the crane and the building alike.

Runway bumpers act similarly to their automobile counterparts—absorbing kinetic energy of the crane's impact. Old-style wood and rubber blocks have given way to contemporary spring and hydraulic bumpers such as the one illustrated in Fig. 15.14. ANSI standard B30 and CMAA 70 require that bridge bumpers be designed to resist the force resulting from the crane hitting the stop at 40 percent of the rated speed.

Runway stops, as the name implies, are intended to stop a moving crane. While proprietary bumpers are commonly selected by their suppliers, design of crane stops belongs to the design professional. For monorails and underhung cranes, a short piece of angle attached to the web of the runway beam may be adequate. For top-running bridge cranes, however, a heavy bracket bolted to the top of the runway girder is needed. The bracket either has an attached

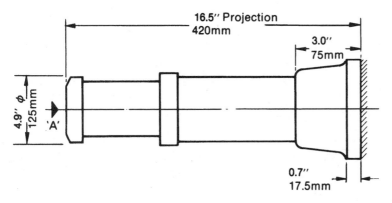

Figure 15.14 Hydraulic bumper. (*Gantrex Corporation.*)

bumper or is designed to come in contact with a bumper installed on the crane's end truck.

Eventually, of course, the force on the stop must be resisted by the building structure and its bracing. In the absence of specific crane and bumper data, this force may be estimated as the greater of 10 percent of the unloaded crane weight or twice the design tractive force, as suggested by AISC *Design Guide* 7 for interior cranes. The 10 percent formula is also found in MBMA *Design Practices*.

15.7 How to Select and Specify a Building Crane

The layout of new industrial buildings, especially those with cranes, is usually governed by the equipment they contain. It makes sense therefore to select a satisfactory crane coverage for the process equipment first and then determine the dimensions of the metal building system, rather than the other way around. The process outlined below roughly follows that of ACCO's *Crane Planning Guide for Metal Buildings*.[9]

The first step involves determination of the required hook coverage—width, length, and height to be serviced by the crane. While the plan dimensions are governed by the equipment layout, the vertical coverage (lift) depends on the size of the lifted items, plus an allowance for the height of any floor-mounted equipment to be cleared and for spreaders or other under-hook devices.

The second step is to determine the type, capacity, and service classification of the crane(s), based on the information contained in this chapter or other guides. For example, one electrically operated 25-ton double-girder top-running bridge crane of CMAA class B in combination with two 5-ton class A jib cranes may be needed. The selected type of movement (hand-pushed or electric) should be satisfactory not only for the short term but for any probable future operational changes as well. If in doubt, it is better to invest in some extra crane capacity from the beginning.

The third step deals, finally, with the building dimensions. The minimum clear span of the building is determined by adding dimensions EG, B, and C to that of the hook coverage (Fig. 15.15). The minimum clear height of the building is computed by adding dimensions A and R to the hook lift; it is measured to the lowest point of the roof, be it bottom of the frame rafters or a suspended sprinkler pipe. All these dimensions are included in Figs. 15.6, 15.8, and 15.9.

In the fourth step, the runway beams, their supports, and bracing methods are selected and designed.

Last, a configuration and the exterior dimensions of the metal building are determined from the interior dimensions computed in step 3 and information contained in Chap. 3. It is prudent to select a slightly larger building to allow for some variability of designs among manufacturers.

The contract documents should spell out who supplies the items likely to fall in the "gray area" of responsibility such as runway beams, rails, and runway

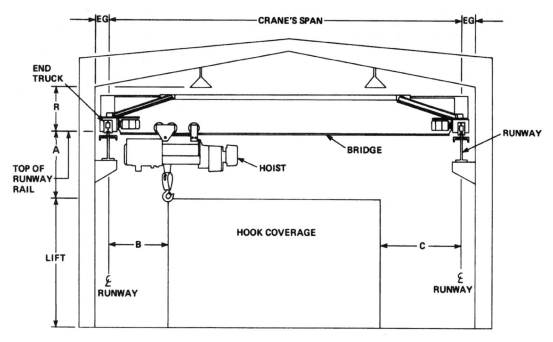

Figure 15.15   Hook coverage. (*FKI Industries, Inc.*)

stops. If not provided by the crane supplier, and if specifically called for by contract, these items could become a responsibility of the metal building manufacturer. In that case, all the information about the crane, its wheel loads, supports, etc., should be provided in the contract documents.

Unless instructed otherwise, pre-engineered building manufacturers are likely to follow the impact allowances and loading and deflection criteria specified in the MBMA *Manual,* still one of the best sources of crane-related information available. If more stringent design standards are to be followed by the building manufacturer and the crane supplier, the appropriate requirements should be included in the contract documents.

References

1. Specifications for Top Running Bridge & Gantry Type Multiple Girder Electric Overhead Traveling Cranes, CMAA Specification 70, Crane Manufacturers Association of America, Inc., Charlotte, N.C., 1994.
2. "Specifications for Top Running & Under Running Single Girder Electric Overhead Traveling Cranes Utilizing Under Running Trolley Hoist," CMAA Specification 74, Crane Manufacturers Association of America, Inc., Charlotte, N.C., 1994.
3. *Low Rise Building Systems Manual,* Metal Building Manufacturers Association, Cleveland, Ohio, 1986.
4. "Guide for the Design and Construction of Mill Buildings," *AISE Technical Report* 13, Association of Iron and Steel Engineers, Pittsburgh, Pa., 1979.

5. James M. Fisher, "Industrial Buildings—Roofs to Column Anchorage," *AISC Steel Design Guide* 7, Chicago, Ill., 1993.
6. Specification for Structural Steel Buildings, Allowable Stress Design and Plastic Design, American Institute of Steel Construction, Chicago, Ill., 1989.
7. *Underhung Cranes and Monorails,* Monorail Manufacturing Association, Pittsburgh, Pa., 1977.
8. Specifications for Underhung Cranes and Monorail Systems, ANSI MH 27.1-1981, Monorail Manufacturers Association, Pittsburgh, Pa., 1981.
9. *Crane Planning Guide for Metal Buildings,* ACCO Chain & Lifting Products Division, 76 ACCO Drive, Box 792, York, Pa., 1986.
10. *Structural Beam Design Guide and Selection Chart for Overhead Crane Runway System,* ACCO Chain & Lifting Products Division, York, Pa., 1991.
11. Kunming Gwo, "Steel Interchange," *Modern Steel Construction,* March 1995.
12. David T. Ricker, "Tips for Avoiding Crane Runway Problems," *AISC Engineering Journal,* vol. 19, no. 4, 1982.
13. W. M. Weaver, *Overhead Crane Handbook,* Whiting Corporation, Harvey, Ill., 1985.
14. *Manual of Steel Construction, Allowable Stress Design,* 9th ed., American Institute of Steel Construction, Chicago, Ill., 1989.
15. John E. Mueller, "Lessons from Crane Runways," *AISC Engineering Journal,* vol. 2, no. 1, Chicago, Ill., 1965.
16. Richard White and Charles Salmon (eds.), *Building Structural Design Handbook,* Wiley, New York, 1987.
17. Julius P. Van De Pas and James M. Fisher, "Crane Girder Design: An Examination of Design and Fatigue Considerations," *Modern Steel Construction,* March 1996.

16

Avoiding
Construction Problems

16.1 Introduction

Any metal building system, no matter how well designed, may become a continuous source of problems if installed incorrectly. Erection of metal buildings is a specialized field in which the builder's success depends on years of experience with a particular system or manufacturer. There is no single perfect way to construct a metal building, even in theory, as various manufacturers suggest slightly different methods of assembly, and erection techniques of various crews differ.

The objective of this chapter is not to guide design professionals, owners, and facility managers through every minute task of a construction process. Apart from being impractical in this context, it is simply unnecessary for those readers who visit the site only periodically. Instead, our aim is to give but a general idea about how construction of metal buildings should proceed and to describe some common "red flags" that signal trouble. Learning how to tell whether the builder follows good practice—and what the good practice is—is a valuable skill for anybody involved in construction of metal building systems.

16.2 Before Steel Erection Starts

At this point, we continue the discussion about the preconstruction process that began in Chap. 9. By this time, we hope, all the required submittals such as a letter of design certification and the shop drawing approval set have already been reviewed, all colors selected, and the site prepared.

Naturally, some construction—foundations, for example—takes place prior to steel erection. A slab-on-grade, if used, may be placed either before or after the metal building assembly. In conventional construction the slab is normally built after the building is enclosed, but in some pre-engineered buildings the slab is placed first. Such buildings include those with tilt-up walls, where slab-on-grade is needed for wall casting, and those which rely on slab-cast ties for lateral resistance. Some topics in slab-on-grade construction are explored in Chap. 12.

However tight the project schedule might be, steel erection should not begin before the concrete foundations are sufficiently cured: "Green" concrete will not hold anchor bolts and may crack under construction loading. Ideally, concrete should be allowed to cure for 28 days, although in practice this period is often shortened to a week. If time is critical, high-early concrete that reaches the required strength in as little as 3 days can be used.

It might be of some interest to the owner and to the engineer of record to observe the process of delivery, unloading, and temporary storing of the metal building system; a general impression from this observation, favorable or not, will likely be confirmed during construction.

Manufacturers concerned with the quality of their systems do not deliver bundles of unmarked metal and let the builders sort through it all. In fact, some building codes, such as BOCA,[1] require that each structural member, including siding and roofing panels, be identified by the manufacturer—imprinted with the manufacturer's name or logo and the part number or name. Similar language is contained in MBMA *Common Industry Practices*.[2] Unfortunately, complaints about shipments lacking piece marks are quite common. Some manufacturers even resort to use of "universal" (one-size-fits-all) purlins and girts to eliminate job-site confusion.

For the erectors, the way delivery trucks are packed can make a difference. The most useful method of packing is in the reverse order of erection, so that the items needed first are removed from the truck first. This packing method is reserved for those dealers with the foresight to ask for it in their agreements (order documents) with the manufacturers. Otherwise, manufacturers are free to load flatbed trailers in any way they please.

Upon delivery, the builder inspects the shipment. If the inspection discovers that packaged or nested metal components have become wet in transit, the builder is expected to unpack and dry them out to prevent rusting. Then the builder arranges for material storage, however brief. Here is where care, or a lack thereof, will show. Careful builders follow proper procedures during lifting slender cold-formed members which twist and deform easily, keeping in mind that any damage from a rough handling of "iron" will be easily noticeable.

Building components must be stored in accordance with the manufacturer's instructions. Roofing and siding panels are normally kept in a slightly sloped position for drainage, while cold-formed girts and purlins may be stored flat, to eliminate hard points at the supports. Proper dunnage keeps the metal members off the ground rather than allowing them to sink into the mud.[3] Rolls of fiberglass insulation are best kept off the ground and covered.

Experienced builders store the building components in a logical way which helps, rather than hinders, future erection. They also plan the erection process beforehand and know when each building part is needed. The old adage—those who fail to plan, plan to fail—is very true in construction of metal building systems.

16.3 Erection of Main Frames: The Basics

16.3.1 The braced bay

Now that the metal building package is in the builder's hands and the foundations are properly constructed, cured, and inspected, steel erection can begin. The erector can be the builder or a separate subcontractor. Normally, building manufacturers are not involved in the actual construction—unless, of course, they erect the building directly—and do not supervise or inspect the process of steel assembly. However, for critical projects or when the erector's expertise is in doubt, project specifications could require that a competent manufacturer's representative be present at the job site throughout the erection process to make certain that the building is put together properly. Whether an extra cost and perhaps contractual uncertainty of such representation are warranted should be decided carefully because it is the erector who is solely responsible for the means, methods, techniques, and sequences of construction. In any case, the manufacturer should provide erection manuals, erection drawings, and printed instructions.

The most common method of assembly begins with construction of a braced bay which consists of two parallel frames interconnected by girts, purlins, and wall and roof bracing. The braced bay is used as a stable element that the adjacent frames and endwalls can "lean on" during their installation. It is usually located in the second bay from an endwall.

The erection process starts when two adjacent columns are lifted into place by a crane or other equipment and temporarily stabilized by cross bracing. (For small rigid frames, the whole frame could be assembled on the ground and installed as described below.) Next, the columns are interconnected by one or two lines of wall girts which provide some stability and allow the columns to be plumbed (Fig. 16.1).

After the columns are plumbed, the wall bracing is tightened and the two columns at the opposite sidewall are similarly erected. Then the frame rafters that have been preassembled on the ground, where connections are much easier done than in the air, are lifted into place by a crane and bolted to the columns. The bolts are tightened only after the crane boom is repositioned to produce some slack in the cables, allowing the rafters to slightly stretch under their own dead load.[4] The procedure is repeated for the second frame in the braced bay, and the roof bracing is secured.

Next, a few purlins, usually including the peak purlin, are installed at the points where the roof cross bracing is attached to the frames, to form a truss-like roof diaphragm. Installation of column and rafter flange braces at the inside flanges, as shown in Figs. 4.14 and 4.15, completes the braced-bay assembly. The flange braces safeguard against lateral buckling of the frames

Figure 16.1 Assembly of multiple-span rigid frames. The columns at left are stabilized by diagonal bracing and wall girts. (*Photo: Maguire Group Inc.*)

and should not be neglected at this stage. In fact, some erection manuals[5,6] recommend installation of column flange braces even before the rafters are in place.

The endwall framing is erected next. For spans under 60 ft, it can be preassembled on the ground, lifted in place as a unit, and braced by purlins and girts extending to the braced bay. For spans over 60 ft, the endwall framing is erected similarly to the interior frames.[4] The assembly then moves to the adjacent frames, which are laterally supported by the girts and purlins attached to the braced bay. Finally, the opposite endwall framing is erected, followed by a final check and cleanup.

16.3.2   Other frame erection methods

A slightly different method is used for erection of small fully assembled single-span rigid frames. In this method the first frame is installed by a crane, plumbed, and stabilized on both sides by temporary guy wires. After the second frame is brought into place, girts and purlins are installed to brace it to the first

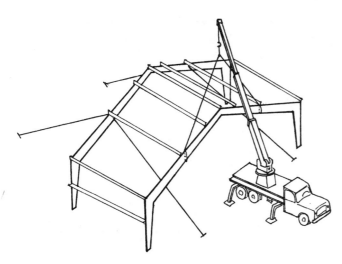

Figure 16.2 Erection of preassembled single-span rigid frames using guy wires.

frame (Fig. 16.2). The permanent wall and roof bracing is secured prior to removing the guy wires for the second frame.

Still another method of construction is used for buildings with tilt-up concrete walls. The tilt-up walls are installed and braced first, so as not to interfere with the frame erection. The best method of wall bracing is by temporary adjustable pipe braces anchored to the concrete slab or to the concrete deadmen cast into the ground for this purpose. Whenever job-built braces are used, a qualified engineer should be retained by the contractor to design the braces and their connections. Poorly planned panel bracing may result in panels lying shattered on the ground.[7]

Once erected and braced, tilt-up walls provide lateral stability to endwalls and to sidewall frame columns. The first intermediate frame can now be connected to the endwall framing by purlins and girts (Fig. 16.3). A similar procedure follows for the other interior frames. Erection of the main steel continues until all permanent roof bracing is in place and all connections are completed.

Regardless of the installation method, fabrication and erection tolerances included in MBMA *Common Industry Practices* apply to most metal buildings. The crane buildings require especially tight tolerances, since sloppy erection may cause excessive forces in the crane runway system and may result in a host of performance and durability problems.

16.3.3 The critical nature of erection bracing

Whichever installation method is preferred by the erectors, it should incorporate proper erection bracing, which may be more substantial than the perma-

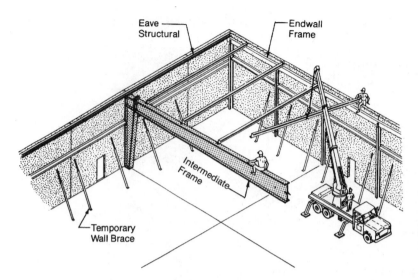

Figure 16.3 Steel erection in a tilt-up building. (*Star Building Systems.*)

nent bracing of the building. Notes the MBMA *Common Industry Practices:* "Bracing furnished by the Manufacturer for the Metal Building System cannot be assumed to be adequate during erection."[2] Indeed, the projected areas of all the exposed roof members might be larger than that of the enclosed building and thus receive more wind loading.

The owner or the engineer of record normally has no way of knowing what kind of bracing is adequate for erection and cannot detect whether a wrong kind is used. However, *some* sort of temporary wall and roof bracing is certainly needed. If there is none, or if only wall bracing is installed, trouble may be on a horizon: Inadequately braced pre-engineered buildings have been known to collapse during erection.

A case in point is documented by Sputo and Ellifritt,[8] who describe collapse of a rigid-frame building in Florida with the clear span of 206 ft. The erectors started installation with a post-and-beam endwall, not bothering to brace it with guy wires. They proceeded to erect the first rigid frame and installed cross bracing only in the sidewalls. The second rigid frame was erected next without any bracing; instead, the erectors seemed to rely on roof purlins that were installed between the two frames and the endwall. No roof cross bracing or frame flange braces were put in place, even though both were specified in the erection drawings. Curiously, a building inspector had noticed the lack of roof bracing during a site visit and had mentioned it to the erectors as a courtesy. No corrective action was taken. The next day, as the third rigid frame was being put in place in a similar fashion, the entire structure collapsed (Fig. 16.4). It was later discovered that the erectors had in their possession neither the erection manual nor the erection drawings and had never before worked

Figure 16.4 Pre-engineered building collapsed during erection. (*Photo: Prof. Duane S. Ellifritt.*)

with a building of that size. The authors conclude: "Inadequate bracing during erection probably contributes to more metal building system collapses than all other factors combined."

16.4 Installation of Girts and Purlins

As we have seen, some girts and purlins necessary for bracing are installed along with the main frames, and the remainder is put in place immediately thereafter. Girts and purlins may be bolted directly to the main framing steel or be attached to it via bearing clips prewelded to the frame, as illustrated in Chap. 5. The details of installation depend on whether girts and purlins are assumed to behave as simple-span or continuous members, on a degree of criticality of web crippling stresses, and on design span and loading.

It is often more cost-effective to raise the purlins onto the roof in bundles rather than one by one. The bundles are placed near the eaves, from where the erectors can move the individual purlins to their intended positions by hand.[5] Purlin bracing, the importance of which was demonstrated in Chap. 5, should be set in place as soon as possible, and definitely before the roofing is installed.

Slender cold-formed C and Z sections are easily distorted during construction. Sagging and twisted girts and purlins, a sad but common result of poor storage and installation practices, do not inspire confidence in the builder's work. Special care should be taken to avoid damaging those members by erection equipment and by careless people using them to support ladders, toolboxes, and similar gear.

Installation of secondary framing around wall and roof openings completes the erection of secondary steel. Some issues in specifying this framing are addressed in Chap. 10.

16.5 Placement of Insulation

Insulation is placed after the secondary steel is installed, but before the cladding. In buildings with fiberglass insulation under through-fastened roofing, the roofing is attached to purlins right through the insulation. The insulation placement usually begins with a 3-ft-wide starter roll being installed near an endwall. The subsequent rolls of normal width (4 to 6 ft) are then attached to purlins with self-drilling screws, which should be of proper length. The insulation blankets 6 in and thicker require longer screws ($1\frac{1}{2}$ or $1\frac{3}{4}$ in) than commonly used for roofing attachment, to avoid squeezing the insulation so tight that the panel gets dimpled.

Insulation placement at the eaves and at the edges of framed openings requires some finesse. Most manufacturers recommend letting the roof insulation overhang the framing edge, removing about 6 in of fiberglass from its facing, and then folding the facing back over the insulation to prevent wicking of moisture (Fig. 16.5).

The attachment of roof blanket insulation is a hazardous job, because no roofing is yet installed to support the workers in case of a fall. Butler Manufacturing Company uses its proprietary Sky-Web* Fall Protection and Insulation Support System, essentially a coated $\frac{1}{2}$-in woven polyester mesh, to provide fall protection at the leading edge of the roof. The mesh has the additional benefits of providing protection from falling objects for the workers below and, in some cases, of supporting the roof insulation.

16.6 Installation of Roof and Wall Panels

Whether roof or wall panels are installed first is a matter of the erector's preference. We slightly favor erecting roof first to allow for any interior finish work to begin and to improve the roof diaphragm action in the partially constructed building (Fig. 16.6).

Installation of roof panels usually starts at one of the endwalls chosen to allow for panel placement in the direction opposite to the prevailing winds; this sequence is intended to decrease the chances of wind-blown water intrusion into the panel sidelaps. The process begins with the roof panels being lifted in bundles by crane and placed directly above the main frames. Each purlin supporting the bundles receives a piece of snug-fitting wood blocking between its top flange and the frame rafter to protect the purlin from distortion.

---

*Sky-Web is a registered trademark of Butler Manufacturing Company.

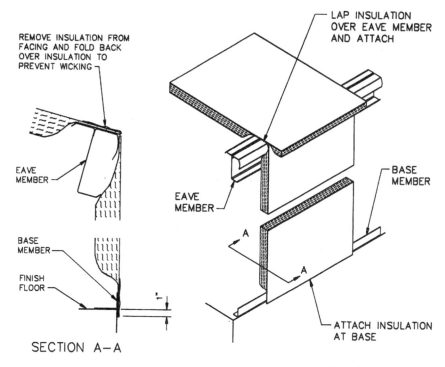

REMOVE INSULATION FROM
FACING AND FOLD BACK
OVER INSULATION TO
PREVENT WICKING

LAP INSULATION
OVER EAVE MEMBER
AND ATTACH

EAVE
MEMBER

EAVE
MEMBER

BASE
MEMBER

BASE
MEMBER

FINISH
FLOOR

A

A

ATTACH INSULATION
AT BASE

SECTION A—A

Figure 16.5 Insulation detail at eave and base. (*Varco-Pruden Buildings.*)

Figure 16.6 The roof of this pre-engineered building is installed before the walls. (*Photo: Maguire Group Inc.*)

The roof panels are laid down from eave to ridge or from lower to higher eave, without fastening. At the ridge line, the panels are held back the distance specified in the manufacturer's details. After further adjustments for the recommended panel endlap splices and low-eave overhang distances, sealants are added if required, and the roofing is fastened down. The action then shifts to the next line of panels, which are installed in a similar fashion, except that the panel seams must now be formed, either by hand or by a mechanical seamer.

One common occurrence during installation of ribbed metal roofing is "growing" and "shrinkage" of the panel width. The erectors can inadvertently make the panel width grow by stepping on the ribs and partially flattening them. Conversely, they may end up shrinking the width by not applying enough pressure while bringing the roofing in contact with the purlins. Such dimensional changes for each panel are small, but the cumulative error may be large enough to be noticeable. A special template or a spacer are useful in such circumstances.[6]

Stepping on the panel ribs is strongly discouraged by the manufacturers, as is walking on partially attached or unattached panels. The safe way to walk on a fully fastened roof is to step on walk boards laid in the panels' flat areas and spanning between the purlins. To prevent slippage, the walk boards should be secured to the roofing. If stepping on the panels is unavoidable, one should attempt to walk directly above the purlins where possible and to stay away from the middle of the flat panel part.

Wall panels are installed similarly to roofing. To minimize visibility of the vertical seams, the panels are best erected in a direction that allows their overlaps to face away from the main viewpoint of the building.

Field cutting of panels, especially of those with the state-of-the-art coatings on galvanized steel, weakens the panels' defense against corrosion by exposing unprotected metal, although galvanizing protects the cut edges to some degree. This reason alone makes factory precut panels preferable to those formed on-site. If unavoidable, panel cutting should be made on the ground, carefully and precisely. The edges should be touched up with a special compound supplied by the manufacturer; all the metal dust and shavings should be promptly removed lest, in humid climates, rust stains appear quickly.

16.7 Some Common Problems During Construction

Disregard of good practices invariably leads to problems with the appearance, function, or longevity of a newly constructed metal building. Some of these mishaps occurring with a disheartening regularity are described below.

16.7.1 Problems with slabs and foundations

Anchor bolts missing or out of alignment. When not placed by a template, anchor bolts are likely to end up in a wrong position and to create a minicrisis involving frantic calls to the engineer and the manufacturer. Some easy fixes such as making new holes in the column base plate and drilling-in new expansion or chemical anchors may solve the problem. Otherwise a new larger base plate or an extension of the existing one may be needed; truly critical cases might even require foundation replacement. All anchor bolts in oversized holes should be supplied with thick washers under the nuts. The washers are typically $\frac{5}{16}$ to $\frac{1}{2}$ in thick; this thickness must be accounted for when the bolt projection is detailed.[10]

The anchor bolts that do not protrude high enough to allow for a proper nut engagement are no less troublesome. The best solution to this common problem is to extend the bolts by welding short pieces of threaded rods; the welded end of each extension piece must be cut at 45° to allow for full-penetration welding. Alternatively, a special threaded coupler can be used for splicing, perhaps necessitating a removal of some base-plate metal and concrete. In any case, a few plate washers are needed to elevate the nut above the connection material.

The usual contractor's proposal of filling the void within the nut with weld will not provide a strong connection and deserves to be rejected, especially for bolts designed for tension loads. A classic example of a failure of anchor bolts plug-welded in this fashion occurred in a Louisiana school during high winds. When the welds and the minimal-length threads suddenly failed, a springlike action reportedly took place and the columns with attached beams were thrown into the air.[9]

Slab cracking or curling. Cracked slabs-on-grade are the perennial source of owners' complaints. Most of the many potential reasons for this cracking have to do with poor construction quality. Drying-shrinkage cracking, which occurs within days of the slab placement, is usually caused by a lack of proper control and construction joints. For example, a popular control-joint detail calls for every other wire of the welded wire fabric to be cut at the control joint locations. The detail will not work if the cutting of wires is not done—a frequent oversight—and the slab is not weakened enough at the joints to induce cracking there. As a result, the slab will crack elsewhere.

Major slab cracking accompanied by settlement indicates an improper subgrade preparation; most other cracks can be traced to inadequate slab curing.

A curling of slab edges usually results from improper detailing and execution of construction joints. It is known, for instance, that keyed construction joints are more likely to curl than doweled joints. The use of underslab vapor barriers without sand cushions (see Fig. 12.27) has also been linked to slab curling. Chap. 12 provides some recommendations on building better slabs-on-grade and on avoiding both slab cracking and curling.

What to do about a cracked or curled slab? The corrective measures could range from doing nothing at all for minor cosmetic cracks and curling to filling the cracks with epoxy or a total slab replacement for critical superflat floor applications.

Improperly placed or missing wall dowels. Reinforcement dowels extending from foundation walls into a slab-on-grade are specified for several reasons. Most commonly, the dowels are needed to support the slab by helping it span over the poorly compacted soil near the wall and to provide lateral support for the top of the wall. These dowels are supposed to be bent in the field, since subgrade compaction cannot take place with the dowels sticking out from the wall.

Unfortunately, regardless of what's shown on the design drawings, dowels are often supplied prebent and then bent in the field again—out of the way. Or dowels are omitted completely and require subsequent drilling and grouting in. Or dowels are too short to be of use. Because of their sheer number, improperly placed dowels can become a significant source of friction between the foundation contractor and the owner.

16.7.2 Problems with metal superstructure

Gaps under column base plates. In conventional construction, column base plates normally bear on grouted leveling plates or are positioned on special leveling nuts. Very large base plates may be installed on a bed of grout separately from the column and welded to it later. In each case, the objective is to ensure a complete bearing under the bearing plates.

In pre-engineered construction, both the leveling plates and the grout are often dispensed with. The MBMA *Common Industry Practices* specifically excludes "grouting or filling of any kind under columns" from the metal building erector's work. The concrete subcontractor is probably long gone when the grouting is needed. Who's to do it?

Not surprisingly, most of the time the pre-engineered columns, which must be plumb within MBMA tolerances, are placed directly on top of concrete piers or foundation walls which are not perfectly level. As a result, the column base plates bear on concrete only at one edge, with a small gap under the rest of the plate (Fig. 16.7). If this gap is not filled with grout or shims, the concrete may crack or the column base may slightly deform and settle under load, causing a rattle-producing "play" between the base plate and the anchor-bolt nut above it. Base plates of building columns which support cranes, and therefore require stability and precision of installation, should always be grouted, even by the owner's personnel if necessary.

Loose cross bracing. Rod or cable bracing installed in roofs and walls of metal buildings is frequently observed to be loose, bent, or even missing. Such bracing does not fulfill its objective of stabilizing the building and ensuring that it can resist external loads without excessive deformations.

A cumulative movement of the structure, which can occur before the bracing is taut enough to be effective, can crack skylights and windows, jam doors, and disrupt operations of the structure-supported equipment—much as an excessive story drift would. The excessive "play" can also damage brittle wall finishes and lead to perceptible rattles and vibrations. Fortunately, this common erection deficiency is easy to spot and to correct.

Figure 16.7   A gap under the column base plate.

Member bracing missing or not properly secured. The importance of properly installed purlin and girt bracing has been emphasized in Chap. 5. Flange bracing for column and rafter interior flanges, where required by design, is equally important. Still, both these kinds of member bracing are regularly found to be missing, not properly connected, or installed the wrong way (e.g., a parallel purlin bracing is installed where a cross bracing was specified). Again, this deficiency is usually easy to correct if noticed in time, prior to an application of the interior finishes and insulation.

16.7.3 Roof leaks

One of the most frequent complaints about metal building systems involves leaky metal roofs. At the design stage, chances of leaks can be greatly reduced if the roofing type and details are properly selected as described in Chaps. 6 and 14. Still, the most common reason for leaks is improper construction. A case in point: Watertightness of standing-seam roofing with trapezoidal seams depends greatly on a proper installation of corrugation closure strips at the eaves. This detail is not perfect even under the best of circumstances. If, however, the end closures are simply omitted, or poorly sealed, leakage is virtually guaranteed, because there is nothing to stop water coming from an overflowing gutter or from an ice dam. Two other examples are unprotected roof penetrations—a notoriously fertile ground for leaks—and an omission of the required sealants. Indeed, as Star Manufacturing Company's *Erection Guide*[5] points out, 99 percent of the leaks can be traced to the following:

- Omission or mislocation of the required bar- and caulk-type sealants in the longitudinal roofing seam cavities
- Failure to install the extra strip of sealant known as pigtail at the four-way panel laps and at the eave connections
- Failure to install tape sealant under the screw heads
- Not caulking between the eave trim and the underside of the roof panels

A presence of the required sealants at these critical locations can be verified by a special "feeler" tool made of thin enough (0.005 in) material to fit inside the roofing seams. Wherever the sealant is determined to be missing, the only certain method of repair is to remove the panel in question and to reapply the sealant.

Such testing is time-consuming and well beyond the expertise of most owners and design professionals to whom few other safeguards of installation quality are available. The most basic and simple precaution—checking the erector's success rate in producing leak-free roofs—should of course be taken prior to signing the contract.

16.7.4 Accepting the job

There could be many more punch-list construction deficiencies, large and small, structural and cosmetic, that are identified by the owners and their design professionals near the end of the job.

Structure-related items are the most serious. In addition to the ones described above, these might include poorly made framing connections such as loose, missing, or improperly tightened bolts and sloppy welds. Some members might be mislocated or have an insufficient length of bearing. Exposed panel fasteners might not be well aligned or properly tightened; applying too little torque to the screws with neoprene washers can leave the penetration unprotected; applying too much can dimple the panel. Identification of these problems is best left to experienced construction inspectors or engineers retained by the owner.

Sloppy fit-up causes appearance problems which may lead not to structural distress but certainly to an emotional one. Door jambs out of plumb, 1-in-wide caulked joints, and sagging gutters are difficult to miss. Field-formed roof and wall panels seem to suffer more than their share of problems with rusting, buckling, oil canning, and poor fit-up.

A convenient job completion checklist used by the builders of Star Building Systems, and similar to the one originally developed by MBMA, is reproduced with permission in Fig. 16.8. Intended mainly for steel erectors, the checklist may prove beneficial to owners and design professionals in their quest to realize the full benefits of metal building systems. A properly constructed metal building system provides an aesthetically appealing, practical, and virtually maintenance-free environment for many years.

JOB COMPLETION CHECK LIST

CUSTOMER'S NAME_____

BUILDING SIZE_____

LOCATION_____

I. STRUCTURAL INSPECTION   **Yes No**

  a. Are all anchor bolts washered and properly nutted?   ____ ____

  b. Is steel plumb, square, and aligned?   ____ ____

  c. Are required brace rods tight with necessary bevel washers? ____ ____

  d. Do base plates align properly with concrete?   ____ ____

  e. Have all connections been properly made and fully bolted according to plans and specifications? ____ ____

  f. Are High Tensile bolts in place and torqued as required? ____ ____

  g. Are eave struts, main building and canopies, straight and level? ____ ____

  h. Are purlins and girts properly made up and in good alignment without roll-over and with all sag rods in place? ____ ____

  i. Have all component parts, ridge sag rods, clip angles, haunch and flange braces, etc., as called for on erection drawings, been properly installed? ____ ____

  j. Is structural primer clean, with shop coat in good condition and any erection marks, burn and/or smoke properly repainted? ____ ____

  k. Have exposed structural members with shop primer been given a field coat? ____ ____

  l. Are headers and jambs for framed openings straight, unwarped and erected plumb and square, are unused holes in the opening filled with bolts? ____ ____

  m. Was any burning or welding, not ordered on erection drawings, necessary? If yes, explain on reverse side. ____ ____

  n. Are any structural components bent, warped, or dented? If yes, identify these areas on erection drawing and return with this form. ____ ____

II. SHEETING AND TRIM

  a. Have roof and wall sheets been properly aligned, lapped and fully fastened? ____ ____

Figure 16.8 Job completion checklist. (*Star Building Systems.*)

SECTION 10. MISC
PAGE NO. 2
DATE 6-1-79
REPLACES 2-3-78

▲ ERECTION GUIDE

JOB COMPLETION CHECK LIST

		Yes	No
b.	Have fasteners been over-driven?	___	___
	Are any fasteners loose?	___	___
	Are any fasteners missing?	___	___
c.	Have bottoms of sheets been dented by improperly aligned base angle?	___	___
d.	Do sheets or trim show ladder scratch or other field damage?	___	___
e.	Are all field-cut sheets cut clean, showing careful workmanship, and properly covered by trim or flashing?	___	___
f.	Is all trim in place, neatly and carefully installed with all laps fastened and all joints and butts closely fitted?	___	___
g.	Is gutter straight; do all laps have sealant and are they adequately fastened?	___	___
h.	Are there any short sections of gutter or trim due to improper field cutting?	___	___
i.	Are down spouts according to plan, properly cut-in and sealed, jointed with the flow, neatly secured and fastened to high rib in vertical line; are bottoms of elbows above finished paving line, or properly lead into underdround drain?	___	___
j.	Have proper fasteners been used in applying trim?		
k.	Has roof been swept, gutteres cleaned, roof and wall sheets cleaned of drill shavings?	___	___
l.	Have scratches in panel and trim been touch-up painted?	___	___

III. WEATHER PROOF

		Yes	No
a.	Has sealant, when required, been properly installed and has loose mastic and paper backing been removed?	___	___
b.	Has closure strip been carefully installed and sealed where required?	___	___
c.	Are there any light leaks?		
d.	Have LTP, roof vents, sheet cuts at openings, windows, thresholds and any other points of possible leakage been carefully and neatly sealed with prescribed material?	___	___
e.	Has loose film been removed from LTP?	___	___

Figure 16.8 (*Continued*) Job completion checklist. (*Star Building Systems.*)

⋀ **ERECTION** | SECTION 10.   **MISC**
★★★ **GUIDE** | PAGE NO. 3
B U I L D I N G S | DATE 6-1-79
| REPLACES 2-3-78

JOB COMPLETION CHECK LIST

		<u>Yes</u>	<u>No</u>
f.	Have four-way laps been checked to insure pigtail was properly installed?	___	___
g.	Have eave pigtails been checked to insure proper location?	___	___
h.	Have mitered eave panels been properly chaulked?	___	___
i.	Are LTP free of cracks (especially at fastening points)?	___	___

IV. INSULATION

a.	Is insulation neatly installed according to standards for lapping and securing?	___	___
b.	If used, has mesh wire been tightly stretched, properly secured, buttlaps hidden behind purlins and side laps made continuous?	___	___
c.	If used, are insulation trim strips properly secured with laps behind purlins?	___	___
d.	Has insulation been folded back at eave strut and around all openings?	___	___
e.	Has exposed insulation at bottom of wall sheets been facing lapped back to prevent wicking?	___	___
f.	Have all punctures or tears in vapor barrier been properly sealed?	___	___
g.	Have the proper length fastener been used for insulation thickness?	___	___

V. ACCESSORIES

a.	Do all accessories having manual or mechanical movement operate freely and properly?	___	___
b.	Do walk doors fit openings and latch properly?	___	___
	Do lock sets operate?	___	___
	Does interlock weather-strip fit?	___	___
	Is glazing complete?	___	___
c.	Do slide doors operate freely; are all guides in place; is latch properly aligned?	___	___

Figure 16.8   (*Continued*) Job completion checklist. (*Star Building Systems.*)

ERECTION GUIDE

SECTION 10. MISC
PAGE NO. 4
DATE 6-1-79
REPLACES 2-3-78

JOB COMPLETION CHECK LIST

Yes No

d. Do overhead doors fit and close; are keepers and lock sets properly adjusted; is tension correct? ___ ___

e. Are all door keys accounted for? ___ ___

f. Do all windows work freely and latch properly? ___ ___

Is all glazing complete? ___ ___

Are latches installed properly? ___ ___

g. Are all vent dampers hooked up and operating properly? ___ ___

OVER-ALL INSPECTION

a. Have original building plans and any changes been fully complied with? ___ ___

b. Are all openings located according to plan? ___ ___

c. Are side wall sheets lapped away from street-front of building or prevailing winds? ___ ___

d. Are all fastener lines straight and in prescribed pattern? ___ ___

e. Are all building lines proper; eave and rake line, openings, ridge and vents? ___ ___

f. Is building identification properly installed? ___ ___

g. Has proper touch-up of color imperfections on sheets and trim been satisfactorily accomplished? ___ ___

h. Have all mud, hand prints or other handling and erection marks been properly removed? ___ ___

i. Has construction site been properly cleaned and cleared of debris? ___ ___

j. Is flashing around openings in roof or between building and other collateral material such as masonry, glass, etc., proper and correct? ___ ___

k. Has pre-engineered building primary or secondary steel been modified to accommodate a field change? If yes, note on erection drawing. ___ ___

Date_____ Signature_____
Erection Foreman
or
Superintendent

Figure 16.8 (*Continued*) Job completion checklist. (*Star Building Systems.*)

References

1. The BOCA National Building Code, Building Officials and Code Administrators International, Inc., Country Club Hills, Ill., 1993.
2. *Low Rise Building Systems Manual,* Metal Building Manufacturers Association, Cleveland, Ohio, 1986.
3. *Metal Building Systems,* Building Systems Institute, Inc., 2d ed., Cleveland, Ohio, 1990.
4. Donna Milner, "Metal Building Basics," *The Journal of Light Construction,* April 1989.
5. *Erection Guide, Star Building Systems,* Star Manufacturing Co., Oklahoma City, Okla., 1989.
6. *Erection Manual,* Steelox Systems, Inc., Mason, Ohio, 1994.
7. "Guidelines for Bracing Tilt-up Walls," *Concrete Construction,* December 1995.
8. Thomas Sputo and Duane S. Ellifritt, "Collapse of Metal Building System During Erection," *Journal of Performance of Constructed Facilities,* vol. 5, no. 4, November 1991, American Society of Civil Engineers.
9. Research at Texas Tech University quoted by Thomas C. Powell in "Steel Interchange," *Modern Steel Construction,* December 1992.
10. David T. Ricker, "Some Practical Aspects of Column Base Selection," *AISC Engineering Journal,* 3d quarter 1989.

Information about Some Pre-engineered Building Manufacturers*

A & M Building Systems

Address:	P.O. Drawer 1450, Clovis, NM 88102
Year Founded:	1964
Founder:	Nels Anderson
President:	Nels Anderson
Plant Location:	Clovis, NM
Company History:	A & M began as a metal building dealer and later developed into a manufacturer of small agricultural buildings. Today, A & M produces buildings for varied uses and design diversities.
Products/Markets:	A & M supplies commercial, industrial, agricultural, and architectural buildings to a 14-state area.

---

*Excerpted from "1995 Who's Who among Pre-engineered Building Manufacturers," *Metal Architecture,* April 1995, to include only MBMA members and Metallic Building Company. Reprinted with permission.

Other MBMA members in 1996 included Crown Metal Buildings, Inc., of Cabot, Ariz.; Dean Steel Buildings, Inc., of Fort Myers, Fla.; Mueller Supply Company, Inc., of Ballinger, Tex.; NCI Building Systems, Inc., of Houston, Tex.; OSI Building Systems of Montgomery, Ala.; and Package Industries, Inc., of Sutton, Mass.

Alliance Steel Inc.

Address:	8600 West Reno, Oklahoma City, OK 73127
Year Founded:	1978
Plant Location:	Oklahoma City, OK
Company History:	Alliance began manufacturing metal buildings and components in Oklahoma. Once a regional producer, the company now services the entire continental United States.
Products/Markets:	From anchor bolts to ridge caps, Alliance manufactures complete metal building systems. Standard R panels, A panels, and standing-seam panels are marketed for the component sector.

American Buildings Co.

Address:	P.O. Box 800, Eufaula, AL 36072
Year Founded:	1947
Founder:	Willard B. Joy
President & CEO:	Robert R. Ammerman
Plant Locations:	Eufaula, AL; El Paso, IL; Carson City, NV; and LaCrosse, VA
Company History:	American Buildings Company began in a small facility under a viaduct in Columbus, GA, and has since grown to become a leading systems building producer.
Products/Markets:	American's complete line of building systems are designed for commercial, industrial, recreational, and agricultural purposes. Wall systems include interlocking panels with no exposed fasteners and an architectural panel system. The company utilizes its own coil coating facility to assure uniform, highly adherent quality finishes. The company has its own roofing and architectural products division plus a heavy fabrication unit in Houston, TX. The company has a nationwide network of approximately 900 authorized builders and 250 preferred roofers. American Buildings is a public company (NASDAQ AmBldgs:ABCO).
Support Services:	Operating on the personal computers of authorized builders and preferred roofings, SPECTRUM and SUMMIT software programs for building systems and roofs do complete proposals and create detail drawings to meet design requirements. Both programs can estimate and produce complete tally sheets.

American Steel Building Co.

Address:	P.O. Box 14244, Houston, TX 77221
Year Founded:	1953
Founder:	Art Crispin

President:	M.B. Tankersley
Plant Location:	Houston, TX
Company History:	American Steel Building Company began as a regional fabricator. Its business has since evolved to include national and international distribution of its product lines. The company is employee-owned and -operated.
Products/Markets:	The company's line includes pre-engineered rigid frame structures, single- and multistory ministorage buildings, Durastone construction panels, standing-seam roofing, bar joists, overhead bridge cranes, industrial pallet rack systems, and other building components.

Behlen Mfg. Co.

Address:	P.O. Box 569, Columbus, NE 68602
Year Founded:	1936
Founder:	Walter Behlen
President & CEO:	A.F. "Tony" Raimondo
Plant Locations:	Columbus, NE; Goshen, IN; Corsicona, TX; and Dublin, GA
Company History:	Established in 1936, Behlen has long been a manufacturer of metal building systems. Behlen focused in the 1980s to establish its ADS Frame Building System in the commercial/industrial metro markets and supplement it with the proprietary S-Span and Ag-Frame building lines. The result was significant growth. Today, Behlen is an employee-owned company.
Products/Markets:	Behlen Building Systems focuses on the urban/metropolitan areas of the country with a highly developed ADS (Advanced Design Frame Building System) supported by the self-supporting S-Span system. The Behlen standing-seam roof system is available for both new and reroof construction. Behlen makes extensive use of computer-aided design (CAD) programs and computer-aided manufacturing (CAM) processes. While Behlen is recognized as a leading regional manufacturer of pre-engineered building systems, its distribution network crosses the nation and circles the globe.
Support Services:	Behlen maintains an experienced staff of registered professional engineers to assist in building design and a knowledgeable team of field sales managers. Engineering manuals are available for both the Behlen Frame Building System and the S-Span building system.

Bigbee Steel Buildings Inc.

Address:	1821 Avalon Avenue, Muscle Shoals, AL 35661
Year Founded:	1962

Founder:	Perry Bigbee
Plant Location:	Muscle Shoals, AL
Company History:	Bigbee Steel Buildings has been a member of the Metal Building Manufacturers Association for more than 30 years and remains privately owned by its founder. Through the years, Bigbee buildings have been constructed in 22 states and seven countries.
Product Line:	Bigbee produces a complete line of buildings in standard and nonstandard designs. Two exterior wall panel configurations are offered. Both high-rib roof panels and standing-seam roof panels are available in 24- or 26-gage steel.

Butler Manufacturing Co.

Address:	P.O. Box 419-917, Kansas City, MO 64141
Year Founded:	1901
Founders:	Charles Butler and Emanuel E. Norquist
President:	Richard Jarman
Plant Locations:	Galesburg, IL; Annville, PA; Laurinburg, NC; Birmingham, AL; San Marcos, TX; Visalia, CA; Kirkcaldy, Scotland; Jeddah, Saudi Arabia; and Guadalajara, Mexico
Company History:	Butler began in 1902 making livestock watering tanks. World War II created a huge demand for quickly assembled metal buildings. Butler provided most of the steel hangars shipped overseas to the Army and Navy during the war. Since WWII, Butler has been the worldwide leader in the production of metal structural, roof, and wall systems. Butler has been the industry's innovator, developing numerous structural, roof, and wall panel systems which have increased the end uses of pre-engineered systems.
Products/Markets:	Butler offers 4 structural systems, 4 roof systems, and 10 wall panel systems. Among the frame styles are a traditional package; a system utilizing long-bay, open-web purlins; a self-bracing long-bay system intended for hardwall construction; and a multistory system. Roof systems include standing-seam profiles, a ribbed panel, and an insulated standing-seam panel. All panel systems are coated with Fluoropon, a full-strength 70% Kynar 500/Hylar 5000.

CBC Steel Buildings

Address:	1700 E. Louise Avenue, Lathrop, CA 95330
Year Founded:	1984
Founders:	Roger Benton and Ron Johnson

President:	Ron Johnson
CEO:	Roger Benton
Plant Location:	Lathrop, CA
Company History:	After 11 years of expansion and growth, CBC has manufactured more than 4000 custom buildings with over $100 million in sales. CBC markets primarily in the 13 western states and Canada through an extensive builder network. Capacity exceeds 500 tons/week.
Products/Markets:	With two fully automated frame bays, CBC can produce up to 7-ft-deep frames with over 300 ft clear span width. The company is a full-service manufacturer, providing all components for a complete building system including a standing-seam roof. Secondary structurals are offered in precoated red oxide or G-90 galvanized and more.

Ceco Building Systems

Address:	P.O. Box 6500, Columbus, MS 39703
Year Founded:	1947
Founder:	C.L. Mitchell
President:	Terrell Landrum
Plant Locations:	Columbus, MS; Rocky Mount, NC; and Mount Pleasant, IA
Company History:	Ceco Building Systems was founded in 1947 as Mitchell Engineering in Columbus, MS. The Rocky Mount, NC, plant was added in 1967, and the Mt. Pleasant, IA, plant was added in 1972. From its start as a regional manufacturer, Ceco has grown to become one of the industry's leading manufacturers.
Company Description:	Ceco markets its products through a network of approximately 600 builders, serving the needs of the local commercial, community, and industrial construction markets. These products include the full range of metal building products including wide-bay and multistory and auxiliary products such as mini-storage buildings and retrofits. Ceco enhances the builder's position in the marketplace by providing support in the form of strong national advertising, direct mail, training, national accounts, and builder computer services. Ceco has developed an exceptionally strong district sales manager group and provides its own erection support service.
Support Services:	Through its presence in a number of architectural publications, Ceco stays in touch with the design community. A design manual and specification disk are available by calling the company's toll-free number. During 1995, a disk with standard Ceco details will be made available. In addition, the manufacturer's district sales manager group is available to assist architects on individual projects.

Chief Industries Inc.

Address:	P.O. Box 2078, Grand Island, NE 68802
Year Founded:	1954
Founder:	Virgil Eihusen
Plant Locations:	Grand Island, NE, and Rensselaer, IN
Company History:	Virgil Eihusen founded Chief Industries as a construction company. In 1966, the company began manufacturing and selling its own line of pre-engineered building systems. Branching out in other manufacturing areas as well, the company has grown to include 1100 employees with 10 divisions and 6 subsidiaries.
Products/Markets:	Chief's Buildings Division supplies standard and specialized pre-engineered building systems for industrial, commercial, institutional, and sports/recreational purposes. Wide-span and multistory systems are available, as well as standing-seam roof systems. Chief and its two manufacturing plants are certified by the American Institute of Steel Construction to have the personnel, organization, equipment, capability, and commitment to fabricate metal building systems of the required quality for Category MB, Metal Building Systems, as set forth by the AISC Quality Certification Program. Chief systems are sold through independent authorized builders.

Garco Building Systems

Address:	P.O. Box 19248, Spokane, WA 99219
Year Founded:	1958
Founder:	Wayne Garceau
President:	Terry Middaugh
Plant Location:	Spokane, WA
Company History:	Beginning in 1958, the company engineered and fabricated the primary and secondary structural steel for buildings, packaging them with panels, trim, and secondary framing from component suppliers. Today, Garco is a fully integrated manufacturer of complete building systems.
Products/Markets:	Garco Building Systems has authorized dealers located in most metropolitan areas throughout the western United States and Canada. The dealers provide full construction services for commercial, industrial, agricultural, and governmental clients. Many of the dealers also provide design/building projects supported by Garco's custom building design services. Component sales include complete wall/roof panel and trim packages. Among the panel systems available is a standing-seam roof system.
Support Services:	Garco's product specification and design manual is available to assist in the use or integration of Garco products. Garco also provides design services through its dealers to assist in the development of projects.

Gulf States Manufacturers Inc.

Address:	P.O. Box 1128, Starkville, MS 39759
Year Founded:	1968
Founder:	Clayton Richardson
President:	Clayton Richardson
Plant Location:	Starkville, MS
Company History:	Gulf States began as a pre-engineered metal building supplier and has evolved into a complete custom metal building system and components supplier. The company was established on principles of exceptional quality and service. Its approach to business has always been conservative, from the design standpoint as well as the financial end.
Products/Markets:	Gulf States produces and supplies complete metal building systems and components, and provides comprehensive customer service. From column base plate to peak flashing, the company provides structural framing, cold form, five wall panels, three roof panels including the Shur-Lok panel, trim, components, and accessories. Gulf States' dealer network covers the eastern half of the United States and a portion of the western half. The International Group has dealers in Central and South America, Korea, Mexico, and the Caribbean, and has shipped buildings into 23 foreign countries. Primary markets are commercial and industrial, offices and retail, municipal and institutional.
Support Services:	Gulf states operates an in-house engineering and design staff, offers efficient and timely sales and estimating services and state-of-the-art AutoCAD services. The company offers its own computer pricing program for building estimates and take-offs.

Inland Buildings

Address:	2141 2nd Avenue S.W., Cullman, AL 35055
Year Founded	1993
Founder:	Dean Grant
President/CEO:	Dean Grant
Executive VP— Sales & Marketing:	Ken C. Williams
Plant Locations:	Cullman, AL; and Waukesha, WI
Company History:	Inland produced its first metal building in 1910, making Inland the oldest name in the metal building business. Inland acquired the Metal Building Division of Steelcraft Industries in Cincinnati, OH. In the late 1960s, Inland moved the buildings division, which was then known as Inland-Ryerson, to Milwaukee and changed the name to Inryco. After selling the business in 1986 to the owners of Sonoco Buildings, the company began a slowly digressing spiral to eventual bankruptcy in 1992. The current company parent organization restarted the business in January 1993 after a bankruptcy asset purchase.

Products/Markets:	Inland Buildings does business in the United States and abroad. Inland currently concentrates its sales efforts in the eastern two-thirds of the United States. The company produces both architectural and ribbed wall panels, as well as the industry's only 9-in architectural ribbed panel. In addition, Inland produces a 24- and 26-gage ribbed roof panel and two types of standing-seam panels offering 18- and 20-in coverages. Roof and wall panels are available in a variety of colors, finishes, and gages.
Support Services:	Inland offers CAD-automated engineering and is capable of generating AutoCAD drawings and details for architects and engineers. For design/build contractors, Inland custom designs and builds every order. Inland works on a daily basis with architects and engineers to develop the best possible building for the given requirements.

Kirby Building Systems

Address:	P.O. Box 390, Kirby Drive, Portland, TN 37148
Year Founded:	1955
Founder:	George F. King
President:	George F. King
Plant Location:	Portland, TN
Company History:	Kirby began as a general contracting firm in Houston, TX. The company began fabrication of its own line of steel buildings in the early 1960s. Distribution through a network of authorized builders soon followed. Today, the company operates under a corporate structure established in 1985, serving the eastern two-thirds of the United States from its centrally located facilities.
Products/Markets:	Kirby manufactures building systems for a variety of markets including commercial, industrial, institutional, and recreational facilities. Kirby also offers a full line of metal building components and retrofit products.

Mesco Metal Buildings

Address:	P.O. Box 20, Grapevine, TX 76099
Year Founded:	1955
Plant Locations:	Grapevine, TX; and Chester, SC
Company History:	When it was formed, Metal Structures Corporation (now known as Mesco) was based in Dallas, TX. In 1957, the company relocated to its present 22-acre site just west of the Dallas/Ft. Worth International Airport. The company opened its second manufacturing facility in Chester, SC, in 1975.

Products/Markets:	Mesco manufactures and markets a low-rise metal building system for commercial, industrial, institutional, recreational, agricultural, and specialty applications. A wide variety of framing systems, roof, and wall panel options (including a standing-seam system) are also available.

Metal Building Products Inc.

Address:	3861 Old Getwell Rd., Memphis, TN 38118
Year Founded:	1986
Founder:	Gary Sims
President:	Gary Sims
Plant Location:	Memphis, TN
Company History:	Gary Sims formed a small metal building fabrication/contracting company in 1971. As well as constructing small buildings, his company performed construction as a metal building dealer for other manufacturers. On January 1, 1986, Gary and Barry Sims moved to Memphis, TN, and founded Metal Building Products Inc. In May of that year, the business moved to a new office and warehouse facility. In 1992, a major expansion project was initiated to accommodate the company's rapid growth. In 1995, a new fabrication shop was erected to increase plant output.
Products/Markets:	MPBI markets pre-engineered buildings for various uses, including commercial, industrial, agricultural, ministorage, and institutional. Primary markets for MBPI includes the southeast, east and midwest United States. The company also exports. MBPI buildings are available in standard configurations or can be customized. The buildings are skinned with screw-down, standing-seam, or concealed fastener panels. MBPI also markets metal building components including zees and cees from 4 to 12 in, roof and wall panels, trim, and accessories. MBPI products are marketed and sold primarily through builders and other metal building industry professionals.
Support Services:	MPBI offers professional and flexible in-house engineering, drafting, and estimating, and a qualified sales staff to assist owners, specifiers, architects, and engineers from project planning to completion.

Metallic Building Company

Address:	7301 Fairview, Houston, TX 77041
Year Founded:	1946
Founders:	Charlie McDaniel and Gilbert Leach
President:	Johnie Schulte

Plant Locations:	Houston, TX; Mattoon, IL; Caryville, TN; Jackson, MS; Hobbs, NM; and Tallapoosa, GA
Company History:	Metallic Building Co. has a rich history that dates back to the mid-1940s. Throughout the years, Metallic Building Co. has been known as Metallic, Marathon Metallic, and Metallic Braden. Some of Metallic Building Co.'s current employees have been with the company since the 1950s, while some of the company's builder dealerships date back to the 1960s.
Products/Markets:	Metallic Building Co. designs and manufactures a full range of metal buildings for industrial, commercial, and institutional uses. Metallic also has specialized designs and products for aircraft hangars, ministorage, and retrofit roofing applications. Metallic's design and fabrication practices are certified by the AISC Quality Assurance program under the MB category. With 3 standing-seam roof systems, 4 wall panels, 11 standard Dura-20 colors, and 8 premium Royal K-70 colors available, Metallic has the solution for most specifications. Metallic markets its products through a nationwide network of authorized builders.

Nucor Building Systems

Address:	2100 Rexford Road, Charlotte, NC 28211
Year Founded:	1986
Founder:	Nucor Corporation
President:	H. Robert Lowe, P.E.
Company History:	While the parent corporation has been in existence for many years, Nucor began an aggressive expansion program in the late 1980s. One such division included Nucor Building Systems. The company's commitment to both its customers and employees, coupled with its philosophy of providing the workforce with the best available technology, assures Nucor customers will receive a high degree of quality and service.
Products/Markets:	Nucor Building Systems is a complete custom-design building manufacturer. Its building packages are based on the latest engineering technology. A full staff of in-house engineers and detailing personnel ensures prompt and efficient service of special design requirements. Nucor is AISC certified. Buildings are sold through a nationwide builder network.

Ruffin Building Systems

Address:	6914 Highway 2, Oak Grove, LA 71263
Year Founded:	1972
Founder:	Shelton Ruffin
President:	Shelton Ruffin

Plant Location:	Oak Grove, LA
Company History:	Incorporated in 1972, Ruffin began operations in a 2000-ft^2 building. In 1981, a $1 million equipment investment resulted in the purchase of rollformers and an automatic welding line. In 1987, Ruffin created a district manager sales force. Today, Ruffin's plant has grown to 210,000 ft^2 with the workforce totaling 120 employees. Sales have reached $24 million, with buildings shipped to 37 states and several foreign countries. In its 22 years of production, Ruffin has manufactured more than 12,000 structures.
Products/Markets:	Ruffin's state-of-the-art computer systems enable it to offer a complete line of customer-designed pre-engineered metal building systems and related components including a ministorage system. By July 1995, a selection of colored architectural panels and a standing-seam roof system will be available. Ruffin products are sold through a network of authorized dealers.
Support Services:	Ruffin's customer service department is available to provide assistance to architects and specifiers.

Southern Structures

Address:	P.O. Box 52005, Lafayette, LA 70505
Year Founded:	1967
Founder:	Alfred Leonpacher
President:	Stefan Leonpacher
Plant Location:	Youngsville, LA
Company History:	Since its creation, Southern Structures has grown steadily from being a local manufacturer to becoming a national supplier with state-of-the-art manufacturing capability. Southern Structures sells through a network of systems builders.
Products/Markets:	Southern Structures primarily distributes in the south, southeast, and midwestern United States, plus Latin America. In addition to manufacturing and marketing complete pre-engineered building systems, the company offers detailed facade, crane, mezzanine, multistory, and retrofit roofing packages.
Support Services:	Southern Structures offers computerized customer design development.

Star Building Systems

Address:	P.O. Box 94910, Oklahoma City, OK 73143
Year Founded:	1927
Founder:	D.H. Roland
President:	Jack Taylor

Company History:	Star Building Systems was formed in 1927 to meet the needs of oil drillers in the historic Oklahoma oil boom. The product at that time was small portable metal buildings located near rig floors to shelter men and equipment. In the late 1930s and during World War II, Star began to meet demands for military warehouses and aircraft hangars. In the 1950s, Star began producing grain storage and sectional steel buildings for utilitarian warehouses and storage sheds. In the 1970s, the market served expanded rapidly, as did the whole metal building industry. Star's market opened up into commercial, industrial, and a broad spectrum of other market segments.
Products/Markets:	Today, Star serves the commercial, industrial, institutional, recreational, agricultural, and retrofit markets, both domestically and internationally. Star products are distributed through more than 650 Star Builders across the country, and in many foreign countries.
Support Services:	Star Building Systems provides services to design professionals to help them specify metal construction projects. Star Building specifications are available in print as well as on disk and are complete specifications for structural framing, roof and wall systems, accessories, and insulation, as well as design loads and erection standards of performance. For architectural firms with Auto-CAD capabilities and modems, Star offers Star-Net Bulletin Board for downloading architectural details on a 24-h basis.

Steelox Systems Inc.

Address:	5412 Courseview Drive, no. 300, Mason, OH 45040
Year Founded:	1929
Founder:	Armco Steel Corporation
President:	Frank B. Chapman
Plant Location:	Washington Court House, OH
Company History:	The business began with the acquisition of Henry Brothers, a metal garage building company. Armco purchased the patents to the Steelox panel in 1934. Known as Armco Building Systems, the company acquired Atlantic Buildings in 1984. In 1986, the assets of Armco Building Systems were sold to Southwestern General Corporation and the name changed to Building Technologies Corporation. Later, BTC, owned by a financial consortium, was sold to its present owners, Kawada Industries USA, and renamed STEELOX SYSTEMS INC.
Products/Markets:	STEELOX SYSTEMS INC. designs, manufactures, and distributes a complete line of metal building and roofing systems for the commercial, industrial, and institutional markets. Its specialty markets include complex metal building systems, small buildings, and new and retrofit roof installation required

by the manufacturing segment. Special emphasis is given to projects requiring weathertight standing-seam roofs and architecturally aesthetic roof and wall coverings.

United Structures of America

Address:	1912 Buschong, Houston, TX 77039
Year Founded:	1980
Founder:	Richard Drake
President:	Richard Drake
Plant Locations:	Houston, TX; and Portland, TN
Company History:	The company began in Houston, TX, in 1980 and opened a second plant in Tennessee in 1985.
Products/Markets:	U.S.A. offers a complete line of pre-engineered metal buildings for industrial, commercial, institutional, hangar, and multistory applications. In addition to standard and custom metal buildings, U.S.A. has a structural steel and retrofit division.

Varco-Pruden Buildings

Address:	6000 Poplar Avenue, Suite 400, Memphis, TN 38119
Year Founded:	1969
Founders:	Clark Prudhon and Jack Hatcher
President:	David M. Gilchrist (as of 1996)
Plant Locations:	Rainsville, GA; Pine Bluff, AR; Turlock, CA; St. Joseph, MO; Kernersville, NC; VanWert, OH; and Evansville, IN
Company History:	Varco-Pruden Buildings was formed in 1969 from a merger between Prudhon Products and Varco Steel. VP is the world's second largest supplier of building systems, with seven plants located throughout the nation to back its network of 1100 builders.
Products/Markets:	Among the primary building components VP manufactures are primary and secondary framing systems, wall systems, and roof systems. VP's family of framing systems include rigid frame, continuous beam frame, unibeam and unibeam lean-to, open-web truss, open-web continuous truss, and a variety of secondary framing systems. Wall systems include Panel Rib, Vee Rib Span Loc, architectural panels, insulated sandwich panels, and nonmetal walls. VP's roof systems include the Standing Seam Roof (SSR), Panel Rib, and PR+.
Support Services:	In response to current technological changes in the industry, VP designed and introduced VP Command, a state-of-the-art computer-aided design system, detail and pricing system, in 1992. Regional, state, and local building codes have been added

to VP Command, assuring accurate designs and price quotes in areas where the codes are unique. VP Command also enables builders to accommodate whatever architectural features a given project requires.

Whirlwind Steel Buildings Inc.

Address:	P.O. Box 75280, Houston, TX 77234
Year Founded:	1955
Founder:	C.O. Sturdivant
President:	Jack Sturdivant
Plant Location:	Houston, TX
Company History:	Whirlwind was founded in 1955 as a manufacturer of power exhaust fans, louvers, and other metal products. In 1963, Whirlwind sold the fan and louver business and has since devoted its resources to the manufacturing of pre-engineered metal buildings and components.
Products/Markets:	Whirlwind designs and manufactures a full line of pre-engineered metal buildings for commercial, industrial, and institutional uses. The company offers a standing-seam roof, 3 wall panels, 17 standard colors, and 8 Kynar500 colors.

B

Properties of Cold-Formed Girts and Purlins

Two sets of tables are presented here. The first set (Tables B.1 to B.4), courtesy of Star Building Systems and Ceco Building Systems, includes section properties of cold-formed girts and purlins produced by those companies and by some other major metal building manufacturers. The second set (Tables B.5 to B.16), includes some cold-formed sections produced by members of Light Gage Structural Institute (LGSI); it is reprinted with permission from LGSI's *Light Gage Structural Steel Framing System Design Handbook* (1995).

These tables include *effective* section properties which can be used directly for preliminary member selection—assuming proper lateral bracing is present, as discussed in Chap. 5.

The exact depth of cold-formed girts and purlins likely to be used by the metal building manufacturer is difficult to predict, because standard member configurations vary among different suppliers. For example, Butler Manufacturing Co. offers 8-, 9.5-, and 11-in Z sections, while Varco-Pruden Buildings makes 6.5-, 8.5-, 11.5-, and 14-in Z sections. Both Star Building Systems and Ceco Building Systems produce 8.5- and 10-in Z sections, and many other manufacturers provide 8- and 10-inch members. LGSI Z sections are available in depths of 3.5, 4, 6, 7, 8, 9, 10, and 12 in.

TABLE B.1

STAR BUILDING SYSTEMS

COLD FORM Z SECTIONS   1986 AISI

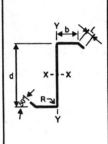

SECTION (d x b)	L (IN)	THK. (IN)	R (IN)	AREA (IN²)	WT./FT (LBS.)	Properties of Section							
						I_x (IN⁴)	I_x DEFL (IN⁴)	S_f (IN³)	S_e (IN³)	r_x (IN)	r_y (IN)	I_{yc} (IN⁴)	I_{xy} (IN⁴)
8.5 x 2.5 Z .057	.769	.057	.219	.83	2.84	8.94	8.94	2.10	1.70	3.27	1.19	.59	2.35
8.5 x 2.5 Z .064	.781	.064	.219	.94	3.18	10.02	10.02	2.36	2.06	3.27	1.19	.67	2.65
8.5 x 2.5 Z .072	.951	.072	.219	1.08	3.66	11.53	11.53	2.71	2.48	3.27	1.26	.85	3.22
8.5 x 2.5 Z .080	.965	.080	.219	1.20	4.07	12.78	12.78	3.01	2.92	3.27	1.26	.95	3.58
8.5 x 2.5 Z .088	.978	.088	.219	1.31	4.47	14.03	14.03	3.30	3.30	3.27	1.27	1.05	3.95
10 x 2.75 Z .075	1.02	.075	.219	1.28	4.34	18.70	18.70	3.74	3.39	3.83	1.35	1.17	4.75
10 x 2.75 Z .084	1.03	.084	.219	1.43	4.86	20.95	20.95	4.19	3.91	3.82	1.35	1.31	5.32
10 x 2.75 Z .100	1.06	.100	.219	1.71	5.80	24.86	24.86	4.97	4.97	3.82	1.36	1.57	6.35

COLD FORM C SECTIONS   1986 AISI

SECTION (d x b)	L (IN)	THK. (IN)	R (IN)	AREA (IN²)	WT./FT (LBS.)	Properties of Section							
						I_x (IN⁴)	I_x DEFL (IN⁴)	S_f (IN³)	S_e (IN³)	r_x (IN)	r_y (IN)	r_o (IN)	C_w (IN⁶)
8.5 x 3.25 C .075	.750	.075	.219	1.18	4.02	13.19	12.52	3.10	2.57	3.34	1.16	4.25	23.20
8.5 x 3.25 C .092	1.00	.092	.219	1.49	5.06	16.53	16.53	3.89	3.43	3.33	1.20	4.34	33.31
10 x 2.75 C .075	1.15	.075	.219	1.28	4.34	18.42	18.42	3.68	3.53	3.79	1.02	4.43	29.50
10 x 2.75 C .084	1.17	.084	.219	1.43	4.86	20.58	20.58	4.12	4.05	3.79	1.02	4.42	33.13
10 x 2.75 C .100	1.21	.100	.219	1.71	5.80	24.38	24.38	4.88	4.88	3.78	1.02	4.42	39.55

COLD FORM EAVE STRUT SECTIONS   1986 AISI

SECTION (d x b)	L (IN)	THK. (IN)	R (IN)	AREA (IN²)	WT./FT (LBS.)	Properties of Section							
						I_x (IN⁴)	I_x DEFL (IN⁴)	S_f (IN³)	S_e (IN³)	r_x (IN)	r_y (IN)	r_o (IN)	C_w (IN⁶)
8.5 x 3.25 ES .075	.750	.075	.219	1.18	4.02	13.19	12.52	3.10	2.57	3.34	1.16	4.25	23.20
8.5 x 3.25 ES .092	1.00	.092	.219	1.49	5.06	16.53	16.53	3.89	3.43	3.33	1.20	4.34	33.31
9.625 ES .084	1.00 & 1.25	.084	.219	1.60	5.43	22.55	20.74	4.28	3.77	3.78	1.51	5.21	56.25
* 10 x 3.25 ES .084	1.204	.084	.219	1.52	5.18	22.72	22.72	4.54	4.22	3.86	1.22	4.75	49.45

* Monticello, Iowa Plant Only

344

TABLE B.2

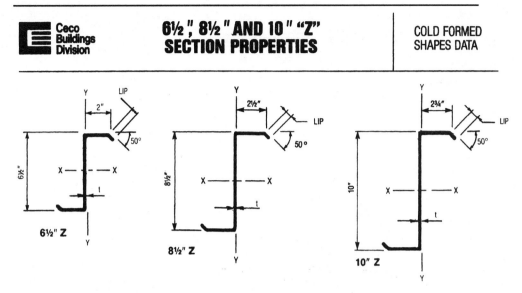

| Ceco Buildings Division | 6½ ", 8½ " AND 10 " "Z" SECTION PROPERTIES | COLD FORMED SHAPES DATA |

MINIMUM YIELD STRESS (Fy) = 55 ksi

							SECTION PROPERTIES									
							Moment of Inertia I_x (in⁴)			Section Modules S_x(in³)		Allow. Moment M_a (ft-kips)	Allow. Shear V_a (kips)	Gross I_y (in⁴)	Radius of Gyr.	
Section	Depth (in)	Flange Width (in)	Lip (in)	Thkns. (in)	Gross Area (in²)	Weight (lb/ft)	Gross	Defl.	Eff.	Gross	Eff.				r_x (in)	r_y (in)
6.5Z58	6.5	2.00	0.620	.058	0.657	2.23	4.13	4.13	3.85	1.269	1.134	3.12	2.74	0.609	2.51	0.96
6.5Z82	6.5	2.00	0.650	.082	0.927	3.15	5.77	5.77	5.77	1.776	1.776	4.88	7.52	0.865	2.50	0.97
8.5Z57	8.5	2.50	0.769	.057	0.834	2.83	8.93	8.65	7.85	2.102	1.698	4.67	1.95	1.183	3.27	1.19
8.5Z64	8.5	2.50	0.781	.064	0.936	3.18	10.01	9.96	9.17	2.356	2.046	5.63	2.76	1.333	3.27	1.19
8.5Z72	8.5	2.50	0.951	.072	1.076	3.66	11.53	11.53	10.96	2.712	2.500	6.88	3.94	1.708	3.27	1.26
8.5Z80	8.5	2.50	0.965	.080	1.195	4.06	12.78	12.78	12.62	3.007	2.945	8.10	5.41	1.905	3.27	1.26
8.5Z88	8.5	2.50	0.978	.088	1.314	4.47	14.03	14.03	14.03	3.300	3.300	9.08	7.21	2.103	3.27	1.27
10Z75	10.0	2.75	1.000	.075	1.277	4.34	18.70	18.70	17.51	3.739	3.366	9.26	3.74	2.301	3.83	1.34
10Z84	10.0	2.75	1.000	.084	1.428	4.85	20.85	20.85	19.96	4.170	3.886	10.69	5.27	2.558	3.82	1.34
10Z100	10.0	2.75	1.062	.100	1.706	5.80	24.86	24.86	24.86	4.971	4.971	13.67	8.92	3.143	3.82	1.36

Section Properties are in accordance with the 1986 Edition with 1989 Addendum of the AISI Specification for the Design of Cold-Formed Steel Structural Members.

For all zees: lip angle 50°; inside radius flange to lip = 0.25 in., flange to web = 0.218 in.

TABLE B.3

Ceco Buildings Division	**8½" AND 10" EAVE STRUT SECTION PROPERTIES**	COLD FORMED SHAPES DATA

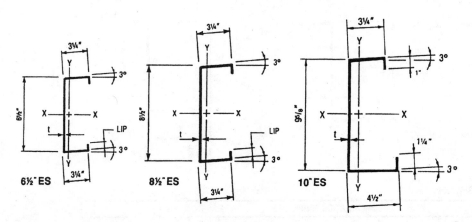

6½" ES   8½" ES   10" ES

MINIMUM YIELD STRESS (Fy) = 55 ksi

SECTION PROPERTIES															
Dimensions						Gross Section Properties								Effective Section Properties	
Section	Depth d (in)	Flange Width b (in)	Lip c (in)	Thkns t (in)	Weight W (lb/ft)	Area A (in²)	Moment of Inertia I_x (in⁴)	Section Modulus S_x (in³)	Radius Of Gyr. r_x (in)	Moment of Inertia I_y (in⁴)	Torsional Constant J (in⁴)	Warping Constant C_w (in⁶)	Polar Radius of Gyr. r_o (in)	Area A_e (in²)	Moment of Inertia I_{xe} (in⁴)	Section Modulus S_{xe} (in³)
6.5E72	6.5	3.25	1.31	.072	3.66	1.08	7.27	2.18	2.60	1.75	.0019	19.54	4.11	0.72	6.71	1.91
6.5E88	6.5	3.25	1.34	.088	4.47	1.31	8.81	2.71	2.59	2.14	.0033	24.16	4.11	1.00	8.49	2.55
8.5E75	8.5	3.25	.75	.075	4.02	1.18	13.18	3.03	3.34	1.58	.0022	21.94	4.20	.65	11.81	2.58
8.5E92	8.5	3.25	1.00	.092	5.06	1.49	16.52	3.80	3.33	2.14	.0042	31.76	4.30	.94	15.46	3.46
10E84	9.625	3.25* 4.25#	1.00 1.25	.084	5.42	1.60	20.62	3.99* 4.62#	1.55							

* = Top Flange in Compression
\# = Bottom Flange in Compression
Section properties are in accordance with the 1986 Edition with 1989 Addendum of the AISI specification for the Design of Cold-Formed Steel Structural Members.
For all eave struts: lip angle 90°; inside radius flange to lip = 0.218 in., flange to web = 0.218 in.

TABLE B.4

6½″ AND 8½″ "C" **SECTION PROPERTIES**	COLD FORMED SHAPES DATA

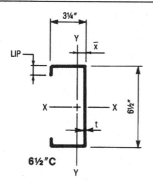

6½″C

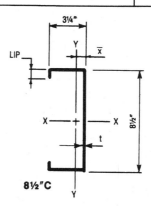

8½″C

MINIMUM YIELD STRESS (Fy) = 55 ksi

SECTION PROPERTIES																	
Dimensions						Gross Section Properties									Effective Section Properties		
Section	Depth d (in)	Flange Width b (in)	Lip c (in)	Thkns t (in)	Weight W (lb/ft)	Area A (in²)	Moment of Inertia I_x (in⁴)	Section Modulus S_x (in³)	Radius of Gyr. r_x (in)	Moment of Inertia I_y (in⁴)	Section Modulus S_y (in³)	Radius of Gyr. r_y (in)	Torsional Constant J (in⁴)	Warping Constant C_w (in⁶)	Polar Radius of Gyr. r_o (in)	Area A_e (in²)	Moment of Inertia I_{xe} (in⁴)	Section Modulus S_{xe} (in³)
6.5C72	6.5	3.25	1.31	.072	3.66	1.08	7.27	2.24	2.60	1.76	0.86	1.28	.0019	19.60	4.12	0.72	6.72	1.96
6.5C88	6.5	3.25	1.34	.088	4.47	1.31	8.81	2.71	2.59	2.14	1.06	1.28	.0034	24.16	4.11	1.00	8.49	2.55
8.5C75	8.5	3.25	0.75	.075	4.03	1.18	13.19	3.10	3.34	1.59	0.68	1.16	.0022	22.02	4.21	0.65	11.73	2.58
8.5C92	8.5	3.25	1.00	.092	5.06	1.49	16.53	3.89	3.33	2.14	0.95	1.20	.0042	31.87	4.30	0.94	15.36	3.45

Section properties are in accordance with the 1986 Edition with 1989 Addendum of the AISI Specification for the Design of Cold-Formed Steel Structural Members.

For all cees: lip angle 90°: inside radius flange to lip = 0.218 in., flange to web = 0.218 in.

TABLE B.5

LIGHT GAGE STRUCTURAL INSTITUTE

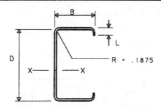

R = .1875

SECTION PROPERTIES

SECTION				AXIS X - X			AXIS Y - Y			
D x B	GA	WEIGHT LB/FT	AREA IN²	Ix IN⁴	Se IN³	Rx IN	Iy IN⁴	Sy IN³	Ry IN	L IN
4 X 2 C	16	1.89	.56	1.39	.68	1.58	.34	.27	.78	.84
	14	2.36	.69	1.71	.85	1.57	.42	.33	.78	.87
	13	2.89	.85	2.06	1.03	1.56	.51	.41	.78	.90
	12	3.32	.98	2.34	1.17	1.55	.59	.46	.77	.93
4 X 2.5 C	16	2.06	.61	1.58	.69	1.63	.55	.35	.95	.75
	14	2.57	.76	1.98	.90	1.62	.68	.43	.95	.78
	13	3.14	.92	2.39	1.18	1.61	.83	.53	.95	.81
	12	3.62	1.06	2.72	1.36	1.60	.95	.61	.95	.84
4 X 4 C	16	2.68	.79	1.94	.80	1.71	1.71	.71	1.47	.75
	14	3.34	.98	2.49	1.04	1.71	2.12	.89	1.47	.78
	13	4.09	1.20	3.12	1.34	1.69	2.59	1.08	1.47	.81
	12	4.71	1.38	3.75	1.60	1.69	2.98	1.25	1.47	.84
6 X 2.5 C	16	2.48	.73	4.04	1.19	2.37	.64	.37	.93	.75
	14	3.08	.91	5.07	1.55	2.36	.79	.46	.93	.78
	13	3.77	1.11	6.15	2.03	2.35	.96	.56	.93	.81
	12	4.34	1.28	7.03	2.34	2.35	1.11	.64	.93	.84
6 X 4 C	16	3.10	.91	4.84	1.35	2.50	1.98	.76	1.47	.75
	14	3.86	1.14	6.18	1.76	2.50	2.46	.94	1.47	.78
	13	4.72	1.39	7.76	2.26	2.49	3.01	1.15	1.47	.81
	12	5.43	1.60	9.35	2.69	2.48	3.46	1.33	1.47	.84
7 X 2.5 C	16	2.68	.79	5.80	1.47	2.73	.67	.37	.92	.75
	14	3.34	.98	7.28	1.91	2.72	.83	.46	.92	.78
	13	4.09	1.20	8.85	2.50	2.71	1.02	.57	.92	.81
	12	4.71	1.38	10.12	2.89	2.70	1.17	.65	.92	.84
7 X 4 C	16	3.31	.97	6.85	1.62	2.88	2.09	.77	1.46	.75
	14	4.12	1.21	8.78	2.17	2.87	2.60	.96	1.46	.78
	13	5.04	1.48	11.03	2.77	2.87	3.18	1.18	1.46	.81
	12	5.80	1.71	13.27	3.29	2.86	3.66	1.36	1.46	.84
8 X 2 C	16	2.68	.79	7.09	1.76	3.00	.41	.27	.72	.75
	14	3.34	.98	8.78	2.20	2.99	.51	.34	.72	.78
	13	4.09	1.20	10.68	2.67	2.98	.62	.41	.72	.81
	12	4.71	1.38	12.23	3.06	2.97	.71	.47	.72	.84
8 X 2.5 C	16	2.89	.85	7.94	1.75	3.08	.70	.38	.91	.75
	14	3.60	1.06	9.98	2.30	3.07	.87	.47	.90	.78
	13	4.41	1.30	12.14	3.00	3.06	1.06	.58	.90	.81
	12	5.07	1.49	13.90	3.47	3.05	1.22	.66	.90	.84

NOTES
1 Properties and allowables are computed in accordance with the 1986 edition of the AISI specifications.
2 Ix is for deflection determination.
3 Se is for bending.
4 Sy and Iy are for full section.

TABLE B.6

LIGHT GAGE STRUCTURAL INSTITUTE

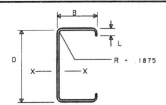

R = .1875

SECTION ALLOWABLES

SECTION						BEARING (KIPS)			
D x B	GA	Ma	Ae	Oc	Va	END		INTERIOR	
		KIP-FT	IN²		KIPS	F	V	F	V
4 X 2 C	16	1.91	.508	1.92	4.24	.539	.088	1.267	.145
	14	2.41	.668	1.92	6.02	.969	.128	2.084	.192
	13	2.91	.849	1.92	7.29	1.605	.173	3.255	.245
	12	3.30	.976	1.92	8.32	2.246	.210	4.413	.289
4 X 2.5 C	16	1.96	.543	1.92	4.24	.539	.088	1.267	.145
	14	2.55	.727	1.92	6.02	.969	.128	2.084	.192
	13	3.34	.924	1.92	7.29	1.605	.173	3.255	.245
	12	3.85	1.063	1.92	8.32	2.246	.210	4.413	.289
4 X 4 C	16	2.26	.539	1.92	4.24	.539	.088	1.267	.145
	14	2.96	.752	1.92	6.02	.969	.128	2.084	.192
	13	3.81	1.015	1.92	7.29	1.605	.173	3.255	.245
	12	4.54	1.214	1.92	8.32	2.246	.210	4.413	.289
6 X 2.5 C	16	3.37	.525	1.92	3.44	.502	.082	1.205	.138
	14	4.38	.747	1.92	6.58	.918	.121	2.004	.185
	13	5.73	.970	1.92	9.85	1.537	.165	3.153	.237
	12	6.63	1.165	1.92	13.04	2.164	.202	4.295	.281
6 X 4 C	16	3.83	.535	1.92	3.44	.502	.082	1.205	.138
	14	5.01	.755	1.92	6.58	.918	.121	2.004	.185
	13	6.42	1.036	1.92	9.85	1.537	.165	3.153	.237
	12	7.63	1.259	1.92	13.04	2.164	.202	4.295	.281
7 X 2.5 C	16	4.16	.522	1.92	2.91	.484	.079	1.174	.135
	14	5.40	.752	1.92	5.66	.892	.117	1.964	.181
	13	7.06	.981	1.92	9.85	1.503	.162	3.103	.234
	12	8.18	1.182	1.92	13.04	2.123	.198	4.236	.277
7 X 4 C	16	4.59	.534	1.92	2.91	.484	.079	1.174	.135
	14	6.14	.756	1.92	5.66	.892	.117	1.964	.181
	13	7.87	1.041	1.92	9.85	1.503	.162	3.103	.234
	12	9.33	1.266	1.92	13.04	2.123	.198	4.236	.277
8 X 2 C	16	4.96	.501	1.92	2.53	.466	.076	1.143	.131
	14	6.20	.673	1.92	4.90	.867	.114	1.923	.177
	13	7.54	.886	1.92	9.03	1.469	.158	3.052	.230
	12	8.63	1.075	1.92	13.04	2.082	.195	4.176	.273
8 X 2.5 C	16	4.95	.498	1.92	2.53	.466	.076	1.143	.131
	14	6.50	.739	1.92	4.90	.867	.114	1.923	.177
	13	8.48	.976	1.92	9.03	1.469	.158	3.052	.230
	12	9.83	1.179	1.92	13.04	2.082	.195	4.176	.273

NOTES
1 Properties and allowables are computed in accordance with the 1986 edition of the AISI specifications.
2 Ma is the allowable moment for sections that are supported laterally for their full length.
3 Oc is the safety factor for axial load.
4 Ae is the reduced area for axial load.
5 Oc and Ae are based on $KL_y = KL_T = 1.0$ ft. and KL_x as follows: D=4, 12 ft.; D=6, 16 ft.; D=7,8, and 9, 18 ft.;
  D=10 and 12, 20 ft.
6 F and V are coefficients used to determine the web crippling strength. Refer to sample calculations.

TABLE B.7

 # LIGHT GAGE STRUCTURAL INSTITUTE

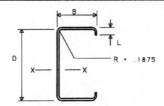

SECTION PROPERTIES

SECTION				AXIS X - X			AXIS Y - Y			
D x B	GA	WEIGHT LB/FT	AREA IN²	Ix IN⁴	Se IN³	Rx IN	Iy IN⁴	Sy IN³	Ry IN	L IN
8 X 3.5 C	16	3.31	.97	8.71	1.84	3.20	1.57	.64	1.27	.75
	14	4.12	1.21	11.23	2.51	3.19	1.96	.79	1.27	.78
	13	5.04	1.48	14.63	3.18	3.19	2.40	.97	1.27	.81
	12	5.80	1.71	17.23	3.68	3.18	2.76	1.12	1.27	.84
8 X 4 C	16	3.51	1.03	9.15	1.87	3.25	2.18	.78	1.45	.75
	14	4.38	1.29	11.93	2.59	3.25	2.72	.98	1.45	.78
	13	5.36	1.58	14.98	3.31	3.24	3.33	1.20	1.45	.81
	12	6.16	1.81	18.01	3.93	3.23	3.83	1.38	1.45	.84
9 X 2.5 C	16	3.10	.91	10.37	1.97	3.42	.72	.38	.89	.75
	14	3.86	1.14	13.21	2.71	3.41	.90	.48	.89	.78
	13	4.72	1.39	16.08	3.53	3.40	1.10	.58	.89	.81
	12	5.43	1.60	18.42	4.09	3.39	1.26	.67	.89	.84
10 X 2 C	16	3.10	.91	11.72	2.24	3.66	.43	.27	.69	.75
	14	3.86	1.14	15.13	3.03	3.65	.54	.34	.69	.78
	13	4.72	1.39	18.43	3.69	3.64	.66	.42	.69	.81
	12	5.43	1.60	21.13	4.23	3.64	.75	.48	.69	.84
10 X 2.5 C	16	3.31	.97	13.04	2.19	3.75	.74	.39	.87	.75
	14	4.12	1.21	17.00	3.15	3.75	.93	.48	.87	.78
	13	5.04	1.48	20.71	4.09	3.74	1.13	.59	.87	.81
	12	5.80	1.71	23.74	4.75	3.73	1.30	.68	.87	.84
10 X 3.5 C	14	4.64	1.36	18.69	3.27	3.90	2.10	.81	1.24	.78
	13	5.67	1.67	24.60	4.33	3.89	2.58	1.00	1.24	.81
	12	6.53	1.92	28.98	5.01	3.89	2.97	1.15	1.24	.84
10 X 4 C	14	4.89	1.44	19.65	3.33	3.96	2.92	1.01	1.43	.78
	13	5.99	1.76	25.13	4.50	3.96	3.58	1.23	1.43	.81
	12	6.89	2.03	30.17	5.32	3.95	4.12	1.42	1.43	.84
12 X 2.5 C	14	4.64	1.36	25.57	3.74	4.40	.97	.49	.84	.78
	13	5.67	1.67	32.25	5.32	4.40	1.19	.60	.84	.81
	12	6.53	1.92	37.00	6.17	4.39	1.36	.69	.84	.84
12 X 3.5 C	14	5.15	1.52	27.85	3.96	4.58	2.22	.83	1.21	.78
	13	6.30	1.85	38.23	5.47	4.58	2.72	1.02	1.21	.81
	12	7.25	2.13	44.57	6.48	4.57	3.13	1.17	1.21	.84
12 X 4 C	14	5.41	1.59	29.30	4.05	4.66	3.09	1.03	1.39	.78
	13	6.62	1.95	38.18	5.58	4.65	3.78	1.26	1.39	.81
	12	7.62	2.24	46.25	6.86	4.65	4.35	1.45	1.39	.84

NOTES
1 Properties and allowables are computed in accordance with the 1986 edition of the AISI specifications.
2 Ix is for deflection determination.
3 Se is for bending.
4 Sy and Iy are for full section.

TABLE B.8

LIGHT GAGE STRUCTURAL INSTITUTE

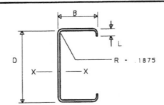

SECTION ALLOWABLES

SECTION						BEARING (KIPS)			
D x B	GA	Ma KIP-FT	Ae IN²	Oc	Va KIPS	END		INTERIOR	
						F	V	F	V
8 X 3.5 C	16	5.22	.521	1.92	2.53	.466	.076	1.143	.131
	14	7.11	.732	1.92	4.90	.867	.114	1.923	.177
	13	9.03	.978	1.92	9.03	1.469	.158	3.052	.230
	12	10.44	1.269	1.92	13.04	2.082	.195	4.176	.273
8 X 4 C	16	5.31	.521	1.92	2.53	.466	.076	1.143	.131
	14	7.36	.739	1.92	4.90	.867	.114	1.923	.177
	13	9.41	1.021	1.92	9.03	1.469	.158	3.052	.230
	12	11.15	1.229	1.92	13.04	2.082	.195	4.176	.273
9 X 2.5 C	16	5.57	.494	1.92	2.23	.448	.073	1.112	.128
	14	7.67	.725	1.92	4.33	.841	.111	1.883	.173
	13	9.99	.973	1.92	7.96	1.434	.154	3.001	.226
	12	11.58	1.177	1.92	12.16	2.041	.191	4.117	.269
10 X 2 C	16	6.32	.500	1.92	1.99	.429	.070	1.081	.124
	14	8.53	.673	1.92	3.87	.815	.107	1.843	.170
	13	10.40	.889	1.92	7.11	1.400	.151	2.951	.222
	12	11.91	1.082	1.92	10.87	2.000	.187	4.058	.265
10 X 2.5 C	16	6.20	.496	1.92	1.99	.429	.070	1.081	.124
	14	8.91	.729	1.92	3.87	.815	.107	1.843	.170
	13	11.58	.979	1.92	7.11	1.400	.151	2.951	.222
	12	13.43	1.186	1.92	10.87	2.000	.187	4.058	.265
10 X 3.5 C	14	9.28	.725	1.92	3.87	.815	.107	1.843	.170
	13	12.28	.965	1.92	7.11	1.400	.151	2.951	.222
	12	14.19	1.255	1.92	10.87	2.000	.187	4.058	.265
10 X 4 C	14	9.45	.732	1.92	3.87	.815	.107	1.843	.170
	13	12.76	1.015	1.92	7.11	1.400	.151	2.951	.222
	12	15.09	1.217	1.92	10.87	2.000	.187	4.058	.265
12 X 2.5 C	14	10.59	.714	1.92	3.19	.764	.101	1.763	.162
	13	15.03	.975	1.92	5.87	1.332	.143	2.849	.214
	12	17.44	1.183	1.92	8.96	1.917	.179	3.939	.258
12 X 3.5 C	14	11.21	.712	1.92	3.19	.764	.101	1.763	.162
	13	15.51	.981	1.92	5.87	1.332	.143	2.849	.214
	12	18.36	1.227	1.92	8.96	1.917	.179	3.939	.258
12 X 4 C	14	11.48	.719	1.92	3.19	.764	.101	1.763	.162
	13	15.83	1.000	1.92	5.87	1.332	.143	2.849	.214
	12	19.45	1.254	1.92	8.96	1.917	.179	3.939	.258

NOTES
1 Properties and allowables are computed in accordance with the 1986 edition of the AISI specifications.
2 Ma is the allowable moment for sections that are supported laterally for their full length.
3 Oc is the safety factor for axial load.
4 Ae is the reduced area for axial load.
5 Oc and Ae are based on $KL_y = KL_T = 1.0$ ft. and KL_x as follows: D=4, 12 ft.; D=6, 16 ft.; D=7,8, and 9, 18 ft.; D=10 and 12, 20 ft.
6 F and V are coefficients used to determine the web crippling strength. Refer to sample calculations.

TABLE B.9

LIGHT GAGE STRUCTURAL INSTITUTE

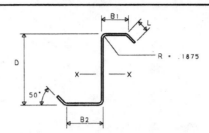

SECTION PROPERTIES

D X B1 X B2	SECTION	GA	WEIGHT LB/FT	AREA IN²	AXIS X - X			AXIS Y - Y			L IN
					Ix IN⁴	Se IN³	Rx IN	Iy IN⁴	Sy IN³	Ry IN	
3.5 X 1.5 X 1.5	3.5 X 1.5 Z	16	1.58	.47	.76	.50	1.38	.38	.19	.91	.74
		14	1.97	.58	.94	.62	1.37	.48	.24	.91	.77
4 X 2.125 X 2.375	4 X 2.5 Z	16	2.06	.61	1.54	.72	1.62	1.12	.39	1.36	.90
		14	2.57	.76	1.97	.95	1.61	1.40	.48	1.36	.92
6 X 2.125 X 2.375	6 X 2.5 Z	16	2.48	.73	3.95	1.23	2.37	1.12	.38	1.24	.90
		14	3.08	.91	5.05	1.63	2.36	1.40	.48	1.24	.92
		13	3.77	1.11	6.13	2.00	2.35	1.73	.59	1.25	.95
		12	4.34	1.28	7.01	2.29	2.34	2.01	.68	1.25	.97
7 X 2.125 X 2.375	7 X 2.5 Z	16	2.68	.79	5.67	1.52	2.72	1.12	.38	1.19	.90
		14	3.34	.98	7.25	2.01	2.72	1.40	.48	1.19	.92
		13	4.09	1.20	8.82	2.47	2.71	1.73	.59	1.20	.95
		12	4.71	1.38	10.10	2.83	2.70	2.01	.68	1.20	.97
8 X 2.125 X 2.375	8 X 2.5 Z	16	2.89	.85	7.78	1.80	3.07	1.12	.38	1.15	.90
		14	3.60	1.06	9.94	2.41	3.06	1.40	.48	1.15	.92
		13	4.41	1.30	12.11	2.97	3.06	1.73	.59	1.16	.95
		12	5.07	1.49	13.87	3.41	3.05	2.01	.68	1.16	.97
8 X 2.625 X 2.875	8 X 3 Z	16	3.10	.91	8.49	1.85	3.14	1.79	.52	1.40	.90
		14	3.86	1.14	10.99	2.43	3.13	2.24	.65	1.40	.92
		13	4.72	1.39	13.56	3.10	3.12	2.76	.80	1.41	.95
		12	5.43	1.60	15.54	3.79	3.12	3.20	.92	1.41	.97
8 X 3.125 X 3.375	8 X 3.5 Z	16	3.31	.97	8.93	1.89	3.20	2.67	.68	1.66	.90
		14	4.12	1.21	11.45	2.55	3.19	3.35	.85	1.66	.92
		13	5.04	1.48	14.95	3.20	3.18	4.13	1.05	1.67	.95
		12	5.80	1.71	17.21	3.73	3.18	4.77	1.20	1.67	.97
9 X 2.125 X 2.375	9 X 2.5 Z	16	3.10	.91	10.31	2.03	3.41	1.12	.38	1.11	.90
		14	3.86	1.14	13.16	2.84	3.41	1.40	.48	1.11	.92
		13	4.72	1.39	16.05	3.51	3.40	1.73	.59	1.12	.95
		12	5.43	1.60	18.39	4.02	3.39	2.01	.67	1.12	.97

NOTES
1 Section properties and allowables are computed in accordance with the 1986 edition of the AISI specifications.
2 Ix is for deflection determination.
3 Se is for bending.
4 Sy and Iy are for full section.

TABLE B.10

LIGHT GAGE STRUCTURAL INSTITUTE

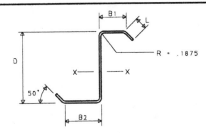

R = .1875

50°

SECTION ALLOWABLES

D X B1 X B2	SECTION	GA	Ma KIP-FT	Ae IN²	Oc	Va KIPS	BEARING (KIPS) END F	END V	INTERIOR F	INTERIOR V
3.5 X 1.5 X 1.5	3.5 X 1.5 Z	16	1.42	.000	.00	4.18	.428	.070	1.195	.137
		14	1.76	.000	.00	5.15	.793	.104	1.966	.181
4 X 2.125 X 2.375	4 X 2.5 Z	16	2.03	.000	.00	4.24	.539	.088	1.267	.145
		14	2.70	.000	.00	6.02	.969	.128	2.084	.192
6 X 2.125 X 2.375	6 X 2.5 Z	16	3.51	.000	.00	3.44	.502	.082	1.205	.138
		14	4.64	.000	.00	6.58	.918	.121	2.004	.185
		13	5.70	.000	.00	9.85	1.537	.165	3.153	.237
		12	6.52	.000	.00	13.04	2.164	.202	4.295	.281
7 X 2.125 X 2.375	7 X 2.5 Z	16	4.33	.000	.00	2.91	.484	.079	1.174	.135
		14	5.71	.000	.00	5.66	.892	.117	1.964	.181
		13	7.03	.000	.00	9.85	1.503	.162	3.103	.234
		12	8.05	.000	.00	13.04	2.123	.198	4.236	.277
8 X 2.125 X 2.375	8 X 2.5 Z	16	5.11	.000	.00	2.53	.466	.076	1.143	.131
		14	6.86	.000	.00	4.90	.867	.114	1.923	.177
		13	8.46	.000	.00	9.03	1.469	.158	3.052	.230
		12	9.69	.000	.00	13.04	2.082	.195	4.176	.273
8 X 2.625 X 2.875	8 X 3 Z	16	5.26	.000	.00	2.53	.466	.076	1.143	.131
		14	6.92	.000	.00	4.90	.867	.114	1.923	.177
		13	8.82	.000	.00	9.03	1.469	.158	3.052	.230
		12	10.79	.000	.00	13.04	2.082	.195	4.176	.273
8 X 3.125 X 3.375	8 X 3.5 Z	16	5.37	.000	.00	2.53	.466	.076	1.143	.131
		14	7.24	.000	.00	4.90	.867	.114	1.923	.177
		13	9.09	.000	.00	9.03	1.469	.158	3.052	.230
		12	10.62	.000	.00	13.04	2.082	.195	4.176	.273
9 X 2.125 X 2.375	9 X 2.5 Z	16	5.79	.000	.00	2.23	.448	.073	1.112	.128
		14	8.08	.000	.00	4.33	.841	.111	1.883	.173
		13	9.98	.000	.00	7.96	1.434	.154	3.001	.226
		12	11.44	.000	.00	12.16	2.041	.191	4.117	.269

NOTES
1 Properties and allowables are computed in accordance with the 1986 edition of the AISI specifications.
2 Ma is the allowable bending moment for bending about axis x-x for sections that are supported laterally for their full length.
3 F and V are coefficients used to determine the web crippling strength. Refer to sample calculations.

TABLE B.11

LIGHT GAGE STRUCTURAL INSTITUTE

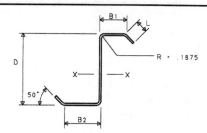

R = .1875

50°

SECTION PROPERTIES

D X B1 X B2	SECTION	GA	WEIGHT LB/FT	AREA IN²	AXIS X - X			AXIS Y - Y			L IN
					Ix IN⁴	Se IN³	Rx IN	Iy IN⁴	Sy IN³	Ry IN	
10 X 2.125 X 2.375	10 X 2.5 Z	16	3.31	.97	13.04	2.27	3.75	1.12	.38	1.07	.90
		14	4.12	1.21	16.95	3.30	3.74	1.40	.48	1.08	.92
		13	5.04	1.48	20.68	4.07	3.73	1.73	.59	1.08	.95
		12	5.80	1.71	23.71	4.67	3.73	2.01	.67	1.08	.97
10 X 2.625 X 2.875	10 X 3 Z	16	3.51	1.03	14.10	2.33	3.83	1.79	.52	1.31	.90
		14	4.38	1.29	18.60	3.18	3.82	2.24	.65	1.32	.92
		13	5.36	1.58	22.96	4.22	3.82	2.76	.80	1.32	.95
		12	6.16	1.81	26.33	5.14	3.81	3.20	.92	1.33	.97
10 X 3.125 X 3.375	10 X 3.5 Z	16	3.72	1.09	14.57	2.39	3.90	2.67	.68	1.56	.90
		14	4.64	1.36	19.30	3.27	3.90	3.35	.85	1.57	.92
		13	5.67	1.67	25.14	4.35	3.89	4.13	1.04	1.57	.95
		12	6.53	1.92	28.95	5.07	3.88	4.77	1.20	1.58	.97
12 X 2.125 X 2.375	12 X 2.5 Z	14	4.64	1.36	25.94	3.92	4.40	1.40	.48	1.02	.92
		13	5.67	1.67	32.21	5.29	4.39	1.73	.58	1.02	.95
		12	6.53	1.92	36.97	6.08	4.39	2.01	.67	1.02	.97
12 X 2.625 X 2.875	12 X 3 Z	14	4.89	1.44	28.58	3.86	4.50	2.24	.65	1.25	.92
		13	5.99	1.76	35.51	5.47	4.49	2.76	.80	1.25	.95
		12	6.89	2.03	40.75	6.64	4.48	3.20	.92	1.26	.97
12 X 3.125 X 3.375	12 X 3.5 Z	14	5.15	1.52	29.13	3.99	4.58	3.35	.85	1.49	.92
		13	6.30	1.85	38.66	5.39	4.57	4.13	1.04	1.49	.95
		12	7.25	2.13	44.54	6.55	4.57	4.77	1.20	1.50	.97

NOTES
1 Section properties and allowables are computed in accordance with the 1986 edition of the AISI specifications.
2 Ix is for deflection determination.
3 Se is for bending.
4 Sy and Iy are for full section.

TABLE B.12

LIGHT GAGE STRUCTURAL INSTITUTE

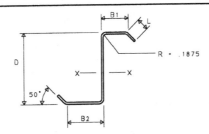

R = .1875

SECTION ALLOWABLES

D X B1 X B2	SECTION	GA	Ma KIP-FT	Ae IN²	Oc	Va KIPS	BEARING (KIPS) END F	END V	INTERIOR F	INTERIOR V
10 X 2.125 X 2.375	10 X 2.5 Z	16	6.45	.000	.00	1.99	.429	.070	1.081	.124
		14	9.37	.000	.00	3.87	.815	.107	1.843	.170
		13	11.58	.000	.00	7.11	1.400	.151	2.951	.222
		12	13.28	.000	.00	10.87	2.000	.187	4.058	.265
10 X 2.625 X 2.875	10 X 3 Z	16	6.62	.000	.00	1.99	.429	.070	1.081	.124
		14	9.03	.000	.00	3.87	.815	.107	1.843	.170
		13	12.01	.000	.00	7.11	1.400	.151	2.951	.222
		12	14.63	.000	.00	10.87	2.000	.187	4.058	.265
10 X 3.125 X 3.375	10 X 3.5 Z	16	6.80	.000	.00	1.99	.429	.070	1.081	.124
		14	9.30	.000	.00	3.87	.815	.107	1.843	.170
		13	12.37	.000	.00	7.11	1.400	.151	2.951	.222
		12	14.41	.000	.00	10.87	2.000	.187	4.058	.265
12 X 2.125 X 2.375	12 X 2.5 Z	14	11.16	.000	.00	3.19	.764	.101	1.763	.162
		13	15.06	.000	.00	5.87	1.332	.143	2.849	.214
		12	17.29	.000	.00	8.96	1.917	.179	3.939	.258
12 X 2.625 X 2.875	12 X 3 Z	14	10.97	.000	.00	3.19	.764	.101	1.763	.162
		13	15.56	.000	.00	5.87	1.332	.143	2.849	.214
		12	18.88	.000	.00	8.96	1.917	.179	3.939	.258
12 X 3.125 X 3.375	12 X 3.5 Z	14	11.35	.000	.00	3.19	.764	.101	1.763	.162
		13	15.33	.000	.00	5.87	1.332	.143	2.849	.214
		12	18.62	.000	.00	8.96	1.917	.179	3.939	.258

NOTES
1 Properties and allowables are computed in accordance with the 1986 edition of the AISI specifications.
2 Ma is the allowable bending moment for bending about axis x-x for sections that are supported laterally for their full length.
3 F and V are coefficients used to determine the web crippling strength. Refer to sample calculations.

TABLE B.13

LIGHT GAGE STRUCTURAL INSTITUTE

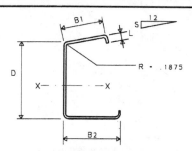

R = .1875

SECTION PROPERTIES
SINGLE SLOPE

SECTION				AXIS X - X			AXIS Y - Y			
D X B1 X B2 X S	GA	WEIGHT LB/FT	AREA IN²	Ix IN⁴	Se IN³	Rx IN	Iy IN⁴	Sy IN³	Ry IN	L IN
6 X 4 X 4 X 1	16	3.10	.91	5.13	1.31	2.63	1.96	.73	1.47	.74
7 X 4 X 4 X 1	16	3.31	.97	7.19	1.57	3.00	2.06	.75	1.46	.74
	14	4.12	1.21	9.20	2.10	2.99	2.57	.93	1.46	.76
8 X 3.375 X 5 X 1	16	3.72	1.09	10.26	2.05	3.39	3.01	.84	1.66	1.05
	14	4.63	1.36	13.00	2.82	3.38	3.75	1.05	1.66	1.08
	12	6.52	1.92	18.65	4.22	3.38	5.26	1.47	1.66	1.13
8 X 4 X 4 X 1	16	3.51	1.03	9.53	1.82	3.36	2.16	.76	1.45	.74
	14	4.38	1.29	12.41	2.51	3.36	2.69	.95	1.45	.76
	12	6.16	1.81	18.76	3.81	3.35	3.78	1.34	1.45	.82
10 X 4 X 4 X 1	14	4.89	1.44	20.26	3.24	4.07	2.89	.98	1.42	.76
	12	6.89	2.03	31.10	5.17	4.06	4.07	1.37	1.42	.82
6 X 4 X 4 X 4	16	3.10	.91	5.70	1.20	2.83	1.84	.68	1.42	.71
7 X 4 X 4 X 4	16	3.31	.97	7.91	1.45	3.19	1.94	.70	1.41	.71
	14	4.12	1.21	10.12	1.92	3.19	2.41	.86	1.41	.74
8 X 3.375 X 5 X 4	16	3.72	1.09	11.23	1.98	3.54	2.91	.80	1.63	1.03
	14	4.63	1.36	14.21	2.65	3.54	3.62	1.00	1.63	1.05
	12	6.52	1.92	20.37	3.93	3.53	5.08	1.41	1.63	1.11
8 X 4 X 4 X 4	16	3.51	1.03	10.33	1.68	3.54	2.03	.71	1.40	.71
	14	4.38	1.29	13.43	2.31	3.54	2.52	.88	1.40	.74
	12	6.16	1.81	20.33	3.51	3.54	3.54	1.23	1.40	.79
10 X 4 X 4 X 4	14	4.89	1.44	18.58	3.01	4.23	2.71	.91	1.37	.74
	12	6.89	2.03	28.20	4.78	4.23	3.80	1.27	1.37	.79

NOTES
1 Properties and allowables are computed in accordance with the 1986 edition of the AISI specifications.
2 Ix is for deflection determination.
3 Se is for bending.
4 Sy and Iy are for full section.

TABLE B.14

LIGHT GAGE STRUCTURAL INSTITUTE

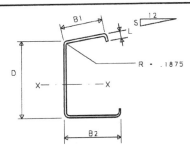

R = .1875

SECTION ALLOWABLES
SINGLE SLOPE

SECTION						BEARING (KIPS)			
						END		INTERIOR	
D x B1 X B2 X S	GA	Ma	Ae	Oc	Va	F	V	F	V
		KIP-FT	IN²		KIPS				
6 X 4 X 4 X 1	16	3.71	.000	1.92	3.43	.502	.082	1.204	.138
7 X 4 X 4 X 1	16	4.46	.000	1.92	2.90	.484	.079	1.173	.135
	14	5.95	.000	1.92	5.64	.892	.117	1.963	.181
8 X 3.375 X 5 X 1	16	5.82	.000	1.92	2.52	.465	.076	1.142	.131
	14	8.01	.000	1.92	4.89	.866	.114	1.923	.177
	12	11.97	.000	1.92	13.04	2.081	.194	4.175	.273
8 X 4 X 4 X 1	16	5.16	.000	1.92	2.52	.465	.076	1.142	.131
	14	7.12	.000	1.92	4.89	.866	.114	1.923	.177
	12	10.81	.000	1.92	13.04	2.081	.194	4.175	.273
10 X 4 X 1	14	9.19	.000	1.92	3.86	.815	.107	1.842	.170
	12	14.66	.000	1.92	10.84	1.999	.187	4.057	.265
6 X 4 X 4 X 4	16	3.40	.000	1.92	3.40	.501	.082	1.203	.138
7 X 4 X 4 X 4	16	4.11	.000	1.92	2.88	.483	.079	1.172	.135
	14	5.45	.000	1.92	5.60	.891	.117	1.961	.181
8 X 3.375 X 5 X 4	16	5.62	.000	1.92	2.50	.465	.076	1.141	.131
	14	7.54	.000	1.92	4.86	.865	.114	1.921	.177
	12	11.17	.000	1.92	13.04	2.079	.194	4.172	.273
8 X 4 X 4 X 4	16	4.78	.000	1.92	2.50	.465	.076	1.141	.131
	14	6.54	.000	1.92	4.86	.865	.114	1.921	.177
	12	9.97	.000	1.92	13.04	2.079	.194	4.172	.273
10 X 4 X 4 X 4	14	8.53	.000	1.92	3.84	.814	.107	1.840	.170
	12	13.57	.000	1.92	10.77	1.997	.187	4.054	.265

NOTES
1 Properties and allowables are computed in accordance with the 1986 edition of the AISI specifications.
2 Ma is the allowable moment for sections that are supported laterally for their full length.
3 Oc is the safety factor for axial load.
4 Ae is the reduced area for axial load.
5 Oc and Ae are based on KL$_y$ = KL$_T$ = 1.0 ft. and KL$_x$ as follows: D=4, 12 ft.; D=6, 16 ft.; D=7,8, and 9, 18 ft.;
   D=10 and 12, 20 ft.
6 F and V are coefficients used to determine the web crippling strength. Refer to sample calculations.

TABLE B.15

LIGHT GAGE STRUCTURAL INSTITUTE

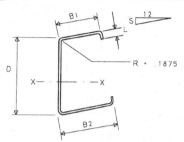
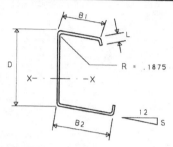

SECTION PROPERTIES
DOUBLE SLOPE AND HIGH SIDE

SECTION				AXIS X - X			AXIS Y - Y			
D X B1 X B2 X S	GA	WEIGHT	AREA	Ix	Se	Rx	Iy	Sy	Ry	L
		LB/FT	IN²	IN⁴	IN³	IN	IN⁴	IN³	IN	IN
6 X 4 X 4 X 1	16	3.10	.91	4.77	1.33	2.57	1.94	.73	1.46	.75
7 X 4 X 4 X 1	16	3.31	.97	6.69	1.59	2.94	2.05	.74	1.45	.75
	14	4.12	1.21	8.55	2.11	2.94	2.56	.93	1.45	.78
8 X 3.375 X 5 X 1	16	3.72	1.10	9.81	2.07	3.29	2.96	.83	1.64	1.06
	14	4.64	1.36	12.47	2.86	3.29	3.68	1.03	1.64	1.09
	12	6.53	1.92	17.91	4.29	3.28	5.17	1.45	1.64	1.15
8 X 4 X 4 X 1	16	3.52	1.03	8.98	1.84	3.31	2.15	.76	1.44	.75
	14	4.38	1.29	11.59	2.52	3.30	2.67	.94	1.44	.78
	12	6.17	1.81	17.54	3.82	3.29	3.76	1.33	1.44	.84
10 X 4 X 4 X 1	14	4.90	1.44	19.27	3.26	4.01	2.88	.97	1.41	.78
	12	6.90	2.03	29.32	5.16	4.00	4.05	1.37	1.41	.84
6 X 4 X 4 X 4	16	3.10	.91	4.78	1.33	2.60	1.72	.63	1.37	.77
7 X 4 X 4 X 4	16	3.31	.97	6.73	1.58	2.97	1.82	.64	1.37	.77
	14	4.12	1.21	8.61	2.09	2.96	2.26	.80	1.36	.80
8 X 3.375 X 5 X 4	16	3.72	1.10	9.65	2.08	3.24	2.56	.75	1.53	1.09
	14	4.64	1.36	12.31	2.69	3.24	3.17	.93	1.52	1.12
	12	6.53	1.92	17.75	4.03	3.23	4.43	1.30	1.52	1.18
8 X 4 X 4 X 4	16	3.52	1.03	9.09	1.81	3.33	1.90	.66	1.36	.77
	14	4.38	1.29	11.63	2.47	3.32	2.36	.82	1.35	.80
	12	6.17	1.81	17.77	3.78	3.31	3.31	1.14	1.35	.87
10 X 4 X 4 X 4	14	4.90	1.44	19.42	3.17	4.02	2.55	.84	1.33	.80
	12	6.90	2.03	29.56	5.06	4.02	3.57	1.17	1.33	.87

NOTES
1 Properties and allowables are computed in accordance with the 1986 edition of the AISI specifications.
2 Ix is for deflection determination.
3 Se is for bending.
4 Sy and Iy are for full section.

TABLE B.16

LIGHT GAGE STRUCTURAL INSTITUTE

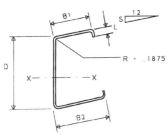

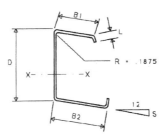

SECTION ALLOWABLES
DOUBLE SLOPE AND HIGH SIDE

SECTION						BEARING (KIPS)			
						END		INTERIOR	
D x B1 X B2 X S	GA	Ma KIP-FT	Ae IN²	Oc	Va KIPS	F	V	F	V
6 X 4 X 4 X 1	16	3.76	.000	1.92	3.44	.502	.082	1.205	.138
7 X 4 X 4 X 1	16	4.53	.000	1.92	2.91	.484	.079	1.174	.135
	14	5.99	.000	1.92	5.66	.892	.117	1.964	.181
8 X 3.375 X 5 X 1	16	5.89	.000	1.92	2.53	.466	.076	1.143	.131
	14	8.13	.000	1.92	4.90	.867	.114	1.924	.177
	12	12.17	.000	1.92	13.04	2.082	.195	4.176	.273
8 X 4 X 4 X 1	16	5.21	.000	1.92	2.53	.466	.076	1.143	.131
	14	7.15	.000	1.92	4.90	.867	.114	1.924	.177
	12	10.84	.000	1.92	13.04	2.082	.195	4.176	.273
10 X 4 X 4 X 1	14	9.26	.000	1.92	3.87	.815	.107	1.843	.170
	12	14.64	.000	1.92	10.87	2.000	.187	4.058	.265
6 X 4 X 4 X 4	16	3.78	.000	1.92	3.46	.503	.082	1.206	.138
7 X 4 X 4 X 4	16	4.48	.000	1.92	2.93	.484	.079	1.175	.135
	14	5.92	.000	1.92	5.69	.893	.117	1.965	.181
8 X 3.375 X 5 X 4	16	5.92	.000	1.92	2.54	.466	.076	1.144	.131
	14	7.64	.000	1.92	4.93	.867	.114	1.924	.177
	12	11.44	.000	1.92	13.04	2.083	.195	4.178	.273
8 X 4 X 4 X 4	16	5.14	.000	1.92	2.54	.466	.076	1.144	.131
	14	7.01	.000	1.92	4.93	.867	.114	1.924	.177
	12	10.72	.000	1.92	13.04	2.083	.195	4.178	.273
10 X 4 X 4 X 4	14	9.00	.000	1.92	3.88	.816	.107	1.844	.170
	12	14.37	.000	1.92	10.91	2.001	.187	4.059	.266

NOTES
1 Properties and allowables are computed in accordance with the 1986 edition of the AISI specifications.
2 Ma is the allowable moment for sections that are supported laterally for their full length.
3 Oc is the safety factor for axial load.
4 Ae is the reduced area for axial load.
5 Oc and Ae are based on KLy = KLT = 1.0 ft. and KLx as follows: D=4, 12 ft.; D=6, 16 ft.; D=7,8, and 9, 18 ft.;
   D=10 and 12, 20 ft.
6 F and V are coefficients used to determine the web crippling strength. Refer to sample calculations.

Typical Specification for Metal Building Systems

This sample specification is included to help readers incorporate many items discussed throughout the book into their contract documents. It is based on U.S. Army Corps of Engineers Guide Specification for Military Construction (CEGS) Section 13121, with numerous changes and additions by the author. The specification is assumed to be a part of contract documents that include other specification sections and drawings with additional loading information.

This specification is intended to be used for custom-designed buildings with eave heights greater than 20 ft or frame spans exceeding 80 ft. It might be too restrictive for smaller pre-engineered buildings that do not require exceptional durability of construction materials and finishes.

The specification covers a typical single-story metal building system with gable rigid-frame structure containing cranes. Delete any requirements not applicable to your building, and add the new ones if needed (add, for example, temperature loading when warranted). Some specific provisions that call for the specifier's attention are:

1. Section 1.2.2 requires the manufacturer to be AISC-certified for category MB. Evaluate this requirement if good manufacturers who are not so certified are available in your area.

2. Section 1.2.3 calls for a manufacturer's representative to be present during construction. This CEGS requirement may or may not be needed, as discussed in Chap. 16.

3. Lateral drift provisions of Sec. 1.3.4 need to be specified. Refer to the discussion in Chap. 11.

4. Downgrade the requirements for warranty and for metal-panel finishes if polyvinylidene fluoride coating is not desired.

Section 13121   Metal Building Systems

PART 1   GENERAL

1.1.1   RELATED DOCUMENTS

A. Drawings and general provisions of the Contract, including General and Supplementary Conditions and Division 1 Specification Sections, apply to this Section.

1.1.2   SUMMARY

A. Furnish and install a single-story, pre-engineered metal building of the length, width, eave height, and roof pitch indicated. The work shall include:
1. Exterior walls covered with field-assembled wall panels attached to framing members using exposed fasteners. Endwalls are not expandable.
2. Roof system with standing-seam roofing and fiberglass insulation.
3. Parapet, mansard, and interior metal walls as shown on the drawings.
4. Framing around overhead doors.
5. Eave gutters and downspouts.
6. Framing, supports, and roof curbs for rooftop-mounted HVAC equipment.

B. Related sections:
1. Section 07720 - Roof Accessories.
2. Section 08110 - Metal Doors and Frames.
3. Section 08700 - Builder's Hardware.
4. Section 09900 - Painting.
5. Section 14630 - Cranes, Electrical (...).

1.1.3   REFERENCES

The publications listed below form a part of this specification to the extent referenced. The publications are referred to in the text by basic designation only.

ALUMINUM ASSOCIATION (AA)

AA Standards   (1993) Aluminum Standards and Data

AA Manual   (1986) Aluminum Construction Manual Series—Section 1 Specifications for Aluminum Structures

AMERICAN ARCHITECTURAL MANUFACTURERS ASSOCIATION (AAMA)

AAMA Specification   (1993) Voluntary Specifications for Aluminum and Poly (Vinyl Chloride) (PVC) Prime Windows and Sliding Glass Doors

AMERICAN INSTITUTE OF STEEL CONSTRUCTION (AISC)

AISC QC Program   (1990) AISC Quality Certification Program Description

AISC Code of Standard Practice   (1992) Code of Standard Practice for Steel Buildings and Bridges

AISC Specification for Structural Joints   (1986) Allowable Stress Design Specification for Structural Joints Using ASTM A 325 or ASTM A 490 Bolts

AISC Specification   (1989) Specification for Structural Steel Buildings—Allowable Stress Design and Plastic Design

AMERICAN IRON AND STEEL INSTITUTE (AISI)

AISI Manual   (1989) Cold-Formed Steel Design Manual

AMERICAN SOCIETY FOR TESTING AND MATERIALS (ASTM)

ASTM A 36   (1994a) Carbon Structural Steel

ASTM A 53   (1993a) Pipe, Steel, Black and Hot-Dipped, Zinc-Coated Welded and Seamless

ASTM A 325   (1994) Structural Bolts, Steel, Heat Treated, 120/105 ksi Minimum Tensile Strength

ASTM A 463   (1994) Steel Sheet, Aluminum-Coated by the Hot-Dip Process

ASTM A 490   (1993) Heat-Treated Steel Structural Bolts, 150 ksi Minimum Tensile Strength

ASTM A 500   (1993) Cold-Formed Welded and Seamless Carbon Steel Structural Tubing in Rounds and Shapes

ASTM A 501   (1993) Hot-Formed Welded and Seamless Carbon Steel Structural Tubing

ASTM A 529   (1994) High-Strength Carbon-Manganese Steel of Structural Quality

ASTM A 570   (1992; R 1993) Steel, Sheet and Strip, Carbon, Hot-Rolled, Structural Quality

ASTM A 572   (1994b) High-Strength Low-Alloy Columbium-Vanadium Structural Steel

ASTM A 588   (1994) High-Strength Low-Alloy Structural Steel with 50 ksi (345 MPa) Minimum Yield Point to 4 in. (100 mm) Thick

ASTM A 606	(1991a; R 1993) Steel, Sheet and Strip, High-Strength, Low-Alloy, Hot-Rolled and Cold-Rolled, with Improved Atmospheric Corrosion Resistance
ASTM A 607	(1992a) Steel, Sheet and Strip, High-Strength, Low-Alloy, Columbium or Vanadium, or Both, Hot-Rolled and Cold-Rolled
ASTM A 653	(1994) Steel Sheet, Zinc-Coated (Galvanized) or Zinc-Iron Alloy-Coated (Galvannealed) by the Hot-Dip Process
ASTM A 792	(1993a) Steel Sheet, 55% Aluminum-Zinc Alloy-Coated by the Hot-Dip Process, General Requirements
ASTM B 117	(1994) Salt Spray (Fog) Testing Apparatus
ASTM B 209	(1993) Aluminum and Aluminum-Alloy Sheet and Plate
ASTM C 518	(1991) Steady-State Heat Flux Measurements and Thermal Transmission Properties by Means of the Heat Flow Meter Apparatus
ASTM C 553	(1992) Mineral Fiber Blanket Thermal Insulation for Commercial and Industrial Applications
ASTM C 612	(1993) Mineral Fiber Block and Board Thermal Insulation
ASTM C 1289	(1986; R 1993) Bitumen Content
ASTM D 522	(1993a) Mandrel Bend Test of Attached Organic Coatings
ASTM D 714	(1987; R 1994) Evaluating Degree of Blistering of Paints
ASTM D 968	(1993) Abrasion Resistance of Organic Coatings by Falling Abrasive
ASTM D 1308	(1987; R 1993) Effect of Household Chemicals on Clear and Pigmented Organic Finishes
ASTM D 1654	(1992) Evaluation of Painted or Coated Specimens Subjected to Corrosive Environments
ASTM D 2244	(1993) Calculation of Color Differences from Instrumentally Measured Color Coordinates
ASTM D 2247	(1994) Testing Water Resistance of Coatings in 100 Percent Relative Humidity
ASTM D 2794	(1993) Resistance of Organic Coatings to the Effects of Rapid Deformation (Impact)
ASTM D 3359	(1995) Measuring Adhesion by Tape Test
ASTM D 3841	(1992) Glass-Fiber-Reinforced Polyester Plastic Panels
ASTM D 4214	(1989) Evaluating the Degree of Chalking of Exterior Paint Films
ASTM D 4397	(1991) Polyethylene Sheeting for Construction, Industrial, and Agricultural Applications
ASTM E 84	(1994a) Surface Burning Characteristics of Building Materials
ASTM E 96	(1994) Water Vapor Transmission of Materials
ASTM E 1042	(1992) Acoustically Absorptive Materials Applied by Trowel or Spray
ASTM G 23	(1995) Operating Light-Exposure Apparatus (Carbon-Arc Type) with and without Water for Exposure of Nonmetallic Materials

AMERICAN SOCIETY OF CIVIL ENGINEERS (ASCE)

ASCE 7   (1995) Minimum Design Loads for Buildings and Other Structures

AMERICAN WELDING SOCIETY (AWS)

AWS D1.1   (1996) Structural Welding Code—Steel

ASSOCIATION OF IRON AND STEEL ENGINEERS (AISE)

AISE 13   (1979) Guide for the Design and Construction of Mill Buildings, AISE Technical Report No. 13

FEDERAL SPECIFICATIONS (FS)

FS HH-I-1972/GEN   (Basic; Am 1; Notice 1) Insulation Board, Thermal, Faced, Polyurethane or Polyisocyanurate

FS HH-I-1972/1   (Basic; Notice 1) Insulation Board, Thermal, Faced, Polyurethane or Polyisocyanurate

MATERIAL HANDLING INSTITUTE (MHI)

MHI CMAA 70   (1994) Electric Overhead Traveling Cranes

METAL BUILDING MANUFACTURERS ASSOCIATION (MBMA)

MBMA Manual   (1986; Errata; Supple 1990) Low Rise Building Systems Manual

NORTH AMERICAN INSULATION MANUFACTURERS ASSOCIATION

NAIMA 202   (1992) Standard for Flexible Fiberglass Insulation Systems in Metal Buildings

SHEET METAL & AIR CONDITIONING CONTRACTORS' NATIONAL ASSOCIATION (SMACNA)

SMACNA Manual   (1993) Architectural Sheet Metal Manual

STEEL DOOR INSTITUTE (SDOI)

SDOI Standard   (1991) Standard Steel Doors and Frames

STEEL WINDOW INSTITUTE (SWI)

SWI Guide   (1989) The Specifier's Guide to Steel Windows

UNDERWRITERS LABORATORIES (UL)

UL 580   (1994; Rev through Apr 1995) Tests for Uplift Resistance of Roof Assemblies

1.2   GENERAL

1.2.1   Building Configuration

Building shall have vertical walls and gable [single-slope] roof. Roof slope shall be as indicated. Building shall be [single-span] [multiple-span] rigid frame [truss] structure. Eave height shall be measured from the top of the finished floor to the intersection of the insides of the roof and sidewall sheets. The clear height between the finished floor and the bottom of the roof steel and other critical dimensions shall be as indicated.

1.2.2   Manufacturer

Metal building shall be the product of a recognized metal building systems manufacturer who has been in the practice of manufacturing metal buildings of the size and complexity of the building shown on the contract drawings, for a period of no less than 5 years. The manufacturer shall be chiefly engaged in the practice of designing and fabricating metal building systems. The manufacturer shall have an AISC Quality Certification, category MB in accordance with AISC QC Program.

1.2.3   Manufacturer's Representative

A representative of the metal building manufacturer, who is familiar with the design of the building supplied and experienced in the erection of metal buildings similar in size to the one required under this contract, shall be present at the job site during construction from the start of the structural framing erection until completion of the installation of exterior covering to assure that the building meets the specified requirements.

1.2.4   Installer

Erector shall have specialized experience in the erection of metal building systems for a period of at least 5 years, and at least 2 years experience in the erection of metal buildings of the size and complexity shown and provided by the selected manufacturer.

1.3   DESIGN REQUIREMENTS

1.3.1   Design Conditions

As a minimum, design shall conform to the requirements of _____ Building Code _____ Edition, hereafter called The Code. Loading criteria as set out by MBMA Manual shall apply only where specifically stated by this Specification.

1.3.1.1   Dead Load.   The dead load shall consist of the weight of the structural frame and all other materials of the building system.

1.3.1.2 Collateral Loads. Collateral load of _____ Pa (_____ pounds per square foot) shall be applied to the entire structure to account for the weight of additional permanent materials other than the building system, such as sprinklers, mechanical systems, electrical systems, and ceilings. This allowance does not include the weight of hung equipment weighing 23 kilograms (50 pounds) or more. Equipment loads of 222 N (50 pounds) or more shall be investigated and the structure (frame, purlins, girts) shall be strengthened as required. The Contractor is responsible for providing the Building Manufacturer the magnitude and approximate location of all concentrated loads greater than 222 N (50 pounds).

1.3.1.3 Roof Live Loads. Roof live loads shall be determined and applied in accordance with ASCE 7.

1.3.1.4 Roof Snow Loads. The design roof snow loads shall be as shown in the Contract Drawings.

1.3.1.5 Wind Loads. Wind load on main framing shall be computed and applied in accordance with The Code and information contained in the Contract Drawings. Wind pressures on parts and components shall be computed and applied in accordance with ASCE 7.

1.3.1.6 Auxiliary (Crane) Loads. Auxiliary (crane) loads such as superimposed loads resulting from craneways shall be as shown in the Drawings. The Contractor shall verify that the auxiliary loads shown in the Drawings exceed the loads imposed by the equipment supplied.

1.3.1.7 Concentrated Loads. Concentrated loads shall be applied at locations indicated on the drawings.

1.3.1.8 Seismic Loads. Seismic loads shall be computed for seismic zone _____ in accordance with The Code using data indicated in the Drawings.

1.3.1.9 Impact Loads. Impact loads due to [monorails] cranes shall be applied as indicated in AISC Specification [MBMA Manual] [AISE 13].

1.3.1.10 Load Combinations. The load combinations contained in The Code or in MBMA Manual, whichever produces more stringent loading, shall be used in the design.

Dead load shall be weight of all permanent construction materials including collateral loads. Dead load for combination (Dead + Wind) shall not include collateral loads. For buildings with cranes, additional load combinations specified in MBMA Manual shall be included. [All load combinations including snow shall consider both balanced and unbalanced conditions.]

1.3.2 Foundation Requirements

The depth of the building columns at the base shall not exceed the pier and wall sizes indicated on the Drawings.

1.3.3 Framing and Structural Members

Structural steel members and their connections shall be designed in accordance with AISC Specification. Structural cold-formed steel framing members and their connections shall be designed in accordance with AISI Manual. Framed openings shall be designed to structurally replace the covering and framing displaced. The allowable live load deflection of roof elements not supporting ceilings shall not exceed 1/180th of the span. The allowable live load deflection of roof members supporting plastered ceilings shall not exceed 1/360th of the span, and of roof members supporting other ceilings—1/240th of the span. The framing shall have sufficient rigidity to resist ponding. Members with openings in their webs shall be designed with consideration of the additional stresses which will result due to the openings. Deflections of the steel framing above and along the side of rolling door openings shall be limited to a maximum of $\frac{1}{2}$ of the allowable movement in the telescoping top roller of the doors to ensure proper operation of the doors. Pinned connections shall be assumed at all column bases.

1.3.4 Drift Provisions

Lateral deflections, or drift, at the roof level in relation to the slab-on-grade shall be calculated based on a 50-year mean recurrence interval and shall not exceed [h/125] _____ or the working limits of the architectural and mechanical details.

1.3.5 Bracing for Girts and Purlins

Lateral bracing for girts and purlins shall be provided by bracing members attached to the member flanges and by approved through-fastened roofing. Standing-seam roofing shall not be considered adequate bracing. Whenever net wind uplift is present, lateral bracing shall be provided for both flanges. Flange bracing shall be located as indicated and as required by analysis but, as a minimum, shall be spaced at quarter points of the member's span.

1.3.6 Exterior Covering

Steel covering shall be designed in accordance with AISI Manual. Aluminum covering shall be designed in accordance with the AA Standards. Section modulus and moment of inertia of aluminum sheet shall be determined for actual cross section dimensions by the conventional methods for actual design stresses and by effective width concept for deflection in accordance with AA Manual. Maximum deflection for wall and roof panels under full dead and live and/or wind loads shall not exceed 1/180th of the span between supports. The design analysis shall establish that the roof when deflected under dead plus live or snow loads will not result in a negative slope. Maximum deflections shall be based on panels continuous across two or more supports. In addition to the loads indicated above, the roof decking shall be designed for a 890 N (200-pound) concentrated load at midspan on a 300 mm (12-inch) wide section of deck. Wall panels are not permitted for diaphragms or shear walls.

1.3.7   Gutters and Downspouts

Gutters and downspouts shall be designed according to the requirements of SMACNA Manual for storms which should be exceeded only once in 5 years and with adequate provisions for thermal expansion and contraction. Supports for gutters and downspouts shall be designed for the anticipated loads.

1.3.8   Louvers

Louvers shall be fixed-blade adjustable type designed for a minimum net open area of ____ square meters (____ square feet), to be rainproof, and to resist vibration when air is passed at the rate of ____ cubic meters per second ____ cubic feet per minute).

1.3.9   Ventilators

1.3.9.1   Circular Ventilators. Circular roof ventilators shall be gravity directional stationary revolving type, designed for a minimum capacity of ____ cubic meters of air per second ____ cubic feet of air per minute) for each ventilator, based on a wind velocity of 8 kilometers per hour (5 miles per hour) and an exterior-interior temperature differential of 5 degrees C (10 degrees F) and without screens in place.

1.3.9.2   Continuous Ventilators. Continuous roof ventilators shall be ridge mounted gravity type designed for a minimum capacity of ____ cubic meters of air per second (____ cubic feet of air per minute) for each 3-m (10-foot) section based on a wind velocity of 8 kilometers per hour (5 miles per hour) and an exterior-interior temperature differential of 5 degrees C (10 degrees F) and without screens in place.

1.3.10   Cranes

The crane loads shall be obtained from the crane manufacturer and shall be applied to the design of the crane runways and supports as per Section 1.3.1.9 and as shown in the Drawings. The cranes, [girders,] rails, end trucks, switches, stops, and bumpers shall be provided by the crane manufacturer as specified in Section 14630 CRANES ELECTRIC OVERHEAD TRAVELING, TOP RUNNING AND UNDERHUNG 30 TON MAX. Cranes with a capacity greater than 18000 kg (20 tons) shall be supported by crane columns, which shall not be tied to the building frame by rigid bracing, but shall have their own independent method of supporting the crane longitudinal forces. The connections which carry the crane lateral forces into the building frames shall be designed and constructed to allow for the difference in required erection tolerances between the building frame and the crane support members. [Provide all built-up framing supporting suspended cranes with double-sided welds for at least 10 feet on each side of the hanger.]

1.3.11 Prefabricated Roof Curbs

Prefabricated roof curbs are specified in Section 07720, furnished and installed under this Section.

1.4 SUBMITTALS

The following shall be submitted in accordance with Section 01300 SUBMITTALS:

1.4.1 Design Calculations

Complete calculations for the building framing, cold-formed members, roofing and siding, as one package with the shop drawings, signed by a Registered Professional Engineer licensed in the State of _____ . Structural calculations shall include all required loading cases and load combinations used in the design and the resulting member forces, reactions, deflections and story drift. The magnitude of maximum column reactions on foundations from all critical load combinations shall be separately tabulated. Critical load conditions used in the final sizing of the members shall be emphasized. The design analysis shall include the name and office phone number of the designer to answer questions during the shop-drawing review.

1.4.2 Shop Drawings

1.4.2.1. Submit complete erection drawings showing anchor bolt settings, sidewall, endwall and roof framing, transverse cross sections, covering and trim details and accessory installation details to clearly indicate proper assembly of building components. Show vertical elevations for bottom of column base plates.

1.4.2.2. Indicate standard designations, configurations, sizes, spacings, and locations of girts and purlins including bridging and connections.

1.4.2.3. Provide drawings of all standard and modified steel connections and details indicating all shop-welded, field-welded and bolted connections for approval.

1.4.2.4 Sheet Metal Accessories. Provide layouts at $\frac{1}{4}$-inch scale. Provide details of ventilators, louvers, gutters, downspouts, and other sheet metal accessories at $1\frac{1}{2}$-inch scale showing profiles, methods of joining and anchorages.

1.4.2.5. Indicate wall and roof system dimensions, panel layout, general construction details, anchorages and methods of anchorage and installation.

1.4.2.6 Manufacturer's Installation Instructions. Indicate preparation requirements and assembly sequence.

1.4.2.7. Fabrication shall not start until shop drawings have been approved.

1.4.3 Statements

1.4.3.1 Qualifications. Qualifications of the manufacturer, including proof of AISC Certification Category MB, qualifications of the manufacturer's representative who will provide job site assistance, and qualifications and experience of the building erector.

1.4.3.2 Letter of Design Certification. Submit, one week prior to bid date, Letter of Design Certification from the metal building manufacturer stating that the metal building design and metal roof system design is based on complete set of the Contract Drawings and Specifications and that the building furnished complies with the specified loading requirements, signed by a Registered Professional Engineer licensed in the State of _____ . The letter shall state the building dimensions, the design criteria, governing building codes including year, the procedures used, and shall attest to the adequacy and accuracy of the design. The letter shall specifically describe design dead, live, snow, seismic, collateral, wind, and concentrated loads; live-load reductions; load combinations; and methods of load application.

1.4.3.3 Contractor's Certificate. One week prior to bid date, submit Contractor's certificate certifying that the Contractor complies with specified requirements and is a manufacturer's currently authorized dealer of the system to be furnished.

1.4.3.4 Installer's Certificate. One week prior to bid date, submit certificate that the building system and roof system Installer has been regularly engaged in the installation of building systems of the same construction for the period of time specified.

1.4.3.5 Mill Certificates. Mill certification for structural bolts, structural steel, wall and roof covering.

1.4.3.6 Roofing System Certificate. Submit certification verifying that the metal roofing system has been tested by UL 580 and awarded a designation of

Class 90. Submit certification that the metal roofing system has been tested and approved by the U.S. Army Corps of Engineers Guide Specification 07416.

1.4.4 Samples

1.4.4.1 Accessories.   One sample of each type of flashing, trim, closure, cap and similar items. Size shall be sufficient to show construction and configuration.

1.4.4.2 Roof and Wall Covering.   One piece of each type and finish (exterior and interior) to be used, 230 mm (9 inches) long, full width. The sample for factory color finished covering shall be accompanied by certified laboratory test reports showing that the sheets to be furnished are produced under a continuing quality control program and that a representative sample consisting of not less than 5 pieces has been tested and has met the quality standards specified for factory color finish.

1.4.4.3 Fasteners.   Two samples of each type to be used, with statement regarding intended use.

1.4.4.4 Insulation.   One piece of each type to be used, and descriptive data covering installation.

1.4.4.5 Gaskets and Insulating Compounds.   Two samples of each type to be used and descriptive data.

1.4.4.6 Sealant.   One sample, approximately 0.5 kg (1 pound) and descriptive data.

1.5 DELIVERY AND STORAGE

Materials shall be delivered to the site in a dry and undamaged condition and stored out of contact with the ground. Materials other than framing and structural members shall be covered with weathertight coverings and kept dry. Storage accommodations for roof and wall covering shall provide good air circulation and protection from surface staining.

1.6 WARRANTY

The metal building system shall be warranted against water leaks arising out of or caused by ordinary wear and tear by the elements and against failure of the factory-applied exterior finish on metal roof and wall panels for a period of 20 years after the date of Substantial Completion.

1.7 EXTRA MATERIALS

Maintenance Stock: Furnish at least 5 percent excess over required amount of nuts, bolts, screws, washers, and other required fasteners for each metal building. Pack in cartons labeled to identify the contents and store on the site where directed.

PART 2   PRODUCTS

2.1   IDENTIFICATION OF BUILDING COMPONENTS

Each piece or part of the assembly shall be clearly and legibly marked to correspond with the shop drawings.

2.2   FRAMING AND STRUCTURAL MEMBERS

2.2.1   Materials

Steel 3 mm ($\frac{1}{8}$ inch) or more in thickness shall conform to ASTM A 36, ASTM A 529, ASTM A 572, ASTM A 570, or ASTM A 588. Uncoated steel less than 3 mm (1/8 inch) in thickness shall conform to ASTM A 570, ASTM A 606, or ASTM A 607. Galvanized steel shall conform to ASTM A 653, G 90 coating designation, 1.143 mm (0.045 inch) minimum thickness. Aluminum-zinc coated steel shall conform to ASTM A 792, AZ 55 coating designation, 1.143 mm (0.045 inch) minimum thickness. Structural pipe shall conform to ASTM A 53, ASTM A 500, or ASTM A 501. Holes for bolts shall be made in the shop. All structural framing not receiving other finishes shall be shop primed with fast-curing, lead-free, universal primer, selected by the manufacturer for resistance to normal atmospheric corrosion, compatibility with finish paint systems, and capability to provide a sound foundation for field-applied topcoats despite prolonged exposure.

2.2.2   STRUCTURAL FRAMING

2.2.2.1 Rigid Frames and Endwall Framing. Fabricate built-up members in accordance with MBMA manual, hot-rolled members in accordance with AISC Specification. Design all column bases as pinned.

2.2.2.2 Secondary Framing

a. Roof Purlins, Eave Struts, Sidewall and Endwall Girts: C- or Z-shaped sections fabricated from 16-gage (0.0598-inch) minimum thickness shop-painted cold-formed steel.

b. Flange and Sag Bracing: Angles fabricated from 16-gage (0.0598-inch) minimum thickness, shop-painted.

c. Base or Sill Angles: Fabricate from 14-gage (0.0747-inch) minimum thickness cold-formed galvanized steel sections.

d. Secondary endwall structural members, except columns and beams, shall be the manufacturer's standard sections fabricated from 14-gage (0.0747-inch) minimum thickness cold-formed galvanized steel.

2.2.2.3 Wind Bracing. Provide adjustable wind bracing using threaded steel rods. [Locate wind bracing only where indicated on the Drawings.]

2.3 ROOF AND WALL COVERING

Panels shall be either steel or aluminum and shall have a factory color mill finish. Length of sheets shall be sufficient to cover the entire length of any unbroken roof slope or the entire height of any unbroken wall surface. Design provisions shall be made for thermal expansion and contraction consistent with the type of system to be used.

2.3.1 Roof Panels

Manufacturer's standard 360° "Pittsburgh" double-lock standing seam roof panel designed for mechanical attachment of panels to roof purlins using a concealed clip fastener. Form roof panels two foot wide with two major corrugations two inches high, 24 inches on center, between and perpendicular to the major corrugations. Provide 16-gage (0.0598-inch) minimum thickness panel clips.

2.3.2 Wall Panels

Wall panels shall have configurations for overlapping adjacent sheets or interlocking ribs for securing adjacent sheets. Wall covering shall be fastened to framework using exposed or concealed fasteners.

2.3.3 Steel Covering

Zinc-coated steel conforming to ASTM A 653, G 90 coating designation; aluminum-zinc alloy coated steel conforming to ASTM A 792, AZ 55 coating; or aluminum-coated steel conforming to ASTM A 463, Type 2, coating designation T2 65. Panels shall be 24 gage (0.024-inch) thick minimum, except that when the mid-field of the roof is subject to design wind uplift pressures of 2.87 kPa (60 psf) or greater or the steel covering is used as a diaphragm the entire roof system shall have a minimum thickness of 22 gage (0.030-inch). Prior to shipment, mill finish panels shall be treated with a passivating chemical and oiled to inhibit the formation of oxide corrosion products. Panels that have become wet during shipment but have not started to oxidize shall be dried, retreated, and re-oiled.

2.3.4 Aluminum Covering

Alloy conforming to ASTM B 209, temper as required for the forming operation, minimum 0.813 mm (0.032 inch) thick.

2.3.5 Factory Insulated Panels

Insulated wall panels shall be factory-fabricated units with insulating core between metal face sheets, securely fastened together and uniformly separated with rigid spacers, facing of steel or aluminum of composition and gage specified for covering, constructed in a manner that will eliminate condensation on interior of panel. Panels shall have a factory color mill finish. Insulation shall be compatible with adjoining materials; nonrunning and nonsettling; capable of retaining its R-value for the life of the metal facing sheets; and unaffected by extremes of temperature and humidity. The assembly shall have a flame spread rating not higher than 25 and smoke developed rating not higher than 50 [100] when tested in accordance with ASTM E 84. The insulation shall remain odorless, free from mold, and not become a source of food and shelter for insects. Units shall be not less than 200 mm (8 inches) wide and shall be in one piece for unbroken wall heights up to 7.32 m (24 feet).

2.3.6 Factory Color Finish

Wall and roof panels shall have a factory applied polyvinylidene fluoride finish on the exposed side. The exterior finish shall consist of a baked-on fluoropolymer enamel topcoat with an appropriate prime coat. Color shall be the manufacturer's standard color most nearly matching the color indicated on the Drawings. The exterior coating shall be a nominal 0.025 mm (1 mil) thickness consisting of a polyvinylidene fluoride topcoat of not less than 0.018 mm (0.7 mil) dry film thickness and the paint manufacturer's recommended primer of not less than 0.005 mm (0.2 mil) thick. The interior color finish shall consist of [the same coating and dry film thickness as the exterior] [a backer coat with a dry film thickness of 0.013 mm (0.5 mil)] [a 0.005 mm (0.2 mil) thick prime coat.] The exterior color finish shall meet the test requirements specified below.

2.3.6.1 Salt Spray Test. A sample of the sheets shall withstand a salt spray test for a minimum of 1000 hours in accordance with ASTM B 117, including the scribe requirement in the test. Immediately upon removal of the panel from the test, the coating shall receive a rating of not less than 8F, few No. 8 blisters, as determined by ASTM D 714; and a rating of [6, 3 mm ($\frac{1}{8}$ inch)] [10, no edge creep] failure at scribe, as determined by ASTM D 1654.

2.3.6.2 Formability Test. When subjected to testing in accordance with ASTM D 522, the coating film shall show no evidence of fracturing to the naked eye.

2.3.6.3 Accelerated Weathering, Chalking Resistance and Color Change. A sample of the sheets shall withstand a weathering test a minimum of 1000 hours in accordance with ASTM G 23, using a Type EH apparatus with cycles of 60 minutes radiation and 60 minutes condensing humidity. The coating shall withstand the weathering test without cracking, peeling, blistering, loss of adhesion of the protective coating, or corrosion of the base metal. Protective coating that can be readily removed from the base metal with tape in accordance with ASTM D 3359, Test Method B, shall be considered as an area indicating loss of adhesion. Following the accelerated weathering test, the coating shall have a chalk rating not less than No. 8 in accordance with ASTM D 4214 test procedures, and the color change shall not exceed 5 CIE or Hunter Lab color difference (delta E) units in accordance with ASTM D 2244. For sheets required to have a low gloss finish, the chalk rating shall be not less than No. 6 and the color difference shall be not greater than 7 units.

2.3.6.4 Humidity Test. When subjected to a humidity cabinet in accordance with ASTM D 2247 for 1000 hours, a scored panel shall show no signs of blistering, cracking, creepage or corrosion.

2.3.6.5 Impact Resistance. Factory-painted sheet shall withstand direct and reverse impact in accordance with ASTM D 2794 equal to 1.5 times metal thickness in millimeters (mils) expressed in Newton-meters (inch-pounds) with no loss of adhesions.

2.3.6.6 Abrasion Resistant Test. When subjected to the falling sand test in accordance with ASTM D 968, the coating system shall withstand a minimum of 50 liters of sand before the appearance of the base metal. The term "appearance of base metal" refers to the metallic coating on steel or the aluminum base metal.

2.3.6.7 Pollution Resistance. Coating shall show no visual effects when immersion tested in a 10 percent hydrochloric acid solution for 24 hours in accordance with ASTM D 1308.

2.3.7 Accessories

Flashing, trim, metal closure strips and curbs, fascia, caps, diverters, and similar metal accessories shall be not less than the minimum thickness specified for covering. Accessories shall be compatible with the system furnished. Exposed metal accessories shall be finished to match the covering building finish. Molded closure strips shall be bituminous-saturated fiber, closed-cell or solid-cell synthetic rubber or neoprene, or polyvinyl chloride premolded to match configuration of the covering and shall not absorb or retain water.

2.4 FASTENERS

Fasteners for steel wall and roof panels shall be of aluminum or stainless steel fasteners for exterior application and galvanized or cadmium-plated fasteners for interior applications. Fasteners for aluminum wall and roof panels shall be aluminum or corrosion resisting steel. Fasteners for structural connections shall provide both tensile and shear strength of not less than 3.34 kN (750 pounds) per fastener. Exposed roof fasteners shall be gasketed or have metal-backed gasketed washers on the exterior side of the covering to waterproof the fastener penetration. Washer material shall be compatible with the covering; have a minimum diameter of 10 mm ($\frac{3}{8}$ inch) for structural connections; and gasketed portion of fasteners or washers shall be neoprene or other equally durable elastomeric material approximately 3 mm ($\frac{1}{8}$ inch) thick. When wall covering is factory color finished, exposed wall fasteners shall be color finished or provided with plastic color caps to match the covering. Nonpenetrating fastener system using concealed clips shall be manufacturer's standard for the system provided.

2.4.1 Screws

Screws shall be as recommended by the manufacturer to meet the strength design requirements of the panels.

2.4.2 Explosive Actuated Fasteners

Fasteners for use with explosive actuated tools shall have a shank diameter of not less than 3.68 mm (0.145 inch) with a shank length of not less than 13 mm ($\frac{1}{2}$ inch) for fastening panels to steel and not less than 25 mm (1 inch) for fastening panels to concrete.

2.4.3 Blind Rivets

Blind rivets shall be aluminum with 5 mm ($\frac{3}{16}$ inch) nominal diameter shank or stainless steel with 3 mm ($\frac{1}{8}$ inch) inch nominal diameter shank. Rivets shall be threaded stem type if used for other than the fastening of trim. Rivets with hollow stems shall have closed ends.

2.4.4 Bolts

Bolts shall be not less than 6 mm ($\frac{1}{4}$ inch) diameter, shouldered or plain shank as required, with proper nuts.

2.5 GUTTERS AND DOWNSPOUTS

Gutters and downspouts shall be fabricated of aluminum, zinc-coated steel or aluminum-zinc alloy coated steel and shall have the same manufacturer's standard factory finish as roof and wall covering. Minimum uncoated thickness of materials shall be 0.457 mm (0.018 inch) for steel and 0.813 mm (0.032 inch) for aluminum. All accessories necessary for the complete installation of the gutters and downspouts shall be furnished. Accessories shall include gutter straps, downspout elbows, downspout straps and fasteners fabricated from metal compatible with the gutters and downspouts.

2.6 LOUVERS

Louvers shall be fabricated of aluminum, zinc-coated steel, or aluminum-zinc alloy coated steel; shall have manufacturer's [standard factory color] [mill] finish; and shall be furnished with bird insect screens. Minimum uncoated thickness of materials shall be 1.219 mm (0.048 inch) for steel and 1.626 mm (0.064 inch) for aluminum. Manually operated louvers shall be designed to be opened and closed from the operating floor.

2.7 CIRCULAR ROOF VENTILATORS

Circular roof ventilators shall be fabricated of aluminum or zinc-coated steel; shall have manufacturer's standard factory color mill finish, and shall be furnished with bird insect screens and chain or cable operated dampers. Minimum uncoated thickness of materials shall be 0.457 mm (0.018 inch) for steel and 0.813 mm (0.032 inch) for aluminum. Ventilators shall be designed to provide rigid weathertight construction upon installation, free from vibration and movement.

2.8 CONTINUOUS ROOF VENTILATORS

Continuous roof ventilators shall be fabricated of aluminum, zinc-coated steel, or aluminum-zinc alloy coated steel, shall have manufacturer's standard factory color mill finish, and shall be furnished with bird insect screens and chain or cable operated dampers. Minimum uncoated thickness of materials shall be 0.457 mm (0.018 inch) for steel and 0.813 mm (0.032 inch) for aluminum. Ventilators shall be furnished in 2.4 m to 3.1 m (8 to 10 feet) long sections braced at mid-length.

2.9 SKYLIGHTS

Skylight panels shall be fabricated of glass-fiber reinforced polyester plastic panels conforming to ASTM D 3841, Type ____ , Grade ____ , Class ____ weighing not less than 2.4 kg per square meter (8 ounces per square foot). Size and color of skylight panels shall be as indicated.

2.10 TRANSLUCENT WALL PANELS

Translucent wall panels shall be manufacturer's standard conforming to ASTM D 3841, Type ____ , Grade ____ , Class ____ weighing not less than 2.4 kg per square meter (8 ounces per square foot). Size and color of translucent wall panels shall be as indicated.

2.11 DOORS

2.11.1 Hinged Doors

Hinged doors and frames shall conform to the requirements of Section 08110 METAL DOORS AND FRAMES. Exterior doors shall have top edges closed flush and sealed against water penetration. Hardware shall be as specified in Section 08700 BUILDERS' HARDWARE.

2.11.2 Sliding Doors

Sliding doors shall be of the metal framed or self-framing metal type. Covering shall be of same material and finish as the wall covering, except heavier gage material shall be used if required to provide rigidity. All hardware necessary for the complete installation of the doors shall be furnished. Accessories shall include galvanized steel track, brackets, permanently lubricated dual wheel trolley hangers, operating handle, slide bolt latch assembly permitting padlocking from either inside or outside of building and rubber or elastomeric weatherstripping.

2.11.3 Overhead Doors

Overhead doors shall be industrial type of standard manufacture, fabricated of 0.864 mm (0.034 inch) or heavier galvanized or aluminum-zinc alloy coated steel. All hardware necessary for the complete installation of the doors shall be furnished. Accessories shall include galvanized steel track, brackets, lifting handles, torsion-spring mechanism, ball bearing rollers, cylinder lock, and weatherstripping. Doors shall be manually operated, except doors over 13.4 square meters (144 square feet) in area shall be [chain hoist operated] [electric motor operated].

2.12 WINDOWS

Windows shall be of steel in accordance with SWI Guide or of aluminum in accordance with AAMA Specification. Windows shall be of the type shown, furnished complete with operating and locking hardware, glazing, screened panels, weatherstripping, and framing and fasteners to properly install the windows.

2.13 INSULATION

Insulation shall conform to NAIMA 202. Thermal resistance of insulation shall be not less than the R-values shown. R-values shall be determined at 24 degrees C (75 degrees F) in accordance with ASTM C 518. Insulation shall be a standard product of a manufacturer, factory marked or identified with manufacturer's name or trademark and R-value. Identification shall be on individual pieces or individual packages. Insulation shall have a facing specified in the paragraph VAPOR RETARDER.

2.13.1 Rigid Board Insulation for Use above a Roof Deck

2.13.1.1 Polyurethane or Polyisocyanurate. Polyurethane insulation shall conform to FS HH-I-1972/GEN and FS HH-I-1972/1. Polyisocyanurate insulation shall conform to FS HH-I-1972/1, Class 2 or ASTM C 1289, Type I, Class 2 (having a minimum recovered material content of 9 percent by weight of core material in the polyisocyanurate portion).

2.13.1.2 Mineral Fiber. Insulation shall conform to ASTM C 612.

2.13.2 Blanket Insulation

Blanket insulation shall conform to ASTM C 553.

2.13.3 Spray-On Insulation

Spray-on insulation shall conform to ASTM E 1042, Type II, Class (a), and shall be odorless, fungi resistant, nonshrinking, nonshedding, noncorroding, fibrous glass material. Spray-on insulation shall not contain asbestos. Insulation shall meet the dusting requirements specified in ASTM E 1042 and shall not crack, lose bond or otherwise fail due to deflections, temperature changes or adverse humidity conditions. Spray-on insulation shall be white [_____] color.

2.13.4 Insulation Retainers

Retainers shall be of type, size and design necessary to adequately hold the insulation and to provide a neat appearance. Metallic retaining members shall be nonferrous or have a nonferrous coating. Nonmetallic retaining members, including adhesives used in conjunction with mechanical retainers or at insulation seams, shall have a fire resistance classification not less than that permitted for the insulation.

2.14 WALL LINERS

Wall liners shall be 0.61 mm (0.024 inch) thick minimum for aluminum or 0.46 mm (0.018 inch) thick minimum for steel of composition specified for covering, and formed or patterned to prevent waviness and distortion, and shall extend from the floor to a height of not less than _____ mm (_____ feet) above the floor. Matching metal trim shall be provided at base of wall liner, top of wall liner, around openings in walls and roof and over interior and exterior corners. Wall liners shall have a factory color finish of the quality specified for the roof and wall panels. Colors shall be selected from manufacturer's standard finishes as indicated.

2.15 SEALANT

Sealant shall be an elastomeric type containing no oil or asphalt. Exposed sealant shall cure to a rubberlike consistency. Concealed sealant may be the nonhardening type.

2.16 GASKETS AND INSULATING COMPOUNDS

Gaskets and insulating compounds shall be nonabsorptive and suitable for insulating contact points of incompatible materials. Insulating compounds shall be nonrunning after drying.

2.17   VAPOR RETARDER

2.17.1   Vapor Retarders as Integral Facing

Insulation facing shall have a permeability of 1.15 [_____] ng per Pa-second-square meter (0.02 [_____] perm) or less when tested in accordance with ASTM E 96. Facing shall be white of [reinforced foil with a vinyl finish]; except that unreinforced foil with a natural finish may be used in concealed locations. Facings and finishes shall be factory applied.

2.17.2   Vapor Retarders Separate from Insulation

Vapor retarder material shall be polyethylene sheeting conforming to the requirements of ASTM D 4397. A single ply of 0.25 mm (10 mil) polyethylene sheet; or, at the option of the Contractor, a double ply of 0.15 mm (6 mil) polyethylene sheet shall be used. A fully compatible polyethylene tape shall be provided which has equal or better water vapor control characteristics than the vapor retarder material. A cloth industrial duct tape in a utility grade shall also be provided to use as needed to protect the vapor retarder from puncturing.

PART 3   EXECUTION

3.1   ERECTION

3.1.1   General

Erection shall be in accordance with the approved erection instructions and drawings and with applicable provisions of AISC Code of Standard Practice and MBMA Manual. The completed buildings shall be free of excessive noise from wind-induced vibrations under the ordinary weather conditions to be encountered at the location where the building is erected, and meet all specified design requirements. Framing members fabricated or modified on site shall be saw or abrasive cut; bolt holes shall be drilled. On-site flame cutting of framing members, with the exception of small access holes in structural beam or column webs, shall not be permitted. Improper or mislocated bolt holes in structural members or other misfits caused by improper fabrication or erection shall be repaired in accordance with AISC Code of Standard practice. Exposed surfaces shall be kept clean and free from sealant, metal cuttings, excess material from thermal cutting, and other foreign materials. Exposed surfaces which have been thermally cut shall be finished smooth within a tolerance of 3 mm ($\frac{1}{8}$ inch). Welding of steel shall conform to AWS D1.1; welding of aluminum shall conform to AA Manual. High-strength bolting shall conform to AISC Specification for Structural Joints using ASTM A 325 or ASTM A 490 bolts.

3.1.2   Framing and Structural Members

Anchor bolts shall be accurately set by template while the concrete is in a plastic state. Uniform bearing under base plates and sill members shall be provided using a nonshrinking grout when necessary; moist cure grout for at least 7 days.

Members shall be accurately spaced to assure proper fitting of covering. As erection progresses, the work shall be securely fastened to resist the dead load and wind and erection stresses. The erector shall furnish temporary bracing as needed to assure stability of the structure during construction; permanent building bracing furnished by the manufacturer shall not be assumed to be adequate during erection. Supports for electric overhead traveling cranes shall be positioned and aligned in accordance with MHI CMAA 70. Installation tolerances for framing members shall be $\frac{1}{4}$ inch from level; $\frac{1}{8}$ inch from plumb. Do not field cut or alter structural members without approval of the metal building manufacturer.

3.1.3   Wall Covering and Roof Covering

Wall covering shall be applied with the longitudinal configurations in the vertical position. Roof covering shall be applied with the longitudinal configurations in the direction of the roof slope. Accessories shall be fastened into framing members, except as otherwise approved. Closure strips shall be provided as indicated and where necessary to provide weathertight construction. Improper or mislocated drill holes shall be plugged with an oversize screw fastener and gasketed washer; however, sheets with an excess of such holes or with such holes in critical locations shall not be used. When cutting prefinished sheets, remove all shavings from finish surfaces and touch up the cut edges with a repair compound similar to the shop finish. Stained, discolored or damaged sheets shall be removed from the site. Installation tolerance shall be $\frac{1}{8}$ inch from true position.

3.1.3.1   Lap Type Panels with Exposed Fasteners. End laps shall be made over framing members with fasteners into framing members approximately 50 mm (2 inches) from the end of the overlapping sheet. Side laps shall be laid away from the prevailing winds. Side lap distances, end lap distances, joint sealing, and spacing and fastening of fasteners shall be in accordance with the manufacturer's standard practice insofar as the maximum spacings specified are not exceeded and provided such standard practice will result in a structure which will be free from water leaks and meet design requirements. Spacing of fasteners shall present an orderly appearance and shall not exceed: 200 mm (8 inches) on center at end laps of covering, 300 mm (12 inches) on center at connection of covering to intermediate supports, 300 mm (12 inches) on center at side laps of roof coverings, and 450 mm (18 inches) on center at side laps of wall coverings except when otherwise approved. Side laps and end laps of roof and wall covering and joints at accessories shall be sealed. Fasteners shall be installed in straight lines within a tolerance of 13 mm ($\frac{1}{2}$ inch) in the length of a bay. Fasteners shall be driven normal to the surface and to a uniform depth to properly seat the gasketed washers.

3.1.3.2   Concealed Fastener Wall Panels. Panels shall be fastened to framing members with concealed fastening clips or other concealed devices standard with the manufacturer. Spacing of fastening clips and fasteners shall be in accordance with the manufacturer's written instructions insofar as the maxi-

mum fastener spacings specified are not exceeded and provided such standard practice will result in a structure which will be free from water leaks and meet design requirements. Spacing of fasteners and anchor clips along the panel interlocking ribs shall not exceed 300 mm (12 inches) on center except when otherwise approved. Fasteners shall not puncture covering sheets except as approved for flashing, closures, and trim; exposed fasteners shall be installed in straight lines. Interlocking ribs shall be sealed according to manufacturer's recommendations. Joints at accessories shall be sealed.

3.1.4 Gutters and Downspouts

Gutters and downspouts shall be rigidly attached to the building. Spacing of cleats for gutters shall be 400 mm (16 inches) maximum. Spacing of brackets and spacers for gutters shall be 900 mm (36 inches) maximum. Supports for downspouts shall be spaced according to manufacturer's recommendations. Slope gutters a minimum of $\frac{1}{8}$ inch per foot. [Connect downspouts to storm sewer system.] [Install splash blocks.]

3.1.5 Louvers and Ventilators

Louvers and ventilators shall be rigidly attached to the supporting construction in a manner to assure a rain-tight installation.

3.1.6 Doors and Windows

Doors and windows, including frames and hardware, shall be securely anchored to the supporting construction, shall be installed plumb and true, and shall be adjusted as necessary to provide proper operation. Joints at doors and windows shall be sealed according to manufacturer's recommendations to provide weathertight construction.

3.1.7 Insulation

Insulation shall be installed as indicated and in accordance with manufacturer's instructions. Final appearance of installed insulation shall be free of unsightly sags and wrinkles.

3.1.7.1 Board Insulation with Blanket Insulation. Rigid or semirigid board insulation shall be laid in close contact. If more than one layer of insulation is required, joints in the second layer shall be offset from joints in the first layer. A layer of blanket insulation shall be placed over the rigid or semirigid board insulation to be compressed against the underside of the metal roofing to reduce thermal bridging, dampen noise, and prevent roofing flutter. This layer of blanket insulation shall be compressed a minimum of 50 percent.

3.1.7.2 Blanket Insulation. Blanket insulation shall be installed over the purlins and held tight against the metal roofing. It shall be supported by an integral facing or other commercially available support system.

3.1.7.3 Spray-On Insulation.   Spray-on insulation shall be applied to a uniform thickness and in accordance with manufacturer's published recommendations. Objects attached or adjacent to the wall or ceiling that are to receive sprayed insulation, such as purlins, braces, piping, and conduits, shall be sprayed completely without voids to provide proper thermal seal. Surfaces to be sprayed shall be free of rust, grease, oil, or any material which would prevent good adhesion of the spray-on insulation.

3.1.8 Integral Facing Vapor Retarder

Integral facing on blanket insulation shall have the facing lapped and sealed with a compatible tape to provide a vapor-tight membrane.

3.1.9 Slip Sheet

A building-paper slip sheet shall be laid over the blanket insulation facing to prevent the vinyl facing from adhering to the metal roofing.

3.1.10 Wall Liner

Wall liner shall be securely fastened into place in accordance with the manufacturer's recommendation and in a manner to present a neat appearance.

3.2 DISSIMILAR MATERIALS

Where aluminum surfaces come in contact with ferrous metal or other incompatible materials, keep aluminum surfaces from direct contact by applications to the other materials by one of the following methods:

1. One coat of zinc chromate primer complying with the performance, but not necessarily with the composition requirements of Federal Specification FS TT-P-645, followed by two coats of aluminum paint, SSPC-paint 101.

2. One coat of high build bituminous paint SSPC-Paint 12, applied to a thickness of $\frac{1}{16}$ in over one coat of zinc chromate primer.

3. Mylar tape or neoprene spacer as recommended by manufacturer.

3.3 FIELD PAINTING

Immediately upon detection, abraded or corroded spots on shop-painted surfaces shall be wire brushed and touched up with the same material used for the shop coat. Shop-primed ferrous surfaces exposed on the outside of the building and all shop-primed surfaces of doors and windows shall be painted with two coats of an approved exterior enamel. Factory color finished surfaces shall be touched up as necessary with the manufacturer's recommended touch-up paint. Finish painting is specified in Section 09900, PAINTING.

Index

ABOUT THE AUTHOR

Alexander Newman, P.E., is principal structural engineer with Maguire Group Inc., a national architectural/engineering/planning firm in Foxborough, Massachusetts. With two decades of engineering and management experience, he has worked as project engineer with a consulting engineering firm, design engineer with a light-gauge framing panel manufacturer, and manager of fabrication for a steel fabricator. He has had the responsibility for structural design and specifications for numerous metal building system projects throughout the country. He has also designed many conventional structures, including a Boston Edison switching and conversion station that won the 1990 American Consulting Engineers Council of New England Award for Engineering Excellence. Mr. Newman holds an advanced degree in structural engineering from the Moscow Civil Engineering Institute in Russia, and a master's degree in business administration with high honors from Boston University. He is the author of several articles that have appeared in leading engineering publications and has conducted seminars on metal building systems for design professionals.